The Broca-Wernicke Doctrine

Geert-Jan Rutten

The Broca-Wernicke Doctrine

A Historical and Clinical Perspective on Localization of Language Functions

Springer

Geert-Jan Rutten
Department of Neurosurgery
St Elisabeth-Tweesteden Hospital
Tilburg
The Netherlands

ISBN 978-3-319-54632-2 ISBN 978-3-319-54633-9 (eBook)
DOI 10.1007/978-3-319-54633-9

Library of Congress Control Number: 2017945400

Printed on acid-free paper

This Springer imprint is published by Springer Nature
The registered company is Springer International Publishing AG
The registered company address is: Gewerbestrasse 11, 6330 Cham, Switzerland

Speech activity, which develops with the mastery of language in a social setting, must rest upon a dynamic system of the greatest complexity involving the simultaneous functioning of various brain areas. Thus the problem of the neurologist, who wishes to study the brain mechanisms of speech processes, is to determine how these complex functional systems derive from the dynamic structure of the cortex and understand the role of each area in these functional systems; but he must not "localize" these highly complex "functions" in separate isolate areas of the brain.

It is easy to see the extent to which this systemic conception of the highly complex forms of nervous activity in man conflicts with the mechanistic type of search for "speech centers", "writing centers", or "reading centers". References to such centers give the appearance of scientific explanation, but actually conceal the path to further analysis of underlying mechanisms.

Alexandr Luria, 1947

What This Book Is About

One hundred and fifty years ago, post-mortem studies shaped the way in which we think about brain function. Theories emerged that linked functions to specific anatomical brain regions. History named some of these areas after the persons who allegedly described their functions for the first time, the most famous being Paul Broca and Carl Wernicke.

This classic neurological view of language functions, with its tight coupling of structure and function, is still widely favoured and remains the predominant view in clinical practice. Most neurologists will classify patients with language deficits as having either motor (Broca's) or sensory (Wernicke's) aphasia or a combination of both. Most neurosurgeons will treat left inferior frontal and left posterior temporal areas with the utmost respect and still often refrain from surgery in fear of grave and lasting language deficits. Surprisingly, this clinical view is in stark contrast to compelling evidence from neuroscience. Here, language functions are represented in complex and ever-changing neural networks that span many different brain areas across both hemispheres.

This is a fascinating paradox. Why is it that, despite mounting evidence against the strict and fixed localization of language functions, clinicians—as well as many scientists—continue to refer to 100-plus-year-old language models to explain their patients' deficits or even to plan surgery? I have tried to find answers to this question and have reviewed historical developments in an attempt to understand why these seemingly contradictory views coexist. Fuelled and inspired by observations of patients with brain tumours and ischaemia, this book is largely written from a clinical and neurosurgical perspective. The reader is warned: there will be no revelation of some new scientific truth nor a meticulous overview of current language theories or anatomical–functional underpinnings of language in the brain. Rather, this book is a reflection of personal travels through journals and books. As such, there is a fair amount of randomness to it. Its main goal is to provide clinicians and neuroscientists with the background information that is needed to understand each other's work.

What struck me most is not that the old iconic neurological models are still being taught in curricula and modern textbooks (they should be, because of their monumental historical and educational value). Rather, it is the naive manner in which they are used in today's clinical practice and how they still dominate the decision-making process. And there is another remarkable thing: the ideas of the classic authors are far more sophisticated than we collectively remember. In fact, some of their theories anticipate our current thoughts on functional brain topography. Sadly, however, these were mostly forgotten or ignored.

Personal Experience

When I learned about neurology and neurosurgery, not too long ago, brain scanners were already routinely used in clinical practice. Modern scanners reach submillimetre resolution and provide amazingly detailed information of diseased brain or spine structures. CT and MRI scanners have greatly increased the speed and accuracy of the diagnostic process. Neurosurgeons, like myself, use them on a daily basis to decide if and where to operate in the brain. It seems that this technology has made it very easy to localize the cause of any patient's neurological or functional deficit.

But practice shows otherwise and turns out to be rather different. The neurological deficits that are observed in our patients, and the complaints these patients have, do not bear a very consistent relationship to the abnormalities that are seen (or sometimes not seen) on brain scans. Many brain tumours are asymptomatic for a long period of time and grow to a considerable size before they cause symptoms. This process can take as much as 10 or 15 years. In retrospect, these 'patients' led a normal life with a tumour hidden in their brains. Sometimes ischemia or a tumour is incidentally found on a brain scan, as can happen when a patient's brain is scanned for another reason (for instance, after an accident or during a routine check-up). I gradually came to realize (like many other clinicians and scientists, I presume) that the size and the location of a lesion are not very good predictors for the patient's well-being. This is something that was not taught to me in medical school but is known as a fact to experienced clinicians. Patients also have an impressive ability to recover from brain damage. But where one patient regains normal functionality, another with an apparently similar lesion does not. This is another thing that textbooks don't tell you.

When I started to work as a neurosurgeon in 2007, I had the feeling that things only grew more complex. Despite the use of awake surgical procedures and newly available functional brain imaging techniques, I found it difficult to inform my patients accurately about their neurological and cognitive status after surgery (this is something that still bothers me). What was the actual chance that they would have language or cognitive problems after surgery, and would these deficits be temporary or permanent? And, most importantly, what would be the impact on their social and professional lives? Neurological models simply did not account for these facts. Tumours in classic language areas, for example, can often be removed with minimal or no lasting deficits. But even normal language areas, so it seems, can be surgically damaged without any apparent neurological deficit.

Removal of Classic Language Areas

Take, for example, the case of JP, who at age 23 suddenly developed epileptic seizures. Unfortunately, this was caused by a malignant brain tumour that had infiltrated a large part of his left insula. JP had never had any complaints before, and besides his seizures, there were no neurological or cognitive abnormalities. (He was tested with formal neuropsychological tests that showed normal results.) The insular region is difficult to reach for neurosurgeons, as parts of the frontal, temporal and parietal lobes completely cover its surface. One possible surgical strategy is to resect a part of the overlying normal cortex to create a corridor through which the tumour can be resected. In our case, we had decided prior to surgery that this was our preferred route. The patient was operated on under awake conditions to ensure that this route was safe from a functional point of view. After opening the dura, the cortical surface was probed with electrical stimulation in search of functional disturbances. These indicate the surgical no-go areas. As is often the case, the inferior part of the left precentral gyrus and pars opercularis (the most posterior part of the inferior frontal gyrus) turned out to be involved in speech and language functions. No language errors were inflicted when the pars triangularis was stimulated, something that we had repeatedly seen before in other patients. We therefore decided to make our entrance to the insula here, and although the corticotomy was thus made in a fairly large part of classic Broca's area, it did not result in any language problems. JP kept talking in a normal manner with the neuropsychologist that sat across from him during surgery while his tumour was resected. At intervals, he was also tested with formal language tasks, and he was able to perform all these tasks correctly. Tests were repeated 1 year after surgery. While JP had some mild complaints regarding the handling of information (he tended to avoid crowded areas and busy events), he did well and experienced no language problems. Formal neuropsychological tests were normal and detected no language or other cognitive impairments. These and other cases, as well as the emerging literature on this subject, made me seriously doubt the clinical conviction that language functions are largely confined to predetermined 'eloquent' brain areas and that it is always hazardous to operate within these regions.

Back to the Future

I decided to read the works of Broca and Wernicke to find some answers. Ironically, I turned to 100-plus-year-old documents to find information that was relevant for my own clinical practice. Was I really hoping to find a more precise anatomical definition of Broca's area or the functional consequences of surgery within Wernicke's region? I guess not. The real reason for my search must have been my amazement and scepticism about the dogmatic localist view that was still haunting my own clinical practice. I wanted to find out what kept this naive thinking alive. Much to my surprise, the historical information that I found was so 'new' and thrilling that I kept on reading and collecting documents, gradually working my way through the nineteenth and twentieth centuries. Reading these papers and books changed my thinking about their authors and their so-called classic language models, but above all, it provided me with new insights and an urge to explore language organization in the brain further. I quickly discovered that many of the original documents from nineteenth-century authors are available in facsimile and that these are easily obtained via the Internet. For example, I got my copy of Wernicke's monumental *Der aphasische Symptomencomplex* for only 10 euro. I can only say how surprised I was to discover deep knowledge and insight every time I read this fairly small 70-page monograph. I now think that this book is a must-read for anyone that is interested in aphasia and the anatomical basis of our language system (that is also the reason that so many pages are devoted to it in this book). What happened next was that I lost direction, not only because the forgotten works of Campbell, Brodmann, Penfield, Goldstein, Luria and all those others were easy to get and a joy to read, but in particular because they posed alternative and relevant views for my own understanding of the anatomical basis of the language system. At some point, I decided to narrow my scope and to follow through the question of why the old language models have such a persistent influence in modern clinical practice. I tried to stick to the more clinical opinions and techniques and also began making notes, as it helped me shape my opinion of it all.

Classic Authors with Modern Opinions

I wasn't really surprised to discover that there is little proof for a strict localist concept of functional localization; language functions are not simply confined to a handful of the same well-defined areas in every person, as has been repeatedly put forward in history (but apparently forgotten or ignored over and over again). What did surprise me, however, was that Broca and Wernicke (and many of their contemporaries and successors) postulated interesting ideas and theories that often strongly contrasted with their most famous legacy: the localist language model with its areas of Broca and Wernicke. It appears that this model, in the way that it is usually taught in medical and neuropsychological textbooks, is actually a fairly bad summary of their work. It is certainly not very representative of their scientific opinions and thoughts.

A good example of history repeating is Wernicke. Wernicke is clearly one of the first connectionists. He never proposed localized (temporal) language areas but instead formulated a theory whereby language-related and conceptual knowledge is distributed over many interconnected areas (similar to our current view). Yet history remembers him as a strict localist. Then there is Brodmann, who divided the brain into areas based on their cytological characteristics. His famous maps are still frequently used in neuroscientific studies to denote the anatomical location of brain functions and to communicate functional results across researchers. But Brodmann himself opposed the strict localization of functions and stressed the significant anatomical–functional variability between individuals. I doubt he would have agreed to the use of his histological findings as a map for the localization of functions. Another example of our historical blindness can be found in the works of Penfield. This famous neurosurgeon is remembered for his functional cartography and the iconic 'homunculus' that resulted from his studies with electrocortical stimulation in neurosurgical patients. The awake surgical methodology that he developed in the first half of the previous century is still our current gold standard for localization of brain functions (i.e. the results of any new technique, such as functional MRI, are weighted against those of cortical mapping). What is much less known is that Penfield also formulated new and elaborate views on nervous system functioning, language representation and brain plasticity. He proposed a dynamic and distributed language model, whereby areas within a network could take over functions from damaged areas. And he explicitly warned his readers, just like Brodmann did, not to interpret the simplified graphical images in his work too literally, as they were only meant to aid memory. But, as was the case with many other authors, these nuances were lost over time, and the diagrams and models were the sole survivors.

This book is a tribute to the originality and grandeur of these authors. For that reason, I have included many quotations and graphical images from the original works. To some extent, I have added my own interpretation, but I have always taken care to provide references wherever possible. These references are usually placed at the end of one or more sentences and indicate that some of the findings or opinions of the author(s) are expressed within that part of the text. I think that it is important to know our true scientific history, and to point out the misconceptions that have evolved over time. We should all revert to the original literature once in a while to check the base upon which our current ideas rest. The Broca–Wernicke model is very attractive from an intuitive and heuristic point of view, but it is simply wrong when you study the facts, as was repeatedly demonstrated in the past. Still, it has kept its prominent place in clinical practice over time.

Foremost, I wrote it all down to educate myself. At some point, however, I decided to turn my notes into a book, wishing that others will benefit from it. It is intended for students, clinicians and scientists that are interested in the anatomical basis of language in the brain. I can only hope that it narrows the gap between those that study the brain with modern imaging techniques, and those that treat patients with neurological diseases. Between neuroscientists who describe human behavior in mathematical terms, and neurosurgeons who have to balance the extent of a tumour resection against the so-called quality of life. I sincerely hope that my mosaical historical overview will stimulate the reader to learn more and to explore further the many topics that are only briefly touched upon in this book. I learned a lot myself along the journey and now realize even more strongly that we need to integrate clinical observations and neuroscientific theories if we want better to understand language organization in the brain.

Acknowledgements

Many thanks to Greg Hickok, for his comments on Chaps. 1 and 2; Martijn Jansma, for vivid discussions and co-authorship of Chap. 8; Hanne Clemens, for help with all the copyrights; Sylvana Freyberg from Springer, who trusted me on a Preface; and all my friends from *Neurochirurgie Brabant* who make it a pleasure and privilege to work and live in Tilburg. Finally, my thanks and love go to Stijn and Lidy, who contributed in many different ways to this project.

Contents

Broca and the Birth of Localization Theories

1

In 2009, Craig Bennett and his co-workers put a mature Atlantic salmon in an MRI scanner and showed it a series of photographs [1]. The (dead) salmon had been instructed to determine the emotion of photographed people during MR scanning. Much to everybody's surprise, three activated areas were found exactly in the brain cavity of the salmon, as shown in Fig. 1.1. The results of this experiment were presented at a human brain mapping conference in Toronto [1]. Statistically speaking, the images made a strong point that the salmon was engaged in a cognitive task. The poster, of course, argued differently and pointed to the dangers of modern functional neuroimaging techniques.

Type in 'phrenology' on the Internet and you will quickly find a vast number of beautiful images such as the one shown in Fig. 1.2. This image illustrates the phrenologists' idea that the mind consists of several different 'organs' or 'faculties' that each harbour a specific mental quality. A broad range of human qualities was covered, as variable as benevolence, amativeness, causality or language. Phrenologists associated these mental qualities with specific locations in the brain. This was a radically new concept in an era where philosophical and religious thinking was dualistic

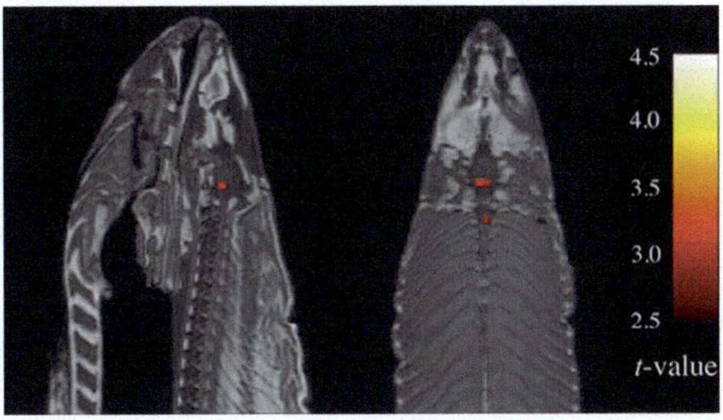

Fig. 1.1 Unexpected results of functional neuroimaging techniques: brain activity in a dead salmon [1]

G.-J. Rutten, *The Broca-Wernicke Doctrine*, DOI 10.1007/978-3-319-54633-9_1

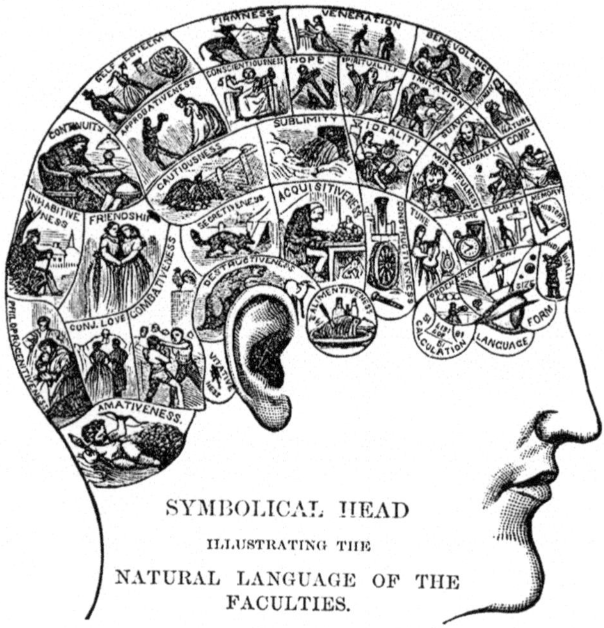

Fig. 1.2 Phrenological representation of brain functions (Figure taken from Samuel Wells' *How to Read Character*, 1888)

in nature and still considered the mind or soul to be undivided, immaterial and immortal. Of course, they were wrong on the details. It is easy for us to criticize the fact that they confined complex behavioural functions to specific brain regions. However, it is often forgotten that the phrenologists paved the way for future neurological and neuropsychological theories about brain function. They were the first to associate brain functions with particular locations in the brain, still the most commonly accepted localist view in clinical practice. The phrenologists were particularly criticized for the fact that they assumed that the strength of a brain function scaled with its physical properties. They considered the size of the organ proportional to its mental power, and the skull to reflect the underlying size of the organ. Samuel Wells (1888) put it as follows in the introduction of his book *How to Read Character* [2]:

> Now, as is the soul which is incarnate in it, so is the brain in texture, size and configuration; and as is the brain, so is its bony encasement, the cranium, on which may be read, in general forms and special elevations and depressions, and with unerring certainty, a correct outline of the intellectual and moral character of man.

It is of course impossible to pick out one point in history where the phrenological ideology started. The fact is that it was ridiculed from the beginning and that the term has often been used with a negative connotation. However, the ghost of phrenology still haunts neuroscience. Examples are easily taken from recent studies that have investigated the relationship between brain structure and function. The results of some of these scientific papers made it to the popular press, and this resulted in remarkable headings such as 'Political Views Reflected in Brain Structure' (www.abcnews.go.com) or 'What's in Voters Heads. Brain Scans Reveal Clues' (www.seattlepi.com). This happened, for example, to the studies of Amodio and colleagues (2007) [3]. They described a relationship between liberalism and brain activity in the anterior cingulate cortex and published this in 2007 in the journal *Nature Neuroscience* under the title 'Neurocognitive Correlates of Liberalism and Conservatism'. Here is their abstract [3]:

> Political scientists and psychologists have noted that, on average, conservatives show more structured and persistent cognitive styles, whereas liberals are more responsive to informational complexity, ambiguity and novelty. We tested the hypothesis that these profiles relate to differences in general neurocognitive functioning using event-related potentials, and found that greater liberalism was associated with stronger conflict-related anterior cingulate activity, suggesting greater neurocognitive sensitivity to cues for altering a habitual response pattern.

Others found that 'this functional correlate of political attitudes has a counterpart in brain structure' [4]. Kanai and colleagues (2011) reported that liberals had a larger area in the anterior cingulate cortex and conservatives had a larger right amygdala [4]. Both these areas are part of the limbic system that deals, amongst others, with emotional processing; the amygdala is specifically associated with responses to fear.

In expert hands, the modern imaging techniques have revolutionized neuroscience. However, these techniques are complex. The measured signals are noisy, and analysis and interpretation can be difficult and are certainly less straightforward than these news headings falsely imply. Quick and dirty conclusions are easily drawn, but not often justified. Of course, the phrenological part of all this lies in the proposed hypotheses and interpretation of the results, not in the technique itself. It is not very difficult to get false-positive findings in individuals or groups with functional imaging techniques; but even with very strict methodological protocols, one can get surprising results, as the fMRI experiment with the salmon showed.

1.1 Gall

Franz Josef Gall (1758–1828) is considered the founding father of phrenology and cranioscopy (the 'reading' of the bumps on the skull). Even in his own time, he received a lot of discredit for his ideas, and he was exiled from Vienna, whereupon he moved to France in the hope of finding a more responsive scientific environment (which he did; he became a well-known figure in his time). Still it is fair to say that Gall made a historical and classical contribution to the concept of cerebral localization. See for a thorough overview Young's brilliant book *Mind, Brain and Adaptation in the Nineteenth Century* [5]. What was revolutionary in itself, and a great contribution to psychology, was the fact that Gall considered behaviour and brain functioning amenable to objective observation. He based his discovery of the 27 faculties on empirical examinations and used several different methods to do this. He collected measurements and casts or skulls of several hundred heads and also made a large number of observations on the crania and behaviour of animals. According to Young (1970), 'he travelled to schools, foundling homes, hospitals, prisons and lunatic asylums, and obtained information on remarkable heads and remarkable talents wherever he could' [5]. As counterproof, Gall also examined individuals with minor qualities for a lack of the corresponding cranial prominence.

Gall and his later followers made a number of suppositions on which their theories were based. Despite the fact that Gall, together with Spurzheim, did extensive and important work in neuroanatomy (e.g. they postulated that white matter served a conduction function and described the anatomical decussation of the pyramids), he considered this work irrelevant for his organology and was convinced that knowledge of functions *preceded* that of the anatomy of the brain [5]. This discrepancy remains something of a mystery, as Gall was well aware of neurological data from victims that had suffered head and brain injuries. He even described such cases himself. Gall stated his main assumptions already in 1798 in a letter to Baron von Retzer, before he did much of his anatomical work [6].

Here are the 'chief principles of phrenology', as formulated in 1868 by another phrenologist, Samuel Wells, 'every one of which is supported by an array of unquestionable facts and susceptible to the clearest proof' [2]. Note that this is almost 50 years after Gall's death and in an era where Broca and Wernicke already based their theories on postmortem examination of damaged brains.

1. The brain is the organ of the mind.
2. Each faculty of the mind has its separate or special organ in the brain.
3. Organs related to each other in function are grouped together in the brain.
4. Size, other things equal, is the measure of power.
5. The physiological conditions of the body affect mental manifestation.
6. Any faculty may be improved by cultivation and may deteriorate through neglect.
7. Every faculty is normally good, but liable to perversion.

Although Wells motivates these principles with a list of 'obvious' facts and findings, he fails to give scientific proof (from our modern point of view). Some arguments are valid from a rational point of view, for instance, when he states that 'Partial injuries to the brain result in suspension of one or more faculties, while others retain their normal activity, which could not be the case if the brain were a single organ' [2].

1.2 Flourens

Opposed to the localist theories of Gall and later Broca and several others, there are the field theories that hold that the brain acts as a single equipotential unit. Jean Pierre Flourens (1794–1867) is considered by many the modern founder of the field theory. He was the first to base his work on experiments with animals from which he systematically removed parts of the brain and observed the resulting disturbed behaviour. In a famous paper from 1824, he describes several of these ablative experiments. Note that in that time little was known of brain anatomy; the only parts that were distinguished were the medulla, the corpora quadrigemina, the cerebellum and the hemispheres. There was no knowledge of white and grey matter.[a] Also, bear in mind that the methods of stimulation and ablation were probably too crude to reveal much information on localization of function [8]. Details of the surgical procedures and behavioural observations are not given to the extent that experiments can be repeated. The following excerpts are all from Flourens' 1824 paper, taken from the book *Readings in the History of Psychology* (1948) [9].

> The entire cranial portion of a young dog was removed. I pushed a needle through the cerebral lobes, cut them in all directions, and also cut through the cerebellum on that side. The animal seemed neither disturbed nor agitated.
>
> I removed both cerebral hemispheres of a pigeon, including the optical layers. The iris retained all of its ability to contract. However, I had only to push through the optic nerves or the corpora quadrigemina to elicit strong and prolonged contractions.
>
> This experiment was repeated on several pigeons. The result was always the same. Consequently the cerebral hemispheres are not responsible for muscular contractions.

Flourens describes several other experiments with rabbits but mainly with pigeons (!) from which he concludes that 'the cerebral lobes are neither the origin of muscular contractions nor are they the origin of the control of movements (...), but it also seems demonstrated that they are the exclusive origin of volition and sensation'. Flourens had noted that without both cerebral lobes pigeons kept intact reflexes but lacked spontaneous movements. Further on he notes:

> One can remove, from the front, or the back, or the top or the side, a certain portion of the cerebral lobes, without destroying their function. A small part of the lobe seems sufficient to exercise these functions.
>
> After certain limits have been surpassed, they are entirely extinguished. The cerebral lobes concur than in their entirety with all of their functions.
>
> Finally as one sensation is lost completely, all of them are. Consequently there is no different origin for any of the faculties nor for any of the sensations.

Interestingly, Flourens also describes recovery of function after damage to the brain. He notes that after some of the removals, the animals regain most or all of their functions.

> We have just seen that is possible to remove a certain portion of the cerebral lobes without destroying their functions completely. However, there is more than that. The lobe can recover these functions in their entirety after having lost them completely.

[a] For more details, see a paper of Tizard [7].

Flourens describes the complete recovery of functions in a pigeon when he stops ablation as soon as the animal 'had lost completely the use of all senses and intellectual functions'. Recovery takes a period of 6 days, in another case it takes 15 days. When he pushes the resection further, he notes that the animals fail to make a complete recovery. Flourens concludes that:

> as long as not too much of the lobes is removed, they may regain in due time the exercise of their functions. Passing certain limits, however, the animal regains them only imperfectly and passing these new limits, it does not recover them at all. Finally, if one sensation comes back, all come back. If one faculty reappears, they all reappear.

Similar recoveries are seen when Flourens removes parts of the cerebellum or corpora quadrigemina. He ends his paper with the conclusion that:

> When one point in the nervous system becomes excited, it excites all others; one point irritated, irritates all. There is community of reaction. Unity is the great reigning principle; it is everywhere; dominates everything. The nervous system is then only one single system.

Flourens related loss of function to the extent of damage, and hereby adopted a holistic concept [8]. His theories were perfectly in line with contemporary philosophical assumptions and remained the dominating view in the first half of the nineteenth century. His experimental work on the brain as a unitary sensorium was used to refute phrenological theories whereby other incompatible observations were disregarded. Tizard refers to two of these observations in a review paper [7]. One was courtesy of Charles Bell (1774–1842), known from several discoveries that bear his name such as Bell's palsy. He indicated that when separate nerve tracts lead to separate areas in the cortex, these areas therefore must have distinct functions. The other was from François Pourfour du Petit (1664–1741) whose experiments were particularly remarkable. This French surgeon first made a number of postmortem observations in patients with a hemiplegia and a contralateral brain lesion. Then he did ablative experiments with dogs and established a relationship between damage in one hemisphere and a consequent paralysis on the opposite side [10]. Almost a century later, Dominique Jean Larrey (1766–1842), another French (military) surgeon and a contemporary of Flourens and Gall, described 12 patients with speech difficulties, most of whom had traumatic brain injuries. His work has been described in a number of books and papers [11]. One of Larrey's patients was a corporal in Napoleon's army who was struck by a musket ball in the left eyebrow at Waterloo, resulting in a depressed skull fracture and brain injury. The soldier, Louis Manez, was left with a right hemiplegia and language deficits. Despite his wound he partially recovered and regained mobility to such an extent that he could return to the army where he became a sergeant instructor. Manez compensated for his remaining naming problems by reading from lists and instruction booklets. It is interesting to know that Larrey consulted with the famous Gall on several of his patients and that Gall included some of these patients in his own works to support these ideas [12]. The skull of Manez was placed in what is now the museum of Natural History in Paris (see Fig. 1.3).

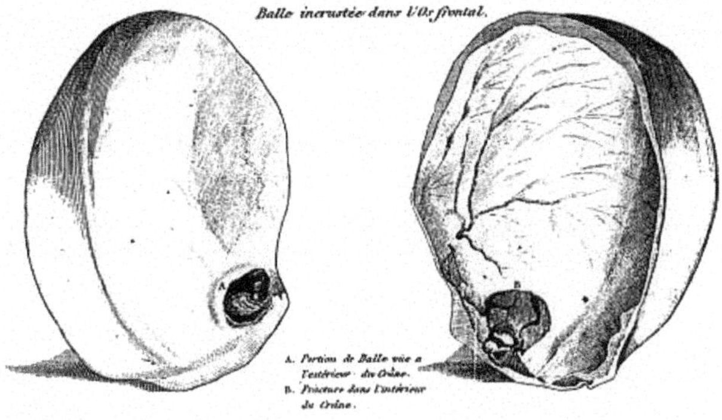

Fig. 1.3 The skull from 24-year-old soldier Louis Manez who suffered a gunshot wound and skull fracture in the left frontal region at Waterloo in 1815. The patient recovered from a hemiparesis but retained most of his language problems. Outside view (*left*) and inside view (*right*) (Figure taken from Jellinek, 2002)

1.3 Bouillaud and Broca

Paul Broca (1824–1880) is credited for the discovery of a language area that now bears his name. In his famous paper from 1861, he located 'the seat of the faculty of spoken language' in the third frontal convolution on the left side of the brain, near the coronary suture [13]. Modern neuropsychology and neurology mark this discovery as the beginning of the era of localization of function in the brain, but the story of Broca's area is much more complex and involves many other researchers. As Young (1990) begins in his chapter on Broca:

> Broca's localization of a centre for 'the faculty of articulate language' was the first localization of function in the hemisphere that met with general acceptance from orthodox scientists. Consequently, Broca is usually credited with priority in initiating the modern doctrine of cerebral localization. This citation has appeared with such regularity that this fact alone gains for it a species of historical truth. However, if one begins to examine his claim to priority, it is difficult to establish with any degree of certainty. His work is part of a continuous consideration of aphasia and cerebral localization that directly stems from Gall and was a live issue throughout the intervening decades. Neither the concept of a faculty of articulate language nor its localization in the frontal lobes was new [5].

Before Broca, several others had already demonstrated patients with speech deficits and focal lesions. According to Whitaker, 'literally hundreds' of case reports had been published in medical or phrenological journals from the 1820s onward, either with or without evidence from postmortem examinations [14]. Therefore, the practising medical doctor must have been aware of the circulating doctrine of a frontal lobe speech centre [15]. One of them was Jean-Baptiste Bouillaud (1796–1881), born a generation before Broca and to become professor of medicine and head of la Charité hospital in Paris. Bouillaud was one of the few remaining followers of Gall's phrenology in the academic world, but advocated a different and more scientific methodology: he was convinced that observation of the brain was always essential, and he sought correlation between clinical symptoms and brain lesions. Gall, as we have seen before, 'explicitly said that inspection of brains and "accidental mutilations" (pathological lesions) played a subordinate role in confirming localizations which he had discovered by his cranioscope methods' [5]. In 1825, Bouillaud published his *Clinical studies—showing that the loss of speech corresponds to a lesion of the anterior lobes of the brain—a confirmation of M. Gall's view regarding the seat of the organ of articulate speech* [16]. Gall had positioned the language faculty on the floor of the orbit in the lower part of the frontal lobes. He supposedly did this after initial observations made in childhood of classmates who could easily learn verbal material by heart and had 'large prominent eyes'. After Gall observed several others of these correlations, he concluded that the area for 'recollection of words or verbal memory' must lie directly behind the eyes and was prominently large in these subjects. Bouillaud was less specific on the exact location of the speech centre and argued on the basis of clinical evidence that loss of speech must be due to a lesion in the frontal lobes; he did not refer to convolutions nor to a specific hemisphere. He was a fierce opponent of theories of equipotentiality and debated amongst others with Flourens.[b]

> No physician who is in the least familiar with clinical studies has failed to observe many defects in locomotor functions produced by an illness of the brain. The inflammation causes spasmodic movements, cerebral compression, more or less widespread paralysis. It is therefore not without a good deal of astonishment that we read in the works of M. Flourens (…) that the brain exerts no immediate and direct influence on the muscular system [17].

Bouillaud proposed not only centres in the brain for movements of the limbs but for all organs that related to muscular movements, such as the tongue and the eyes. He deduced from case studies that 'paralysis of the speech organs can exist independently from other paralysis' [17]. He also made a distinction 'between the ability to produce words as signs of ideas while preserving their memory—and the ability to articulate these same words. There exists, as it were, an internal and external speech the latter is only the expression of the former' [17]. This relates to our current clinical distinction of dysphasia and dysarthria. Bouillaud even presented remarkable in vivo evidence when he had the opportunity to study the brain of a patient with a gunshot wound and a skull defect over the frontal lobes.

> Curious to know what effect it would have on speech if the brain were compressed, we applied to the exposed part a large spatula pressing from above downwards and a little from front to back. With moderate pressure, speech seemed to die on his lips; pressing harder and more sharply, speech not only failed but a few words were cut off suddenly [17].

So why is it that modern neuropsychology considers theories of cerebral localization to begin in 1861 with the observations of Broca and not, for instance, with those of Bouillaud at some earlier time? If one reads Broca's original paper, it is rather remarkable that it made such a historical impact. But, as always, one has to take the circumstances into account, and in 1861, localism was reconsidered more with an open mind than was done in the decades before. In that year, the Societé

[b]In 1848, Bouillaud famously offered 500 francs to anyone who could present him with a patient with a deep lesion in the frontal lobes without speech deficits.

d'Anthropologie (of which Broca was founder and secretary) held a series of debates on localization of function with Gratiolet arguing in favour of holism and Auburtin (pupil and son-in-law of Bouillaud) in favour of localism. During the debate, Auburtin promised to abandon his belief in cerebral localization if anyone could produce a case of loss of speech without a lesion in the anterior lobes of the brain. Precisely at that time, in April 1861, the patient Leborgne had come under the surgical attention of Broca because of a gangrenous infection of his right lower extremity. The 51-year-old Leborgne had already been in the Bicêtre hospital for 21 years because of speech problems and a hemiparesis. Despite his epilepsy, he had been a miller until at the age of 30 'he lost the ability to speak' [13]. Broca was unable to discover retrospectively whether this loss of speech had an acute or a more gradual onset and whether there were accompanying symptoms at the time. Leborgne was known as 'Tan' because the only thing he said in response to questions was 'tan tan', accompanied by 'varying movements with which he was able to express most of his ideas'. He quickly got angry when people did not understand him and was thought to be egoistical, vengeful and mean according to Broca's case report. Ten years after he lost his speech, Leborgne gradually developed a right-sided hemiplegia, starting with the muscles of the arm and gradually spreading to involve his leg and making him bed-ridden. The exact diagnosis has never been made and remains something of a riddle; Broca speaks of a 'chronic and progressive softening'. Broca remarks that the examination of this patient was difficult because of his inability to speak and to move the right side of his body; furthermore, Leborgne was 'in such a perilous state that it would have been cruel to have tortured him with long examinations' [13]. Still, considerable neurological details are given in his paper:

> The tongue was absolutely free; it was not deformed in any way; the patient could move it in any direction and stick it out of his mouth. The two halves of the organ were of the same thickness. The difficulty in swallowing I mentioned, was due to a paralysis, which was beginning at the pharynx, and not due to a paralysis of the tongue, as it was hard to swallow the third time. The muscles of the larynx did not seem to be altered, the timbre of the voice was natural, and the sound the patient produced, to pronounce his monosyllables, was completely pure.
>
> The state of intelligence could not exactly be determined. It is certain that Tan understood almost everything that was said to him; but as he could only express himself by moving his left hand, our dying man could not make himself understood as well as he could understand others. The numerical answers were the ones he did the best with, by opening or closing his fingers. I asked him many times, for how long had he been sick? He would answer sometimes five days, sometimes six days. How many years had he been at the Bicetre? He opened his hands four times in a row and added the rest with a single finger; this came out to twenty-one years, and as we have seen above, this information is absolutely exact.
>
> I showed him my watch two days in a row (…) he could indicate the exact time every time. It is indisputable that this man was intelligent, that he could think, and that he had maintained, to a certain extent, the memory of old things [13].

Broca invited Auburtin to see his patient 'above all in order to know what his diagnosis might be and if he would accept the outcome of this observation as conclusive' [17]. Broca said that he gathered the case history 'with the greatest care because it seemed to serve as a touchstone for the theory of my colleague' [17]. Broca himself had no doubt that from the clinical and neurological information it was clear that there existed a progressive cerebral lesion that had gradually spread but initially had to be 'quite bounded and (…) had not attained the organs of motility, nor the sensory organs' [13]. Sensorimotor functions were at that time not considered to be located on the convolutions, something that was only reserved for 'intellectual' functions such as speech or reasoning. This strengthened Broca in his opinion that the loss of speech was not a motor disorder. Broca considered the striate body the closest motor organ to the anterior lobes and he concluded that 'the probable diagnosis was thus original lesion on the left anterior lobe, then spreading to the striate body on the same side' [left, because the paralysis was on the right] [13]. The goal of the autopsy now was to determine, if possible, the seat of the first lesion. Auburtin took up the invitation and agreed that from the clinical and neurological examination, the lesion must have started on one of the anterior lobes. Because Leborgne died within a week after he was admitted to the surgical ward, the case was inserted in the meeting on April 18.

> M. Broca presented the brain of a fifty-one-year-old man who had died [on the previous day] in his service at the hospital Bicetre. (…) As it is planned to deposit the specimen at the Musee Dupuytren and to publish the complete records in the Bulletin de la Societe Anatomique, only a short resume will be given; the case is quite similar to some of those about which M. Auburtin has talked at the last meeting [17].

Broca himself was initially not very convinced of the principle of localization and made a brief statement without firm conclusions. Later that year, he published the case more extensively in his now famous paper 'Comments regarding the seat of the faculty of spoken language, followed by an observation of aphemia (loss of speech)' [13]. But only 4 years later, in 1865, when he had collected more cases, Broca felt that he could state more convincingly that the left frontal region was responsible for articulated speech [14].

Broca begins his 1861 paper by saying that his observations come in support of the ideas of Bouillaud and acknowledges his clinical–pathological work that locates lesions that can abolish speech in the anterior [i.e. frontal] lobes. The first part of the paper comments on contemporary theories and ideas of language in the brain; Broca explicitly states that language is not a 'simple faculty, dependent on only one cerebral organ' and distinguishes the 'faculty of spoken language' from that of the 'general faculty of language'. The latter faculty defines a constant relationship between 'an idea and a sign, be it a sound, a movement, a picture, or whatever sort of sketch' [13]. In addition to the language organ, Broca speaks of organs that are responsible for emitting and receiving information (given examples are, respectively, the tongue and the ear). Next, he gives the name 'aphemia' to all patients that are missing the faculty to articulate words and have a restricted vocabulary with only a small number of articulated sounds. He notes that these patients have no problems with vocalization per se and can 'easily emit vocal sounds; they move their tongues and lips, producing movements that are much more expansive and energetic than it would take to articulate sounds' [13]. These patients also 'perfectly understand spoken and written language' leading Broca to conclude that the faculty of language itself must be intact [13]. There is a lengthy discussion in the paper about the nature of the faculty for spoken language and its place in the 'cerebral hierarchy'. Broca speaks of two hypotheses and considers aphemia either a 'higher and intellectual disorder' or 'no more than a disorder of locomotion' [13]. He favours the first option but does not rule out the alternative one.

Broca asks himself to what extent the principle of localization, which he considers extremely probable in itself, is applicable. He refers to studies of comparative anatomy in 'normal, abnormal and pathological humans' from which it was known that the highest faculties must have their seat in the frontal convolutions (he names reasoning and thought), 'whereas the convolutions that are located on temporal, parietal and occipital lobes are responsible for emotions, for likes and for passions' [13]. The question of whether every particular faculty is located on a particular convolution seems 'quite unanswerable to me in the current state science is in' [13]. At that time, the convolutions had not received much attention.

> the classical works in anatomy have up till now, [have] not popularized the study of the cerebral convolutions, which thc phrenologists have unfortunately neglected as well. (…) We have let ourselves be dominated by this old prejudice, that the cerebral convolutions have nothing fixed about them, that they are simple folds, randomly produced, comparable to the disordered twists and turns in the loops of the intestine [13].

Broca acknowledges these shortcomings and refers to the more recent anatomical works of Gratiolet and Wagner for a description of the convolutions and their connections [13]. He also criticizes that descriptions of the location of lesions in the literature are usually very crude, being, for instance, 'so many centimetres from the big median fissure or from the fissure of Sylvius'.

The second part of the paper starts with the case history and neurological examination of the patient Leborgne, followed by a very detailed description of the postmortem findings. Broca describes that after removing a thick and vascularized dura mater, a fluid-filled cavity is found at the level of the left fissure of Sylvius, 'with a capacity of holding a chicken's egg' [13]. The cavity ends at the back up to the fissure of Rolando. The convolutions around the cavity show signs of chronic softening and atrophy, as do large parts of the left hemisphere.

> The posterior half of the third frontal convolution is completely destroyed in all its thickness; the second frontal convolution is affected a little less. At least two thirds of its external part has disappeared, and the remaining external third is extremely soft. In the back, the inferior third of the frontal transversal convolution is destroyed in all its thickness up to the fissure of Rolando [13].[c]

'The marginal inferior convolution' [the superior temporal gyrus] has been destroyed, as well as the convolutions of the insula and the anterior part of the striate body. In this way the cavity connects 'by a long opening, half a centimeter

[c] Note that Broca suggests here that the inferior part of the precentral gyrus was also lesioned. This implies that pathology had also affected the ventral premotor cortex, an area that is increasingly recognized in speech production.

long', to the lateral ventricle [13]. Broca remarks that the thalamus had retained its normal size and consistency and then stopped his examination of the deeper parts of the brain, 'in order not to destroy the specimen, which I consider very important to donate to the museum' [17]. Broca must have 'sensed its historical role', Schiller writes in his 1979 biography [17]. Schiller also remarks that there will remain an 'ever so slight doubt' about the identity of the preserved brain [17]. He himself eventually found it on a dusty shelf in the basement of the École de Médecine. With a sense of drama, he notes in his book that the number that refers to the jar differs from that of Broca's numbering in his own 1861 paper (and, as a matter of fact, from the number that is given by Marie in his critical papers of 1906) [18].

Broca was thus well aware of the extensive brain damage, not only at the cortical surface but also extending subcortically, although he was unaware of the exact extension because he stopped his dissection. Still he was very convinced that the damage must have started 'in the centre of the defect' and gradually have spread to other parts of the brain because of the slowly progressing course of the neurological disease with initially only speech disturbances [13]. This is a crucial assumption that is repeatedly stated in the paper. Broca eventually concludes that:

> the original seat of the lesion was in the second or third convolution, most probably the latter. It is thus probable that the faculty of spoken language is located in one or the other of the two convolutions; but we cannot know it yet, as previous observations are silent about the particular state of each convolution, and we cannot even theorize on it, since the principle of localization by convolution does not rest on a certain foundation yet [13].

Broca thus confirmed the opinion of Bouillaud, but indicated that further cases were necessary to know whether the speech centre was confined to a single and fixed convolution or resided in multiple convolutions or even an entire lobe. It seems that he was very lucky with the next case he encountered, that of Lelong, that strongly supported his idea that the third convolution must be important for speech. Broca himself writes that:

> in my second patient the lesion was rigorously occupying the same site as in my first; not only were the same convolutions affected but they were so at the same point, ie, immediately behind that middle third, opposite to the insula, and precisely on the same side (left) [17].

Still there were important differences between both cases. Lelong was 83 when he suddenly collapsed and had lasting and severe speech difficulties; he was left with only a few words and often tried to correct his inappropriate verbal responses with gestures. There was no hemiparesis. Because of a fractured leg, one and a half years after his stroke, Lelong came under the attention of Broca in the Bicêtre hospital. Twelve days later he died. During the autopsy, signs of haemorrhage were found as it was shown that 'microscopically (…) haematin crystals were present to account for the small orange-yellow colored patches in the wall of the lesion' [17]. The lesion was far more restricted than in the case of Leborgne; it only seemed to

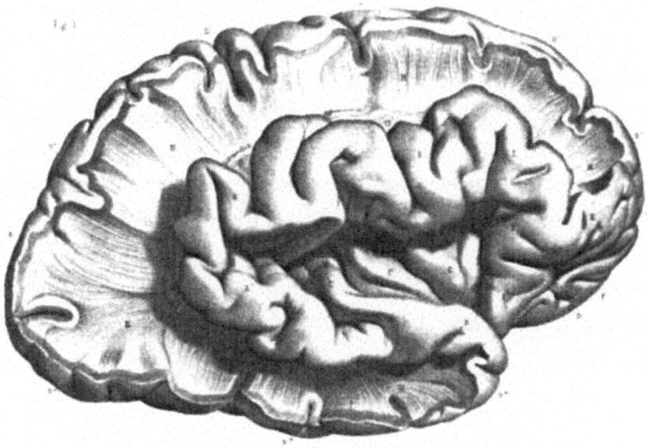

Fig. 1.4 Illustration of the brain by Foville (1844) showing the *circonvolution d'enceinte*, a continuous convolution that wraps around the Sylvian fissure [51] (Figure taken from Schiller, 1992)

occupy two of the frontal convolutions. Due to generalized atrophy, part of the insula was exposed, which seemed unaffected.

In the following 2 years, Broca and his colleagues collected another 16 cases that all showed left hemisphere damage [14]. All but one case involved the third frontal convolution. The exception was an aphemic patient of Charcot, where there was destruction of a left parietal convolution (the supramarginal gyrus) without any signs of frontal pathology. Because of this case Broca refrained from any bold conclusions on the exact cortical location of the speech centre but instead speculated that the seat of articulate speech might extend to the inferior parietal convolution. He referred to anatomical studies that described a continuous convolution around the Sylvian fissure (*circonvolution d'enceinte*, see Fig. 1.4). Again citing from the book of Schiller (1992), Broca takes back these initial speculative thoughts and replies to the Societé Anatomique:

> But this is all hypothetical; we must await further facts. (…) One negative fact does not destroy this series of positive ones; in pathology and especially in cerebral pathology, there is no rule without some exception [17].

1.4 Trousseau and Marie

Of course, 'exceptions' accumulated. Armand Trousseau (1801–1867), who was the first to coin the term 'aphasia' (that, for some reason, quickly replaced 'aphemia'), presented an astounding series of over a hundred patients with left hemisphere lesions. He had noted several cases that were discrepant with Broca's theory which led him to conclude that 'M. Broca's contention is less generally true than that of M. Dax and especially that of M. Bouillaud' [19]. In contrast to this mild response, Pierre Marie (1853–1940) was a lot less diplomatic [18]. He became one of the fiercest opponents of Broca's ideas and in particular of his 'dogma of the third convolution' [15]. Marie was a former intern at Broca's and Charcot's departments and the famous successor of Charcot. His critique appeared in 1906, long after Wernicke had proposed his theory of aphasia and several complicated language diagrams had already been introduced in the literature to explain the role of various brain regions and language disorders. Marie opposed these schemes and was convinced that aphasia 'was one' and could not be subdivided into different syndromes. On examination of the brains of his own patients, he had located the speech-related lesions at the posterior end of the Sylvian fissure (in or near what is now usually termed Wernicke's area). Marie re-examined the brain of Leborgne and was convinced that the original examination was inadequate and that Broca had wrongly inferred that the initial lesion was located on the third convolution. Marie considered Leborgne's aphasia to be the result of the destruction of the superior temporal gyrus and the supramarginal gyrus (the latter was not mentioned by Broca in his papers).

1.5 The Era of CT and MRI

We now know that the lesions in both Broca's patients, Leborgne and Lelong, extended beyond what is now most frequently considered as Broca's area (namely, the posterior part of the inferior frontal gyrus and more specifically its pars triangularis and pars opercularis). In 1980 the brain of Leborgne was scanned with CT by Castaigne and colleagues [20]. Even with the relatively low-quality scans of that time, it was obvious that the brain damage extended beyond the cortex alone. More recently, in 2007, Dronkers and co-workers made high-resolution MR images of both historical brains that allowed them to virtually dissect the brains and clearly visualize both the cortical and subcortical damage [21]. Before elaborating on the details of the MR scans, they started with visual inspection of the brains (Fig. 1.5). Much to their own surprise, they found that the most extensive damage to Leborgne's brain was in the middle third of inferior frontal gyrus (thus not strictly on the posterior half of the gyrus), whereas of Lelong's brain only the posterior part of the pars opercularis was damaged. So already on macroscopical inspection, the cortical lesions had only partly damaged the classic Broca's area and—more importantly—did not completely overlap.

So what did the MRI scans reveal that had not been visible before? In Leborgne's brain, there was (sub)cortical damage in the left inferior frontal gyrus and generally the whole left hemisphere. Imaging yielded specific lesions in the inferior parietal lobe and the anterior part of the superior temporal gyrus. The insula was 'completely destroyed'. There was extensive damage to the basal ganglia and internal capsule, as well as other white matter tracts: the superior longitudinal fasciculus (in particular the arcuate fasciculus), the periventricular white matter and the subcallosal fasciculus.[d] Although the exact functionality of these tracts is currently still under debate, there is nowadays good evidence that they play an important role in normal language function. The right hemisphere was unaffected and could therefore

[d] See also Thiebaut de Schotten (2015), who used an atlas of white matter connections (obtained from diffusion tractography) to estimate more precisely the subcortical damage in the brain of Leborgne [22].

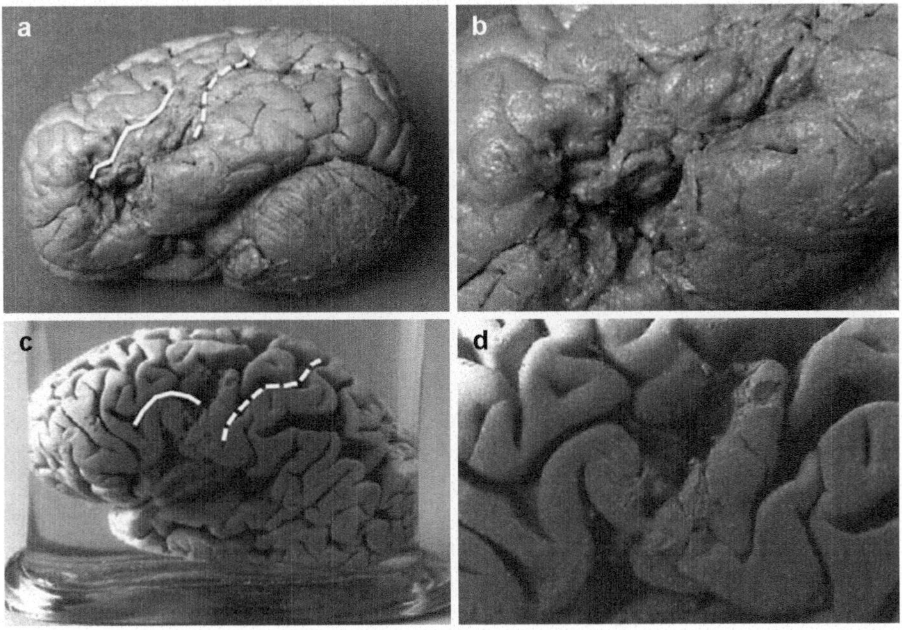

Fig. 1.5 Photographs of the brains of Leborgne (**a**, **b**) and Lelong (**c**, **d**) (Figure adapted from Dronkers et al., 2007 [21]). The central sulcus and the inferior frontal sulcus (IFS) are indicated with, respectively, *dashed and continuous lines*. Note that in the brain of Lelong, the IFS is shorter because of a long anterior horizontal ramus of the Sylvian fissure

serve as an 'excellent comparison' to the damaged left hemisphere. The left hemisphere was up to 50% smaller than the right one, and due to damage to identifying landmarks, it was 'difficult to tell from examining the 3D images or even the brain itself whether the supramarginal and angular gyri are affected' [21]. Of Lelong's brain, only the left hemisphere was preserved. In addition to the generalized atrophy and the damage in the pars opercularis that was known before, MRI showed some small lesions in the superior longitudinal fasciculus and in the white matter pathways of the temporal lobe.

So in these two cases there was overlapping damage in the most posterior part of the inferior frontal gyrus, but even more so in the subcortical areas, specifically in parts of the superior longitudinal fasciculus (SLF). The functional importance of white matter connections within the brain was not yet appreciated in Broca's time and only gradually gained clinical relevance at the end of the nineteenth century with publications from the so-called diagram makers. One may speculate whether the SLF was crucially involved in the 'aphemic' problems of both patients. Probably it was, in conjunction with the inferior frontal damage. Parts of the SLF are known to end in the lower parts of the motor cortex and the inferior frontal gyrus, and the SLF is nowadays related to phonological processing and articulation [23]. Modern lesion-deficit studies indicate that damage to the anterior part of the arcuate fasciculus is a key factor in patients with non-fluent aphasia [24].

The extensive damage found in Broca's historical cases fits our modern concepts that only relatively large areas of cortical and subcortical damage of the left inferior frontal lobe result in significant and lasting language deficits such as those described by Broca. Conversely, when only the classic Broca's area is lesioned, language problems are usually mild and transient and do not resemble Broca's aphasia. These observations stem from more detailed analyses that were initiated in the 1970s by Mohr and others [25]. They gained weight with the introduction of neuroimaging techniques (CT) that allowed for a better identification of the damaged brain areas [26]. This led Mohr to write in 2006 that:

> In retrospect, had Broca emphasized the extent of the lesion topography in his two cases, he might have prevented over a century of controversy [27].

It is important to realize that all lesion-deficit studies implicitly assume that there is an invariant relationship between anatomy and function. Such an assumption, namely, that everyone has the *same anatomical brain*, was already a crucial starting point for Broca's comparative studies. But these assumptions are not true; we know now. Even in healthy subjects, there is a significant intersubject variability in gyral and sulcal patterns. In addition to that, there is good evidence that the cortical representation of functions may change over time due to the influence of pathological processes (such as tumours or arteriovenous malformations). The early CT studies did a better job than the postmortem studies, also because at that time language deficits were disentangled in more detail than had been done before. For example, Alexander and colleagues (1990) studied nine patients with aphasia and lesions in the left frontal operculum [28]. They found that lesions in the lower part of the motor cortex, the operculum and the adjacent and more deeper lying periventricular white matter all three led to a different complex of language disturbances. Several other studies have confirmed that larger lesions, such as are typically observed in stroke patients, damage multiple functional subsystems and that it is therefore virtually impossible to categorize the aphasias by lesion site alone [26, 28, 29]. This led Alexander and colleagues to end their paper with the remark that 'In any case, the analysis of the pathologic anatomy of aphasia is most profitably pursued through investigations of the distributed anatomy of individual performance deficits, not the syndromes' [28].

Newer MRI analysis methods avoid some, but not all, of the limitations of these traditional lesion-deficit studies [30, 31]. Bates and colleagues (2003) used MR scans to study the damaged areas in 101 patients with a left-hemisphere stroke and language impairments (Fig. 1.6). What makes MRI a powerful tool is its high spatial resolution (~ 1 mm) and the fact that the images are digital. Images of different patients can be compared and overlaid once they have been put in similar 'data space' (done by scaling and morphing every brain until it fits a standard brain). These brains can then be virtually cut up into thousands of small cubic elements (voxels), after which statistical analyses can be done. Bates and colleagues were interested in the areas that were most often damaged in patients with disturbances either in language production or comprehension. Surprisingly, these were not the classical areas. In patients with non-fluent speech, it was not Broca's area that

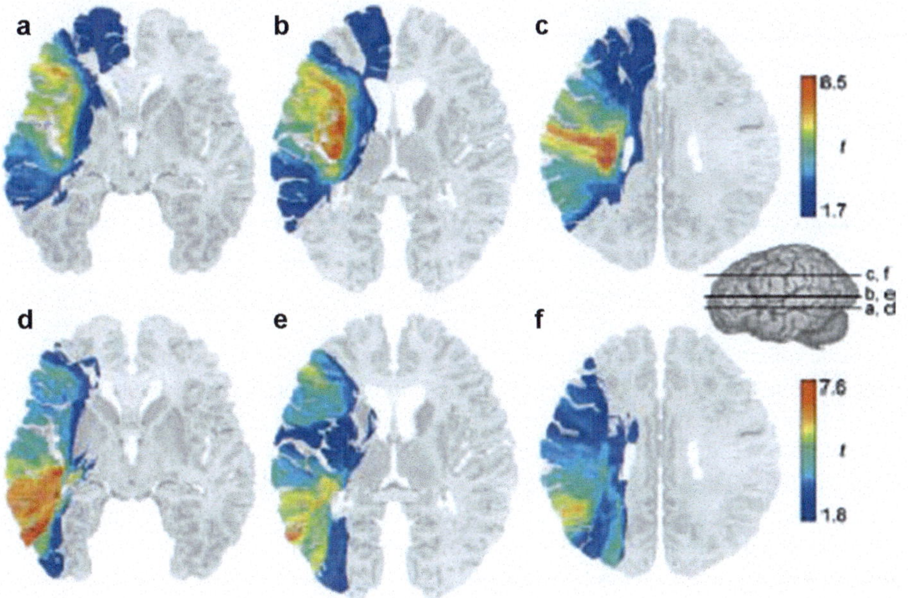

Fig. 1.6 'Voxel-based lesion symptom maps' showing the results of 101 stroke patients with disturbances of fluency (**a–c**) and auditory comprehension (**d–f**). They all had ischaemia of the left hemisphere. Hotspots (*red*) reflect the brain areas that are most frequently damaged in these patients. For fluency this is the anterior part of the insula and the subcortical parietal white matter. For auditory comprehension, this is the middle temporal gyrus, dorsolateral prefrontal cortex and the inferior parietal lobe (Figure taken from Bates et al., 2003)

scored best, but the anterior part of the insula and the parietal subcortex (specifically part of the superior longitudinal fasciculus/arcuate fasciculus). In a similar manner, auditory comprehension was most affected with lesions in the middle temporal gyrus, dorsolateral prefrontal cortex and the inferior parietal lobe, but not typical with lesions in Wernicke's area (see next chapter).

1.6 From Single Words to Sentences

For a long time, language studies focused mainly on processing of single words or objects, in particular in clinical studies. It was as late as the 1970s and 1980s, initiated by the seminal study by Caramazza and Zurif (1976), that theories gained momentum that patients with Broca's aphasia have difficulties with syntax and that this may cause their telegraphic style of speech and the lack of function words and inflections (i.e. agrammatical production) [32–35].

Up until that time Broca's area was thought to be basically a motor speech area. Even the agrammatic speech output of Broca's aphasics was thought, by prominent researchers, to reflect not a syntactic deficit but an economy of effort induced by the difficulty of articulating speech. (taken from www.talkingbrains.org, June 2010).

Interestingly, several authors at the beginning of the twentieth century had suggested that Broca's area was involved in grammar, notably Bonhoeffer (1902) [36], Heilbronner (1906) [37] and Salomon (1914) [15, 38]. Salomon was the first to suggest that agrammatic speakers could have problems with sentence comprehension [38]. When agrammatism regained interest in the 1970s, it was indeed confirmed that Broca's aphasics not only have difficulties with the production of speech but also with the understanding of complex sentences. They have particular problems when they have to derive the meaning of a sentence from only lexical–semantic information. For example, in a multiple-choice examination, patients with Broca's aphasia point to the correct picture when the sentence 'The worm that the bird eats is brown' is shown, because the semantic and pragmatic cues only leave one solution. However, the performance for sentences like 'The girl that the boy chases is tall' is at chance level (examples taken from reference [39]). So in fact these patients can be viewed both as agrammatic speakers and listeners, and this led to a breakdown of the classic production/comprehension dichotomy [39]. Caramazza and Zurif concluded in their 1976 paper:

> It appears that the presumed dissociation between language production and comprehension does not hold for Broca's and conduction aphasics: The present analysis of their comprehension skills suggests that such patients are as impaired in comprehension as they are in production. (…) With respect to neurolinguistic theories, the results are contrary to the view that Broca's aphasics have retained a normal tacit knowledge of their language. The present data together with the previously reported metalinguistic data (Zurif & Caramazza, 1975) suggest that, at least for the Broca's aphasics, brain damage affects a general language processing mechanism that subserves the syntactic component of both comprehension and production. The implication that follows is that the anterior language area of the brain is necessary for syntactic-like cognitive operations [35].

Afterwards, new lesion-deficit studies were published that weakened this claim, as it was found that many patients with Broca's aphasia still had fairly good grammatical insight. The syntactical disorders appeared to be restricted to certain domains. Thus, syntax could not be fully attributed to the area that was damaged in patients with Broca's aphasia [40]. The debate on the nature of the language deficits of patients with Broca's aphasia continues, and in the mean time functional imaging studies have contributed importantly to this discussion (see also Chap. 8) [32, 41–47]. Most scientists nowadays will assume that at least some form of syntactical processing is disabled in patients with Broca's aphasia. However, alternative explanations have also been proposed and are currently under investigation—for instance, that difficulties in sentence processing are in part related to disturbances in phonological working memory.[e]

Thus, modern lesion studies point to the fact that the older postmortem studies are even more wrong than we thought. It needs to be said that these modern studies

[e] For a good introduction to this controversy, see the paper of Rogalsky (2008) [32].

have their own, significant, drawbacks [48]. Despite all this and looking back 150 years, it is clear that lesion studies stood at the foundation of modern neurology and neuroscience. For some reason, the time was right for Broca's observations to gain momentum and acceptance, and he was recognized as the physician and scientist with whom cerebral localization started. What *is* remarkable and above all intriguing is that Broca's area today is still a widely used clinical concept, despite the devastating critique it has suffered over the years. This does not disqualify Broca's personal contributions—on the contrary—but emphasizes the fact that certain dogmas seem very much to prevail in clinical practice [49]. Most neurosurgeons still consider Broca's and Wernicke's areas as 'no-go' areas as opposed to all the other prefrontal or temporal areas that seem to be far more forgiving during surgical procedures [50]. Damasio and Damasio (2000) suggested that the classic neurological models have put a hold on 'the effort to map the language brain':

> The cartoon of the key brain structures required to receive and produce language -Broca's and Wernicke's areas, bridged by the arcuate fasciculus- was so attractive that the evidence in favor of a more varied and complex neural account was all but ignored. The view that Broca and Wernicke are *the* language centers pervades in most textbooks and monographs to our day [29].

As we will see in the next chapter, from Broca's time on, more and more language areas were 'discovered'. The important next step was the recognition that these areas were interconnected and formed a language network. Meynert and Wernicke paved the way for modern neurocognitive theories that consider functions to be sustained by networks, thereby spreading a function over several different brain areas.

References

1. Bennett CM, Baird AA, Miller MB, Wolford GL. Neural correlates of interspecies perspective taking in the post-mortem Atlantic Salmon: an argument for multiple comparisons correction. Organization for Human Brain Mapping; 2009.
2. Wells SR. How to read character: a handbook of physiology, phrenology and physiognomy, illustrated with a descriptive chart; 1888.
3. Amodio DM, Jost JT, Master SL, Yee CM. Neurocognitive correlates of liberalism and conservatism. Nat Neurosci. 2007;10:1246–7.
4. Kanai R, Feilden T, Firth C, Rees G. Political orientations are correlated with brain structure in young adults. Curr Biol. 2011;21:677–80.
5. Young RM. Mind, brain and adaptation in the nineteenth century. New York: Oxford University Press; 1970.
6. Gall FJ. Schreiben über seinen bereits geendigten Prodromus über die Verrichtungen des Gehirns der Menschen und der Thiere, an Herrn Jos. Fr. von Retzer. Der neue Teutsche Merkur. 1798;3:311–32.
7. Tizard B. Theories of brain localization from Flourens to Lashley. Med Hist. 1959;3:132–45.
8. Pearce JM. Marie-Jean-Pierre Flourens (1794-1867) and cortical localization. Eur Neurol. 2009;61:311–4.
9. Dennis W. Readings in the history of psychology. New York: Appleton-Century-Crofts; 1948.
10. Rawson NR. Early steps in cerebral localisation. Newcastle Med J reprint 1927.

11. Jellinek EH. An unlikely aphasiologist: D J Larrey (1766-1842). J R Soc Med. 2002;95:368–70.
12. Gall FJ, Spurzheim JG. Anatomie et physiologie du systeme nerveux en general et du cerveau en particulier avec des observations sur la possibilite de reconnatire plusieurs dispositions intellectuelles et morales de l'homme et des animaux par la configuration de leur tàtes. Paris: Schoell, F.; 1810.
13. Broca P. Remarques sur le siège de la faculté du langage articulé, suivies d'une observation d'aphémie (perte de la parole). Bulletins de la Societe Anatomique de Paris. 1861;6:330–57.
14. Stemmer B, Whitaker HA. Handbook of neurolinguistics. New York: Academic Press; 1998.
15. Tesak J, Code C. Milestones in the history of aphasia: theories and protagonists: Psychology Press; 2008.
16. Bouillaud J. Recherches cliniques propres a démontrer que la perte de la parole correspond a la lésion des lobules antérieurs du cerveau, et a la confirmer l'opinion de M. Gall, sur le siège de l'organs du langage articule. Arch Gen de Med. 1825;8:25–45.
17. Schiller F. Paul Broca: founder of French anthropology, explorer of the brain. New York: Oxford University Press; 1992.
18. Marie P. Revision de la question de l'aphasie: la troisième circonvolution frontale gauche ne joue aucun rôle dans la fonction du language. Semaine Med. 1906;26:241–247.
19. Trousseau A. De I'aphasie. Clinique medicinale de l'Hotel-Dieu de Paris. 1865;2:571–626.
20. Castaigne P, Lhermitte F, Signoret JL, Abelanet R. Description et etude scannographique du cerveau de Leborgne (la decouverte de Broca). Rev Neurol. 1980;136:563–83.
21. Dronkers NF, Plaisant O, Iba-Zizen MT, Cabanis EA. Paul Broca's historic cases: high resolution MR imaging of the brains of Leborgne and Lelong. Brain. 2007;130:1432–41.
22. Thiebaut de Schotten M, Dell' Acqua F, Ratiu P, et al. From Phineas Gage and Monsieur Leborgne to H.M.: revisiting disconnection syndromes. Cereb Cortex. 2015;25:4812–27.
23. Duffau H. The anatomo-functional connectivity of language revisited. New insights provided by electrostimulation and tractography. Neuropsychologia. 2008;46:927–34.
24. Fridriksson J, Guo D, Fillmore P, et al. Damage to the anterior arcuate fasciculus predicts non-fluent speech production in aphasia. Brain. 2013;136:3451–60.
25. Mohr JP, Pessin MS, Finkelstein S, et al. Broca aphasia: pathologic and clinical. Neurology. 1978;28:311–24.
26. Schwartz MF. What the classical aphasia categories can't do for us, and why. Brain Lang. 1984;21:3–8.
27. Mohr JP. Broca's area and Broca's aphasia. In: Grodzinsky Y, Amunts K, editors. Broca's region. Oxford: Oxford University Press; 2006.
28. Alexander MP, Naeser MA, Palumbo C. Broca's area aphasias: aphasia after lesions including the frontal operculum. Neurology. 1990;40:353–62.
29. Damasio AR, Damasio H. Aphasia and the neural basis of language. In: Mesulam MM, editor. Principles of behavioral and cognitive neurology. New York: Oxford University Press; 2000. p. 294–315.
30. Bates E, Wilson SM, Saygin AP, et al. Voxel-based lesion-symptom mapping. Nat Neurosci. 2003;6(5):448–50.
31. Halai AD, Woollams AM, Lambon Ralph MA. Using principal component analysis to capture individual differences within a unified neuropsychological model of chronic post-stroke aphasia: revealing the unique neural correlates of speech fluency, phonology and semantics. Cortex. 2017;86:275–89.
32. Rogalsky C, Matchin W, Hickok G. Broca's area, sentence comprehension, and working memory: an fMRI study. Front Hum Neurosci. 2008;2:14.
33. Goodglass H. Studies on the grammar of aphasics. In: Rosenberg S, Koplin J, editors. Developments in applied psycholinguistics research. New York: MacMillan; 1968.
34. Gleason JB, Goodglass H, Green E, et al. The retrieval of syntax in Broca's aphasia. Brain Lang. 1975;2:451–71.
35. Caramazza A, Zurif EB. Dissociation of algorithmic and heuristic processes in language comprehension: evidence from aphasia. Brain Lang. 1976;3:572–82.

36. Bonhoeffer K. Zur Kenntnis der Rueckbildung motorischen Aphasien. Mitteilungen aus der Grenzgebieten der Medizin und Chirurgie. 1902;10:203–24.
37. Heilbronner K. Ueber Agrammatismus und die Störung der inneren Sprache. Arch Psychiatr Nervenkr. 1906;41:653–83.
38. Salomon E. Motorische Aphasie mit Agrammatismus und sensorisch-agrammatischen Störungen. Monatschtift für Psychiatrie und Neurologie. 1914;35:181–208, 216.
39. Segalowitz SJ (ed.). Language functions and brain organization. Academic Press; 1983
40. Grodzinsky J. The neurology of syntax: language use without Broca's area. Behav Brain Sci. 2000;23:1–71.
41. Benson DF, Ardila A. Aphasia: a clinical perspective. New York: Oxford University Press; 1996.
42. Rogalsky C, Hickok G. Selective attention to semantic and syntactic features modulates sentence processing networks in anterior temporal cortex. Cereb Cortex. 2009;19:786–96.
43. Caplan D, Alpert N, Waters G. Effects of syntactic structure and propositional number on patterns of regional cerebral blood flow. J Cogn Neurosci. 1998;10:541–52.
44. Caplan D, Alpert N, Waters G. PET studies of syntactic processing with auditory sentence presentation. NeuroImage. 1999;9:343–51.
45. Just MA, Carpenter PA, Keller TA, et al. Brain activation modulated by sentence comprehension. Science. 1996;274:114–6.
46. Mazoyer BM, Tzourio N, Frak V, et al. The cortical representation of speech. J Cogn Neurosci. 1993;5:467–79.
47. Vandenberghe R, Nobre AC, Price CJ. The response of left temporal cortex to sentences. J Cogn Neurosci. 2002;14:550–60.
48. Mah YH, Husain M, Rees G, Nachev P. Human brain lesion-deficit inference remapped. Brain. 2014;137(Pt 9):2522–31.
49. Uylings HB, Malofeeva LI, Bogolepova IN, et al. Broca's language area from a neuroanatomical and developmental perspective. In: Brown CM, Hagoort P, editors. The neurocognition of language. Oxford: Oxford University Press; 2000. p. 319–36.
50. Benzagmout M, Gatignol P, Duffau H. Resection of World Health Organization grade II gliomas involving Broca's area: methodological and functional considerations. Neurosurgery. 2007;61:741–52.
51. Foville AL. Traité complet de l'anatomie, de la physiologie et de la pathologie du système nerveux cérébro-spinal. Paris; 1844.

2 Wernicke and Connectionism

Roughly a decade after Broca published his findings on the left frontal speech area, in 1874 Carl Wernicke (1848–1905) published a now famous monograph on different clinical forms of aphasia, *The Symptom Complex of Aphasia: A Psychological Study on an Anatomical Basis* (*German: Der aphasische Symptomencomplex: Eine Psychologische Studie auf Anatomischer Basis*) [1]. Wernicke is particularly remembered for his description of sensory aphasia and for his wiring diagram that eventually grew to become the classical language model in neurology. He noted that patients with a lesion of the left temporal lobe had comprehension difficulties but, as opposed to Broca's aphasics, had fluent speech with a relatively intact vocabulary. As in the case of Broca, the temporal language area was eventually named after Wernicke in honour of his discovery. The posterior part of the superior temporal gyrus is generally considered the core of Wernicke's area, but the area itself has always remained poorly defined in anatomical terms. Wernicke was not the first to write about aphasia with comprehension disorders. Bastian (1869) and Schmidt (1871) did before him, but they did not provide autopsy results or any underlying anatomical framework, and their papers failed to gain the attention of the medical community [2–4].

From what one reads about Wernicke in medical textbooks or on the Internet, it is easy to get the impression that he, like Broca, was a strict localist and that their two language areas nicely complement each other and form a comprehensive and complete language model: Broca's area for the productive/motor part of language and Wernicke's area for the receptive/sensory part of it. However, Wernicke did not plead for strict localization of function to one area; he was in fact one of the first 'connectionists', holding that multiple areas are necessary in the orchestration of a given brain function. He himself wrote in 1874 that:

> only the most elementary psychic functions can be assigned to defined areas of the cortex [examples are given of visual, olfactory and tactile perception] (…) Everything that goes beyond these simplest functions, the association of different impressions into a concept, thinking, consciousness, is an achievement of the fiber tracts which connect the different regions of the cortex to each other, the so-called association system of Meynert [1].

G.-J. Rutten, *The Broca-Wernicke Doctrine*, DOI 10.1007/978-3-319-54633-9_2

2.1 Meynert

Wernicke was not the first to suggest such ideas. Most of the credit is probably due to Theodor Meynert (1833–1892), Wernicke's teacher in neuroanatomy, who made several important contributions and improved histological methods to study the structure and function of the connecting fibre systems [5]. Wernicke repeatedly acknowledges Meynert, for instance, at the beginning of his 1874 monograph: 'the theory put forward here follows almost automatically from the study of Meynert's writings and dissections' [1]. Although the first descriptions of fibre bundles date back to Vicq d'Azyr (1786) and Gall (1810), Meynert described them in detail and differentiated between projection fibres (connecting cortical areas to subcortical parts of the brain) and association fibres (interconnecting cortical areas) [6]. By following the tracts, he was able to deduce that the posterior part of the brain (posterior to the central sulcus) was 'sensory' in function and the anterior part 'motor'. Wernicke adopted the sensorimotor division of Meynert, although he was unsure of the nature of the parietal lobe which he described as an intermediate area of still conflicting functions.[a] Meynert had also published an important work in which he demonstrated that the auditory fibres terminated in the cortex of the Sylvian fissure, and he assumed this area to be a 'sound-field' (German: Klangfeld) [7]. This view was supported by autopsy findings of lesions in the insular and the perisylvian region. Eminent authors and aphasiologists have suggested that Meynert's historical role has been underestimated [8]. Geschwind wrote in 1974 that:

> the significance of Meynert's contribution to the study of aphasia has generally been overlooked. In the discussions of the history of aphasia that I have so far consulted, he is mentioned only as Wernicke's teacher. (…) It is important to appreciate that it was apparently he who first convincingly showed that aphasia could occur in temporal lobe lesions. It was Wernicke's function to complete the cycle started by Meynert by showing how this fact achieved meaning in terms of Meynert's own studies of the brain, and to point out the important fact that this new localization corresponded to a clinical picture different from that seen in lesions of Broca's area [9].

Wernicke wrote his monograph when he was just 26. It was his first work on aphasia and also the one that became his most well known. In particular, his concepts of clinical aphasia and his connectionist thinking made a lasting impact and are still the subject of lively debate. Wernicke made several other important contributions in the field of neurology and psychiatry, but it is not widely known that he continued writing and reviewing articles on aphasia and that he refined and even changed some of his opinions in later works. The basis of his work on aphasia was clearly laid in his monograph, however, and therefore I will describe it in more detail in the rest of this chapter.

[a] In a footnote of his monograph, Wernicke referred to recent findings of Hitzig who had elicited movements while stimulating cortex that was located *posteriorly* to the central sulcus [1].

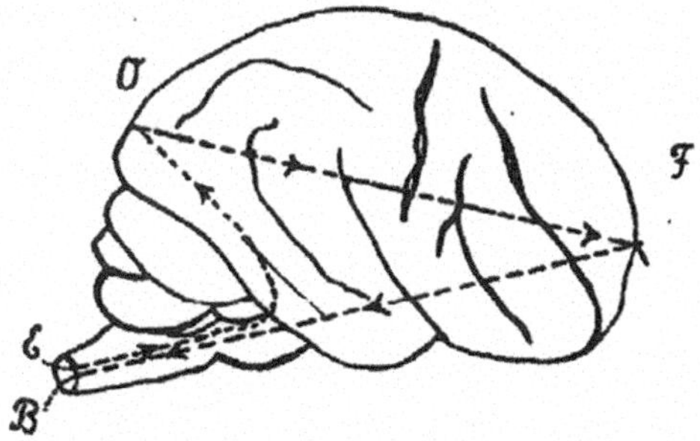

Fig. 2.1 Wernicke's elementary scheme to explain the representation of cognitive functions in the brain. Any sensory stimulus (E, *Empfindung*) can leave a memory image (O, *Erinnerungsbild* or *Empfindungsrest*) in the sensory part of the brain. Movement images (F, *Bewegungsbild*) are in a similar manner represented in the anterior part of the brain and subsequently can lead to movements (B, *Bewegung*). The 'reflex arc' is completed by (variable) interconnections between sensory and motor areas (OF) (Figure taken from Wernicke, 1874 [1])

2.2 The Symptom Complex of Aphasia, Part I

Wernicke's monograph consists of three parts. In the first part (pp. 3–12), a general theory is formulated for the representation of cognitive functions in the brain. Wernicke introduces a scheme that is the basis for all of his language models and is in its most elementary form a 'reflex arc' (see Fig. 2.1). In this model, any sensory stimulus (E, *Empfindung*) is able to leave a memory image (O, *Erinnerungsbild* or *Empfindungsrest*) in the sensory part of the brain. Wernicke assumed that 'molecular changes' in brain cells then consolidate the effects of the brief peripheral stimuli [1]. Bodily movements or changes in musculature leave memory images in a similar manner. These are termed movement images (F, *Bewegungsbild*) and are located in the motor (anterior) part of the brain. Sensory and motor images are connected by associative fibre systems of which the strength of the connections is variable. Wernicke speaks of 'resistance' that is 'lowered' once a trajectory is used more often [1]. This resembles the Hebbian principles that were only to be formulated in the middle of the twentieth century. Wernicke's model also resembles the contemporary view on brain functional organization [10]. Take, in an example from his later work, the way in which Wernicke describes the representation of an object, a bell:

> The memory images of a bell (…) are deposited in the cortex and located according to the sensory organs. These would then include the acoustic imagery aroused by a bell, visual imagery established by means of form and color, tactile imagery acquired by cutaneous sensation, and finally, motor imagery as gained by exploratory movements of the fingers and the eyes.[b]

[b]Quotation taken from the paper of Gage and Hickok (2005); see their detailed comparison of Wernicke's views with current neuroscientific ideas of representation of brain functions [11].

A concept thus arises when different brain areas work in unison. The 'knowledge' of any object (made up by the combined properties of it) is thereby distributed over a number of brain areas and is thus not strictly localized. Because all different functional units are interconnected, it is possible to activate the entire 'network' by activation of part of it. Again quoting Wernicke from the paper of Gage and Hickok (2005):

> Close association between these various memory images has been established by repeated experience of the essential features of bells. As a final result, arousal of each individual image is adequate for awakening of the concept as a whole. In this way a functional unit is achieved. Such units form the concept of the object, in this case a bell.

Any movement image is linked to multiple interconnected memory images that can have different locations in the brain. Vice versa, there is usually more than one movement image 'active' (German: *gemeinschaftlich ins Bewusstsein gerufen*) [1]. Wernicke thus describes a functional topography that is much more complex than is shown in his illustrations. Cognitive functions were to him represented in mosaic-like association systems. The figures that are depicted in his monograph were intended as schematic illustrations to make his point; Wernicke himself referred to them as 'schemes' [12]. Unfortunately, these illustrations have been interpreted far too literally in the history of neuroscience and neurology. It is certainly ironic that Wernicke's language model has been used to advocate a strict localist view on language representation, whereas Wernicke himself had a different and much more sophisticated vision. His illustrations deliberately lack detailed anatomical information, as anyone familiar with Wernicke's work will agree. Wernicke devoted a large part of his scientific career to the study of anatomy and neuropathology and was well trained in anatomical dissections. In his 1874 monograph, for instance, he describes stepwise dissection of the temporal lobe and insula. Figure 2.2 shows some of the illustrations from a paper on comparative anatomy he published just 2 years later, when he was still a resident in the Charité in Berlin. It is of interest, although maybe only from a historical point of view, to quote from the introduction of this 1876 paper where Wernicke still somehow seems to doubt the location of the language centre:

> In recent times, evidence has been presented for the fact that certain functions are bound to certain convolutions or regions. There is no doubt that the first frontal gyrus is the centre for speech movements; it is also likely that the first temporal gyrus is a sensory speech area [13].

In the period 1897–1903, Wernicke published the three-part *Atlas des Gehirns* (brain atlas), together with other eminent authors such as Foerster. The atlas included myelin-stained sections of the brain in coronal, axial and sagittal planes [14]. Wernicke identified several fasciculi, some of which were named after him [6]. Wernicke was convinced that anatomical development and architecture needed to be

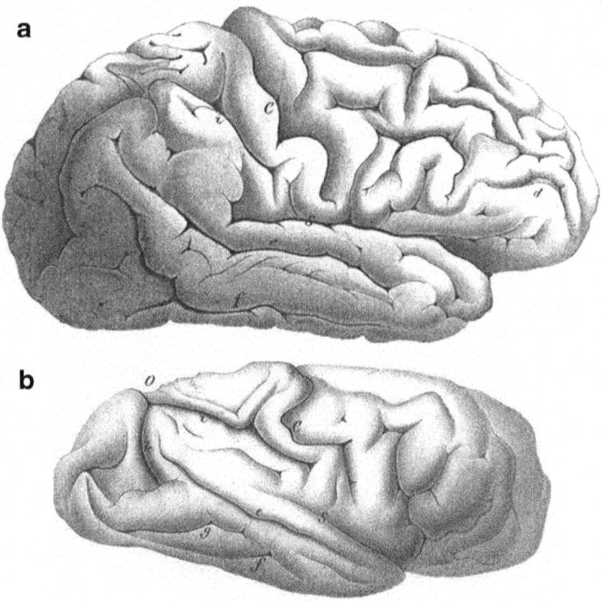

Fig. 2.2 Drawings of the brain of a human (*top*) and a chimpanzee (not to scale) (Figures taken from Wernicke's *The Primitive Convolution of the Human Brain*, 1876 [13])

studied in order to identify normal from abnormal brains and to understand the neural basis of higher cortical functions. He rejected the idea that the convolutional pattern was random or circumstantial.

2.3 The Symptom Complex of Aphasia, Part II

In the second part of the monograph (pp. 12–38), Wernicke describes his theory of aphasia by giving a detailed description of the clinical profiles that result from lesions at five different locations in the psychic reflex arc of his language model (shown in Fig. 2.3A):

> Any interruption of the pathway aal bbl can cause aphasia. The clinical picture will however vary according to the portion of the path affected by the interruption [1].

This essentially forms the foundation of what became the classical neurological model of aphasia. It also set the framework for research on aphasia for at least the first half of the twentieth century. However, when Wernicke's doctoral thesis and his later work is read carefully, it becomes clear that he describes a much more modern view on language representation and organization than he is usually credited and remembered for. The rest of this section will be structured according to the five different lesions that Wernicke described and the subsequent research on each.

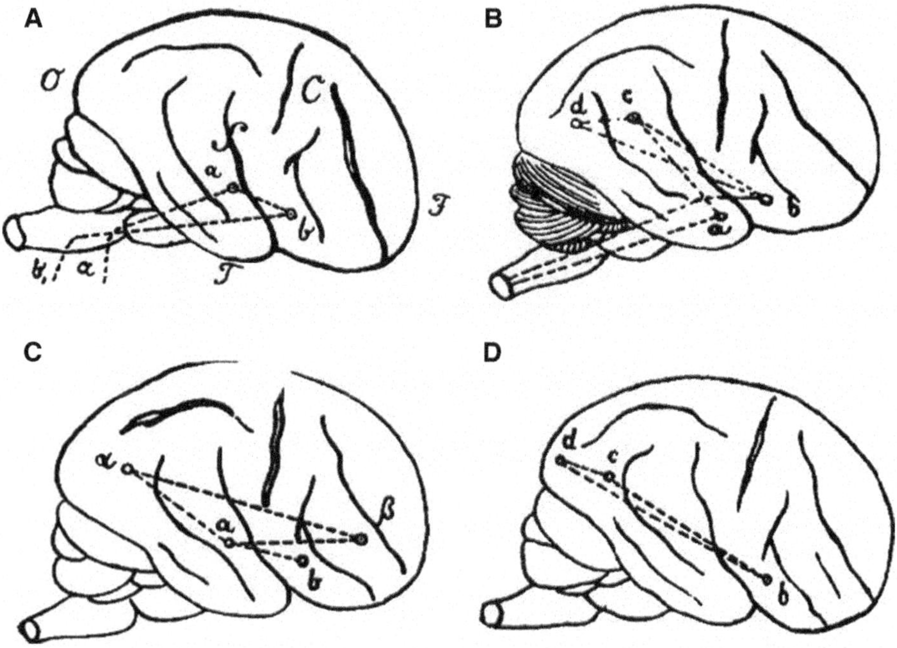

Fig. 2.3 Four different schemes that Wernicke used to explain his theory of aphasia, taken from his 1874 monograph. For unknown reasons these schemes are projected on the right hemisphere. The structure of the schematic brains is similar in all cases, showing three frontal and three temporal convolutions. The central sulcus (C) separates the precentral and postcentral convolution. Parietal and occipital regions are not shown in any detail. Note that the memory areas differ in location between the various models. (*Top left, A*) This is the well-known figure from medical textbooks that essentially forms the basis of classical neurological language models. a is the acoustic nerve that in part terminates in the temporal sensory language area. This area (a) is connected to the frontal motor language area (b). (*Top right, B*) The model has been expanded with memory areas for tactile (c) and optical (d) images. 'Knowledge' of objects or words is diffusely stored in the network that is formed by areas a, c and d. Wernicke is not very specific on locations for these tactile and optical areas, other than that the optical memory areas are located in the posterior parts of the brain. Note also that area a has a different location within the various figures. (*Bottom left, C*) Specific areas for reading (α) and writing (β) have been added. Wernicke postulated that connections $\alpha\beta$ and ab in reality have an intimate anatomical relationship (not clear from the model). (*Bottom right, D*) Language model in subjects that are born deaf-mute, but who were able to acquire speech. According to Wernicke, this proves that there must exist direct connections from tactile (c) and optical (o) areas to the motor speech area b. In deaf-mute subjects, areas c and d form the first part of the 'reflex arc' (Figures taken from Wernicke, 1874 [1])

2.3.1 Lesion of the Acoustic Nerve

A lesion of the afferent tract to *a* (acoustic nerve) will lead to deafness without aphasia, but only, Wernicke writes, if language has already been learned. Children need the auditory input to be able to form auditory memory images. In a first phase, they acquire the words by repeating them. 'Begriff' arises in a later phase, when the word itself has already been mastered. Wernicke asks himself what would happen when a child is deaf in the right ear but still has a normal language development. Assuming that the right acoustic nerve terminates in the left hemisphere, and vice versa (although Wernicke explicitly states that this was still unsure at the time), he hypothesizes that in this case the right temporal lobe shall have the potential to acquire language functions. At a later part in his monograph, Wernicke elaborates on this in a more general manner and states that language is normally a function of the left hemisphere, but that the right hemisphere can take over language functions in case of left hemisphere pathology. In fact, in many of the case reports that are listed in the third part of his monograph, Wernicke describes recovery of aphasic symptoms over time. He thereby repeatedly describes a phenomenon that we now would call 'brain plasticity'. Similar suggestions were also made by Broca.

> In the course of the illness the other hemisphere could have taken over the function of the left temporal lobe, as indeed takes place very rapidly in sensory aphasia [1].

2.3.2 Lesion of the Auditory Memory Centre: 'Wernicke's Aphasia'

A lesion of the centre for auditory images (area *a*) will lead to 'sensory aphasia'. The resulting symptom complex later became known as Wernicke's aphasia, although Wernicke's own description is significantly different from some later definitions. The most important symptom of sensory aphasia, according to Wernicke, is that 'language is not understood, but hearing is demonstrably preserved' [12]. Words will be heard as meaningless noise [15]. The vocabulary is largely intact, but there is a 'confusion of words' [German: Verwechslungen der Wörter]. The term 'paraphasia' was not yet used and was later introduced by Kussmaul in 1877 [16]. According to Wernicke, the explanation for these (mild) speech problems is a defective mechanism via route *ab*. This prevents an unconscious correction of what has been spoken.

> It seems that during normal speech the sound image [Klangbild] is always unconsciously innervated, as is easily understandable in terms of the genesis of language. The sound image is simultaneously hallucinated, as it were, and therefore able to continuously correct the execution of movements [Bewegungsvorstellungen] [1].

Wernicke explains why patients are not deaf when area *a* is lesioned, and he does so by assuming that the acoustic nerve disperses over a wider area of the brain than just area *a* (again demonstrating that Wernicke's illustrations are only simplified schemes).

> Thus that part of the central projection of the acoustic nerve which contains word sounds [Wortklänge] may be destroyed while all noises or all musical tones may still be perceived [1].

Next, Wernicke explains in more detail what happens when area *a* is lesioned:

> If area a1, the cortex of the first temporal convolution, is destroyed, the sound images [Klangbilder] of the names of all possible objects will be extinguished from memory, although the concepts may still remain in their full clarity. For in most cases the sound image is of secondary importance for the concept of the object, whereas the somaesthetic sense images [Gefüls- und Tastsinnsbilder] are of critical importance for it [1].

When I first read this passage, I had difficulty understanding it, because I had never really thought about the difference between *hearing* a word and truly *understanding* its meaning. Just hearing a word does not necessarily imply that you grasp its meaning. Take, as an example, listening to pseudowords, e.g. vonk or bhutap. These words conform to phonological conventions (i.e. they sound like real words) but have no meaning. Therefore, these pseudowords do not trigger the associations in the brain network that normally lead to an understanding of an underlying meaning or concept. So Wernicke's auditory area *a* is needed to hear the words, but is more or less redundant for comprehension [Begriff] because this will only arise when several different memory areas are conjointly activated. Wernicke uses several illustrations (Fig. 2.3B) to visualize this concept and is now able to explain why destruction of one sensory memory area does not lead to a loss of understanding of the concept as a whole, as auditory memory images (*a*) are connected to associated tactile images (*c*) and optical images (*d*).

All three sensory modalities also conjointly determine which word is eventually selected by means of a weighed contribution of their inputs to area *b*. If one of the sensory areas is lesioned, 'innervation' of motor speech centre *b* is determined by the sum of the remaining other areas. This is fundamentally different to the classical interpretation of the model in Fig. 2.3A, whereby a lesion of area *a* would deprive area *b* of any input. Due to a parallel design, the system has inbuilt redundancies and is able to keep functioning despite a lesion of one of its subcomponents. Note that each sensory area has a similar architecture and itself consists of multiple interconnected areas. This is why Wernicke can speak of a 'partial' lesion of a particular sensory language centre, whereby the local extent of the lesion determines the characteristics and the severity of the aphasia [1]. In later publications, these ideas were further worked out, and his schemes also showed more connections between the various memory areas.[c] Wernicke acknowledges that the motor and sensory centres may not be completely independent and that spoken language production relies to some extent on the auditory word form for the purpose of speech monitoring [18]. A careful reading of Wernicke's original work thus reveals that he never proposed the serial and simple language model that later became one of the hallmarks of neurology.

There is a second important assumption that Wernicke makes in order to understand why a lesion of the auditory centre *a* will not induce lasting deficits in language comprehension [Begriff]. Wernicke states that area *a* and tract *ab* are very important in language comprehension, but only in childhood during language acquisition. In this phase, when language is learned, associative connections are formed between auditory and motor images.

> Now the association of sound images with representations of movement essentially has its value in the fact that it makes it possible to learn language [1].

[c] See Fig. 2.4, which appeared in a review of a paper by Lichtheim who introduced a language model that explicitly included a 'concept centre'; for Lichtheim's ideas, see also Chap. 4 [11, 17].

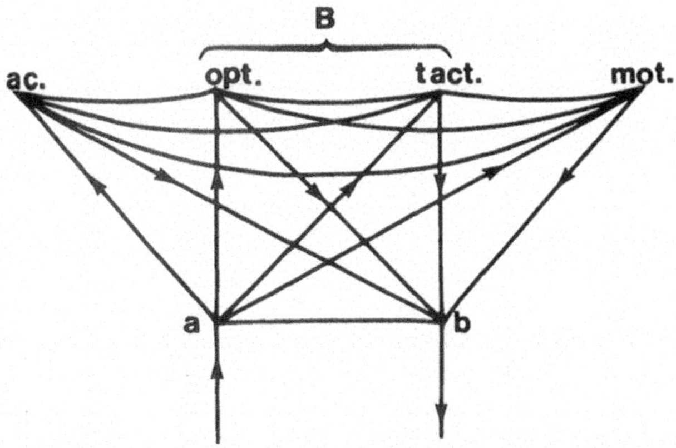

Fig. 2.4 Wernicke's model of how properties of words (ab) and the knowledge of objects [B, *Begriff*] are represented in the brain. B is the conceptual representation of an object. It is a construct of the association of acoustic (ac.), optical (opt.), tactical (tact.) and motor (mot.) images. The representation of the word (ab) that refers to an object consists of its auditory (a) and motor memory images (b) (Figure taken from de Bleser, 1996 [18])

Once these associative connections have been established, the tract *ab* is predominantly used for repetition and is no longer critically involved in comprehension anymore.

> Very soon after we have learned to speak a word, the intention of merely reproducing the sound disappears, making way for the intention of reproducing a specific meaning. The actual sensory images of an object [realen Sinnesbilder eines Gegenstandes], that is, are now able to innervate the representation of movement of a word [Bewegungsvorstellungen des Wortes] directly. (…) Later on, however, this pathway is no longer the one primarily used. Rather, the shorter pathways cb and db are chosen, and the mere existence of the pathway a1 b, without its being intentionally innervated, is sufficient to insure the choice of the correct representation of movement. [1]

So the areas that eventually sustain the information that is necessary to grasp the conceptual knowledge [Begriff] of any object are predominantly located outside the auditory language area *a*. This is stated quite boldly (and abruptly) in the monograph:

> The concept [Begriff] is nothing more than the connection cd. [1]

Wernicke thus argues that once we have learned what a word means, the auditory or visual image of the word itself does not contribute much to the *knowledge* of the object (as it is not an innate property of it). In his own words:

> The spoken and written name of an object is not a new attribute of the object. It is thus clearly different from the actual sensory memory images of the object. Only the latter make up the concept of the object. [1]

Using the example of the bell again:

> The concept of a bell, for example, consists of memory images of visual, tactile, and auditory perceptions that are connected (associated) with each other. These memory images are essential attributes of the bell. The spoken word 'bell', however, has nothing in common with the acoustic impression that a bell makes on us, and there is likewise not the slightest similarity between the written word 'bell' and the image of a bell. (…) It is necessary to keep them apart. [1]

To Wernicke, loss of memory images that critically impair conceptual knowledge [Begriff] leads to asymbolia and not aphasia. He refers to Finkelnburg's asymbolia, which he defines as 'the loss of any memory images essential to the concept of an object' [1]. Wernicke specifically adds that this can affect optical, tactile and also auditory images and also repeatedly stresses that language and intelligence are independent processes. He considers asymbolia a defect of intelligence and therefore not an aphasia:

> Disturbances of the concepts of the things with which we deal in the process of thinking are always disturbances of intelligence; disturbances of language, on the contrary, cause difficulties only in the use of the conventional means [eingeführten Verkersmittel] of representation of the concepts. [1]

For Wernicke, 'comprehension' occurs at two different levels, so to speak. At the more superficial level, there is processing of auditory information and selection of auditory memory images. Deficits at this stage are considered a language comprehension disorder and will lead to a form of sensory aphasia. Then there is 'comprehension' at a more fundamental and conceptual level, where knowledge itself is stored. In modern terms, we would define the asymbolia of Wernicke and Finkelnburg as a form of agnosia: a *not knowing* or *not recognizing* of specific sensory stimuli. In later papers, Wernicke was more explicit on how these different concepts related. In fact, Wernicke was an expert on this topic, as he was the mentor of authors who wrote seminal papers on agnosia (see also Chap. 3). In a paper published in 1886, Wernicke wrote:

> It may be helpful to bear in mind that more precisely two activities must be differentiated in language comprehension. In the first step, the concept of the word is activated, in the second one, the concept of the corresponding object. The process is similar in spontaneous speech, but in the reverse order, so that first the concept of the object arises and then that of the word. [18]

2.3.2.1 A Modern Definition of Wernicke's Aphasia

In Mesulam's *Principles of Behavioral and Cognitive Neurology*, Damasio describes the key features of a modern definition of Wernicke's aphasia [19]. There is fluent and often unintelligible speech, and patients have difficulty with selection of the words (resulting in semantic paraphasias) or with the order of sounds (resulting in phonological paraphasias). In addition to that, there are important problems with auditory comprehension of sentences. Damasio describes that Wernicke's area itself is not the centre where auditory comprehension takes place, in terms quite similar to Wernicke's own formulations in his monograph:

> We see it [Wernicke's area] as a processor of speech sounds which recruits auditory inputs to be mapped as words, and to be used subsequently to evoke concepts. Auditory comprehension in the proper sense occurs later in a chain of events initiated in Wernicke's area, when the concepts that are pertinently associated with a given word's records become activated and attended. The process of auditory comprehension involves numerous cerebral cortices of varied sensory modalities as well well as higher-order cortices distributed over parietal, temporal, and frontal regions. [20]

2.3.2.2 Agraphia and Alexia

Wernicke associated sensory aphasia with agraphia, because he considered writing to be intimately linked to sound and speech: 'Writing is a voluntary movement learned in close dependence on sound and always executed under the guidance of sound' [1]. Whether or not there are reading difficulties—according to Wernicke—depends on the education of the patient. He assumed that unskilled readers needed to hear themselves read aloud in order to comprehend the text, and consequently had an associated alexia when auditory area *a* was damaged. Skilled readers, in contrast, did not need any auditory feedback and thus did not necessarily have alexia. Wernicke states that lesions outside of the sensory language area can also lead to alexia and agraphia, notably lesions of the optical memory areas [1]. These areas are located in the posterior parts of the brain, but their specific location is still unclear:

> We have only a few clues to the anatomical site of the cortical zones that function as the seat of optical and tactile sensory images. [1]

With lesions in this area, Wernicke postulates that alexia and agraphia can also occur independently of a sensory aphasia. Given these circumstances, they can be seen as distinct phenomena and are not considered aphasic problems [1]. Wernicke does acknowledge that in practice lesions are not very selective and that usually more than one memory centre is destroyed, leading to a complex of symptoms. Wernicke is also very clear that there is not such a thing as *one* optical image area (or *one* sensory memory area, as we have seen before). This is further demonstrated in his illustrations (see Fig. 2.3C), where an optical area α (alpha) is shown for letter recognition; Wernicke considered such area to be only 'a part of the entire optical memory area' [1].

In a similar way that connection *ab* is involved in oral language acquisition, optical area α is connected to a frontal motor area β (beta) that is 'the center of writing movements' (Fig. 2.3C) [1]. The connection $\alpha\beta$ associates optical and movement images during the period when writing is learned. According to Wernicke, it is likely that both connections *ab* and $\alpha\beta$ have an intimate anatomical relationship, and it is therefore understandable that lesions of the insular region are usually accompanied by agraphia. He is also not very specific on the location of area β, other than that it must be located in the frontal lobe. In Fig. 2.3C, the area has been positioned in the middle frontal gyrus, in fact very close to the 'writing centre' that was proposed by Exner some years later (in 1881). Exner is generally credited with the discovery of this area which was named after him. It was located in the second frontal gyrus immediately above Broca's area and just anterior to the primary motor cortex controlling the hand. Exner referred to it as the 'graphic motor image centre'. However, the evidence that has been presented so far in the literature for Exner's writing centre is rather thin [21].

Initially, Wernicke shared the ideas of Charcot, Pitres and others, whose language models had independent centres for reading and writing (see also Fig. 4.4) [22]. In later years, Wernicke changed his opinions on alexia and agraphia and eventually rejected a specific writing centre in the second frontal convolution. Wernicke became convinced that reading and writing proceeded letter by letter, whereby letters were represented in the optic areas of both hemispheres and these optic areas had connections with motor areas. Wernicke's opinion on this latter issue was in particular influenced by a single paper of Grashey (1885), who had described a patient (Johann Voit) who had a rather atypical anomia. Voit suffered from a serious head trauma after falling from a staircase. This resulted in a skull fracture, loss of consciousness, impaired vision and cranial nerve palsies. Voit probably also had impairments of memory and intelligence [23]. He made a gradual recovery but was left with language disturbances whereby he instantly forgot spoken words or written letters unless they were permanently repeated. As a result, he had developed a remarkable strategy to name an object: he looked at the object, wrote down the first letter, looked again, wrote down the second letter and so forth until he had (almost) completed the name; only then could he produce it orally. Grashey assumed that his patient had problems holding perceptual information in his memory, and thereby introduced a new aspect to aphasiology, namely, that disturbances in the time course or the storage of mental representations can cause cognitive dysfunction.[d] On the basis of these and other symptoms, Grashey posed the model that is shown in Fig. 2.5. Wernicke was impressed by these findings, as can be read in his review of Grashey's paper:

> In our consideration on isolated disorders of reading and writing, we have tacitly assumed so far that the word concept itself was unimpaired, in contrast to those cases treated in my earlier contribution to the different types of aphasia. It is now necessary to go into more detail about the word concept and its relation to the letter concept. I was given the most essential insights into this by Grashey's article, and I have no hesitation to claim that it constitutes the most significant contribution to aphasiology in the last ten years. Grashey introduces an entirely novel—and his work proves—very fruitful element, namely, the time period necessary for the formation of a word as well as for reading and writing it. All these procedures require a specific amount of time, and aphasia may be the result of such a reduction of memory that objects or concepts arising via association can no longer be named because they cannot be fixed in memory long enough. [25]

The merit of Grashey's work lies in the introduction of *time* and *information capacity* as crucial elements in theories or models of language. However, the evidence of his case was meagre, and the case itself was certainly not without any controversy. Some authors have even suggested that Voit might have been a malingerer.

---

[d] See for a detailed case description and interpretation the original paper [24] and a book chapter by Bartels and Wallesch from 1996 [25].

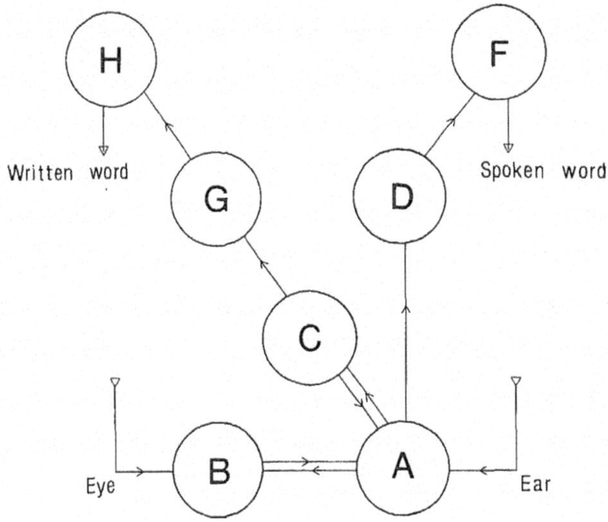

Fig. 2.5 Grashey's model (1885) whereby the symptomatology of the patient Voit led him to conclude that there are separate centres in the brain for naming, reading and writing. The case provided specific arguments that the connection from *A* to *B* was spared, but that the connection in the opposite direction was interrupted [24, 48]. Grashey presented his arguments in a diagram, as he said that this 'has now become general custom'. A: centre of sound images. B: centre of object images. C: centre of symbols, i.e. for cursive/printer letters and words; numbers. D: centre for motor images of speech. F: core of phonatory and articulatory nerves. G: centre of motor images for writing. H: core of motor nerves functioning during writing (Figure and legend taken from Grashey, 1885 [24] and de Bleser, 1996 [18])

Because Grashey's findings were adopted by Wernicke, they had a great influence in the aphasiological community, despite significant criticism in the German and French literature. Due to his assumption of a letter-by-letter basis for reading and writing (in contrast to 'words' as elements in his model for speech and comprehension), Wernicke now emphasized the dependence of written language on oral language: letter chains had to be converted into spoken words by means of inner speech and vice versa. There was no provision for whole-word reading or writing. Wernicke rejected specialized graphemic centres in the optic or motor areas and disagreed with Dejerine's (1891) and Charcot's view of a unilateral left hemispheric centre for reading in the angular gyrus [26, 27]. This was the start of a still ongoing debate, whereby alexia (with or without agraphia) has been attributed to several different (sub)cortical centres [28, 29]. The reader is referred to the extensive literature on this subject [30].

2.3.3 Lesion of Tract *ab*

Wernicke's model predicted another new type of aphasia: conduction aphasia [Leitungsaphasie]. This was the consequence of the proposed subcortical connectivity within the model. Wernicke is considered by many to be the founding father of the disconnection theory. He initially claimed that the fibres that connect the frontal and temporal language areas passed through the insula. In later years, he accepted the view that the connection passes via the arcuate fasciculus that curves around the insula [31]. In patients with conduction aphasia, according to Wernicke, 'the choice of words is disrupted', but both comprehension and speech remain intact [1]. The reason that these patients could still—to some extent—correct their false utterances despite a lesion of pathway *ab* was that there are alternative routes to the frontal motor centre: via the intact auditory language centre and more posteriorly located memory areas. This enables patients to judge their own production and show concern or frustration over their errors. Strictly speaking, such an alternative route is missing in the classical model usually shown in neurological textbooks. Wernicke does not explicitly mention disturbances in 'repetition', which later became the characteristic feature of conduction aphasia [31]. He could have realized this at the time, as it was in later years emphasized by Lichtheim (1885) and Wernicke himself (1910) [15]. Wernicke did note that in these patients there was often an hemiplegia, in line with his ideas that conduction aphasia was the result of lesions located deeper in the insular region. Lichtheim agreed with Wernicke's ideas and confirmed an insular lesion in a patient with conduction aphasia [15]. An interesting historical overview is given by Henderson in his chapter 'Early Concepts of Conduction Aphasia', along with alternative views and critiques of other nineteenth-century authors [15].

2.3.3.1 A Modern Definition and Anatomical Substrate of Conduction Aphasia

Although its existence, and in particular its neurological foundations, have been doubted, conduction aphasia is now well established as an individual type of aphasia. It is usually defined as a language disturbance with relatively fluent spontaneous speech and good comprehension, but poor repetition. There are also abundant (phonological) paraphasias [32]. Patients with conduction aphasia can additionally have impairments in naming, reading or writing or have a mild hemiparesis. According to Damasio (2000):

> Conduction aphasia is caused by damage to one of two regions: (1) the left supramarginal gyrus (Brodmann Area 40, BA40), with or without extension to the white matter underneath the posterior insula, or (2) the left primary auditory cortices (BA 41 and 42), the insula, and the underlying white matter. In either variant most of BA22 is spared. [20]

In the 1960s, Geschwind revived the disconnection syndromes and attributed a major role for the arcuate fasciculus (AF) in conduction aphasia (although he also related it to lesions of the association cortex) [33]. However, current opinion holds that conduction aphasia is probably not caused by a pure white matter lesion (AF or otherwise) and may not even be a disconnection syndrome [20, 34–36]. The posterior part of the left planum temporale (area Spt), for instance, has been hypothesized as an important area for phonological working memory and as the critically damaged area in patients with conduction aphasia [37, 38]; however, discussion continues [39]. Terminology also remains confusing; the AF is generally considered part of a larger tract, the superior longitudinal fasciculus (SLF), but functional and anatomical descriptions of the SLF vary in the literature [40]. The SLF has been proposed to include four different types of connecting fibres, each with specific terminations in frontal and temporoparietal areas [6, 41]. Modern non-invasive fibre tracking methods enable virtual dissection of white matter tracts and have shed a new light on the anatomical and functional connectivity. Bernal and Ardija (2009) pointed out that the rostral part of the AF predominantly ends in the lower part of the precentral gyrus and not so much in the classic Broca's area [32]. More specifically, the AF terminates in the anterior bank of the precentral gyrus; this part is considered premotor cortex (BA 6) and is cytoarchitectonically different from the rest of the gyrus which is primary motor cortex (BA 4). The ventral premotor cortex has been associated with speech programming and seems therefore a suitable target for the AF. This is also in line with the results of electrocortical stimulation of this area in patients who are operated on while awake and where almost invariably a speech arrest is elicited. Bernal and Ardila propose a new model whereby Broca's area, Wernicke's area and the motor cortex (BA 4) are all indirectly connected via a relay station in the ventral premotor cortex/primary motor cortex (BA 6/BA 4). Here is a fragment from their paper, elaborating on the role of the AF (see also Fig. 2.6):

> All things considered, it could be suggested that a model that defines the AF as an accessory aid in transferring information from the temporal lobe to premotor/motor areas, and not directly to Broca's area, may be more congruent with the neuroimaging results and clinical findings. A connection like this may suggest that auditory representation of speech plays a direct role in verbal motor planning. Interestingly, that was the initial hypothesis of Wernicke [42], and it also favours the explanation by Luria [43] who sees the repetition problems of conduction aphasia as an ideomotor verbal apraxia. Moreover, it seems easier to explain the phenomenon of 'conduit d'approche' [ie, successive

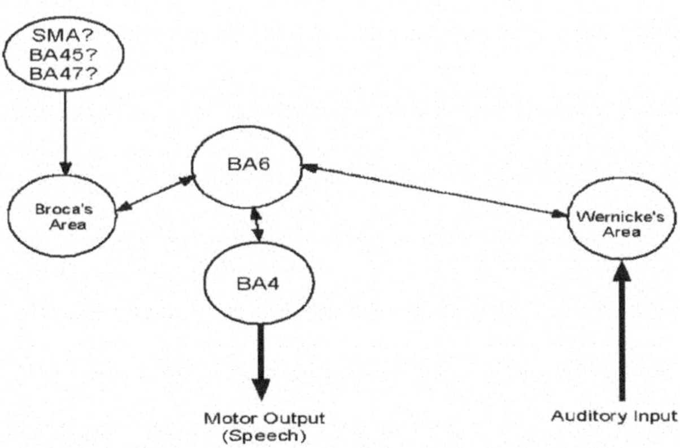

Fig. 2.6 A very recent language model by Bernal and Ardila (2009) whereby Wernicke's and Broca's areas are indirectly connected via the ventral premotor cortex (BA 6). It was demonstrated with MRI-based fibre tractography that the arcuate fasciculus terminates for the larger part in the lower end of the precentral gyrus (i.e. premotor and primary motor cortex) and not in the classic Broca's area (Figure taken from Bernal and Ardija, 2009 [32])

attempts to self-correct mispronunciations] as having a disconnection of the premotor area. Indeed, a premotor disconnection may impair the proper motor sequencing necessary to utter word components (word segments or syllables). As a result, a phonological paraphasia is generated. Since purely language areas are not affected, phonological awareness is not impaired and the patient attempts to correct the output by exhibiting the typical 'conduit d' approche', in the same manner as an apraxic patient tries over and over to find the proper movements to allow him/her to perform a particular action. This contrasts with the truly language phonological paraphasia in which the patient is not usually aware of his/her errors, and therefore, the fluency of the speech is not impacted. The AF, nonetheless, may still connect to Broca's area through a relay station located in the premotor or motor cortex. [32]

2.3.4 Lesion of Movement Centre *b*

In cases of destruction of movement centre *b*, the patient is mute or can utter only a few simple words. Wernicke agreed with Broca's observations after damage to the frontal motor area. Comprehension is not impaired. Whether or not a more extensive frontal lesion (which includes the motor movement centre for writing) will lead to an agraphia is not entirely clear; Wernicke asks himself the question in his monograph, but he does not answer it satisfactorily. Wernicke remarks that movement images for writing are not solely represented in the left hemisphere. 'The left hand is no more awkward in writing, in comparison with the right hand, than it is in any other movement' [1]. He was, however, unaware of callosal connections and bilateral innervation of speech musculature at the time [44].

2.3.5 Lesion of the Efferent Tract *b*

Wernicke assumes that a lesion of tract *b* produces a similar picture of motor aphasia to that when movement centre *b* itself is lesioned, although he remarks that it would be unlikely that any lesion would selectively destroy only this fibre tract. At the time, the lenticular nucleus was thought to be a collection of various nuclei, each innervating different parts of the face or tongue musculature. With a partial lesion, this would result in various dysarthric phenomena (note that the term *dysarthric* was not used at the time).

As a result of circumscribed destruction within the lenticular nucleus it would never be possible for all speech movements to be affected at once. Instead, partial aphasias are produced, which are manifested by paralysis of certain muscles innervated in the course of the speech act, e.g., the muscles of the mouth supplied by the facial nerve. At the same time the other movements of the tongue and the larynx involved in speech can take place normally, so that speech is still intelligible. [1]

Again, these interpretations must be seen against a historical background of incomplete, and in this case also incorrect understanding of the (functional) anatomy.

2.4 The Symptom Complex of Aphasia, Part III

In the third and final part of his monograph (pp. 38–70), Wernicke describes case histories of ten patients with aphasic symptoms. If we broadly characterize these patients according to the four clinical categories of aphasia that follow from Wernicke's model, there are three cases with predominantly motor aphasia (5, 6, 9), two cases of sensory aphasia (1, 2), four cases of conduction aphasia (3, 4, 7, 10) and one case of 'total' aphasia (8). In four cases, post-mortem findings are presented (2, 5, 8, 10), whereby in only two cases (2, 8) there is supportive evidence for Wernicke's theory of a temporal sensory language area. Case 10 was a patient with a brain abscess which did not provide any reliable localizing information. Case 8, Louise Funke, only came to post-mortem after the monograph was finished and was briefly described in an addendum (where Wernicke incorrectly refers to Funke as case 9). The meagre anatomical evidence on which the monograph rests has often been a point of critique. For Head, for instance, who wrote an influential book on aphasia (1926), the evidence presented by Wernicke was wrong and insufficient. He wrote that 'It can only be said that the clinical records are inadequate, or the details of the post-mortem findings unconvincing' [45]. Head even stated that 'Wernicke was completely satisfied with his attempts to deduce the clinical manifestations from hypothetical lesions' [45].

Wernicke, however, was himself very critical of the studies that were available in his own time. He opens the third part of his monograph with the sentence: 'Despite its extent, the clinical literature of aphasia is useful only to a small extent in supporting any theory based on anatomy' [1]. For Wernicke, there were two major problems with the literature of his time. The first problem, so he wrote, was the many subjective descriptions given by observers. They either focused on the symptom of interest (failing to give an accurate description of the complete mental picture) or they did not even list the most important symptom. Such observations were thus often incomplete or invalid. In his monograph, Wernicke repeatedly warns that to an untrained observer, a patient with a sensory aphasia can easily be misinterpreted as a having a psychiatric disorder (e.g. confusion or delusion):

> The fact that these cases [Wernicke refers here to patients with a severe form of sensory aphasia] have until now not been observed or at least not yet been published is the result, not only of the infrequency of such cases but also of the fact that even thoroughly experienced and intelligent physicians interpret this condition as a confusional state, as I myself have had the opportunity to observe. For those who are psychiatrically trained and who know the clinical pictures of confusional states, the diagnosis presents no difficulties whatsoever. [1]

The second problem of the literature was the low quality of the post-mortem examinations. Wernicke stated that the majority of authors were unable to do these investigations themselves and lacked detailed anatomical knowledge (he makes notable exceptions for eminent authors such as Broca, Ogle and Hughlings-Jackson). Hence, he did not trust the case descriptions in the literature. More importantly, he had noted that there had been no publications of clear cases of pure sensory aphasia before. Although Wernicke acknowledged that he had seen few 'material' cases himself, he reasoned that because there existed both patients with pure motor aphasia (i.e. motor symptoms but no sensory symptoms) and patients with pure sensory aphasia (vice versa), there must be two anatomically different language centres. (Note that this argument rests on a double dissociation.) So already on clinical grounds, Wernicke could predict a language area other than that of Broca's.

The first case of sensory aphasia is that of 59-year-old Susanne Adam, who was initially diagnosed with a confusional disorder and consequently transferred to the mental ward of Allerheiligen Hospital, where Wernicke worked. Her speech seemed normal, but she gave completely wrong answers to questions. There was fluent, mildly paraphasic speech with naming errors. Sentences were often structured correct, but comprehension and repetition was severely disturbed. There was also alexia and agraphia. Wernicke gave a lengthy description of her neurological and neuropsychological presentation with a verbatim excerpt from a conversation he had with the patient [1]. Here is a part of it:

> On March 18, 1874 [17 days after the symptoms have started], the following conversation took place, that was taken down word for word. It already shows significant progress.
> 'Good morning, how are you?'
> 'Thank you, I'm quite well.'
> 'How old are you?'
> 'Thank you, I'm fine.'
> 'How old are you?'
> 'Do you mean, what I'm cal, how I hear?' [Meinen Sie, wie ich hei, wie ich höre?]
> 'How old you are is what I wanted to know.'
> 'Well, I don't know exactly what I'm called shear.' [… wie ich so heissen schwiere], (corrected) 'What I am called hear' [wie ich so heissen höre].
> 'Would you like perhaps to give me your hand?'
> 'I really don't know, what I, etc.' (no sign of comprehension)

Over the next weeks, the patient made a good recovery, except for the agraphia which remained. For obvious reasons there was no post-mortem. Wernicke's final notes are from 20 April 1874, 6 weeks after the initial start of the symptoms.

> She has made further progress and now understands almost everything that is repeated to her several times. She still speaks somewhat haltingly, but for the most part correctly, and reads without stumbling. When she is asked to write about some topic of her own choice, only a few words come to her, and she is also not capable of writing words dictated to her. On the other hand she can copy fairly well words that she is given in written form and she copies all the single letters correctly. Agraphia is thus her most striking language disturbance at this time. [1]

Here, in his case notes, Wernicke labels agraphia as a language disorder. In the text of his monograph, he separated reading and writing disorders from aphasia. Perhaps Wernicke refined his ideas later on, when he turned his notes into a book. Although perhaps confusing, it is understandable that even Wernicke had trouble strictly classifying these disorders. Intuitively, they go together. Even today there is no strict and unique nomenclature for all the different disorders that involve hearing, reading and writing.

This case reminded Wernicke 'vividly' of a quite similar case that he had previously encountered, but unfortunately not documented very extensively. This second case was 75-year-old Susanne Rother, who was admitted to Wernicke's hospital on 7 October 1873. She is of key importance for Wernicke's theories, as being the only case of sensory aphasia that was documented with an isolated (left) temporal lesion on post-mortem examination. However, there was not much medical information available, as Wernicke clearly realized. The medical history reports that the patient had suffered a weakness of the left leg for 10 years. Suddenly, on 2 November 1873, her reactions and speech were inadequate [German: Verwirrte Sprache]. The patient was in a generally poor condition; she was bedridden and incontinent.

> Her mental condition was regarded at the time as a confusional state associated with aphasia. She answered all questions directed at her in a completely confused way, and carried out commands either not at all or in a completely confused manner, which at the time gave the impression of apraxia. The attendants thought that she was deaf because of her lack of understanding of what was said to her. Furthermore, she paid little attention to her surroundings, and in keeping with her severe malaise showed little urge to communicate. Her (spontaneously used) vocabulary thus seemed small in contrast to that of the case described above but was nevertheless large enough that motor aphasia (see above) could not be considered. The presence of aphasia could be recognized by her substitutions and distortions of words. Thus she often said correctly, 'Thank you very much' [Ich danke recht herzlich], but at other times, 'I thank you very giving' [Ich danke recht geblich]. 'I am very sick', 'Oh, I'm so cold', 'You are a good man', are expressions she used frequently. The doctor she had just called 'a good man' she soon afterwards called 'my little daughter' or 'my little son', without distinction. [1]

Further details of her neurological status are not given, and it is unknown whether there was also agraphia or alexia. There was no improvement in her mental or physical condition, and on 1 December Susanne Rother died after a protracted course of enteritis. Post-mortem findings are only very briefly described and—as in Broca's cases—not documented by drawings of the brain or the lesion itself. The patient's brain was not preserved and was never re-examined by others, in contrast to the brains of Broca's famous cases Leborgne and Lelong. Wernicke describes a generalized atrophy of the brain with major arteriosclerotic changes in all brain arteries. A thrombus was found in the left Sylvian artery, and a large part of the first temporal convolution was weakened [in einen weissgelben Brei verwandelt]. Insula and basal ganglia were unaffected.

Although the cases of Adam and Rother differed substantially in their clinical course, Wernicke was convinced that they both demonstrate the symptoms of sensory aphasia, whereby patients have a relatively large vocabulary but have definitely lost the ability to comprehend spoken language. He postulates that in Adam there was also a lesion of the left first temporal convolution. In his monograph, Wernicke discusses two points that he thinks are unrelated to the clinical syndrome of sensory aphasia. The first point is the generalized atrophy that was found at the post-mortem of Susanne Rother. Wernicke had observed this atrophy in 'numerous' other post-mortem examinations, and he had never found it to be responsible for any isolated functional deficits. He is very firm in his conclusion that:

> the softening of the left first temporal convolution in the case of Rother was the only brain lesion that could have caused the localized symptom of aphasia that persisted throughout the whole course of the illness, and that the generalized convolutional atrophy was either the result of senescence or, which is considerably more probable, a consequence of the presence of a circumscribed focal lesion. [1]

The second point is the paresis of Rother's left leg. Wernicke assumed a lesion in the right hemisphere but did not find any at the post-mortem examination. Instead a degenerative lesion was found on the left side of the spinal cord. The main evidence that the language problems and the paresis were unrelated was, according to Wernicke, that the paresis had already existed for 10 years.

There is a third case in the monograph (case 8) that Wernicke uses in support of his theory of a temporal sensory language area. It is the case of Louise Funke, 59 years old, who suffered a massive stroke that left her with a right-sided weakness and severe problems with comprehension and speech. She was only able to answer with 'yes', which she did to all questions asked. Spontaneous speech consisted of a repetition of 'yes yes', etc. This case of global aphasia came to post-mortem after the monograph was finished and was published in an addendum. The right hemisphere was intact, but the left perisylvian convolutions demonstrated clear ischemic changes. There was 'an extensive yellow softening which occupied almost the entire first primitive convolutional arc [fast den ganzen I. Urwindungsbogen], thus affecting both banks of the Sylvian fissure'. A thrombosis of the Sylvian artery was found as the cause of the ischemia. As Wernicke did with other cases that showed 'similar' clinical symptomatology, he now compares this case with that of Rother (case 2). He writes that both have in common that the first temporal convolution and the anastomosis with the second temporal convolution are lesioned. Then he states: 'Both suffered from sensory aphasia. Could this agreement be only a coincidence?' This is a very interesting remark, made in a (short) addendum and not in the original monograph; here, Wernicke questions the causality between the clinical and anatomical findings. He does not elaborate on it any further, so it is not clear to me whether he refers specifically to the cases of Funke and Rother or more generally to his theory of sensory aphasia. Obviously, both cases are different on both clinical and anatomico-pathological grounds, so at least I find it is somewhat strange that Wernicke groups them together in the first place. However, we must not forget that in Wernicke's time both cases presented new clinical symptoms (sensory aphasia) and new pathological findings (left temporal lesion). Thereby, Wernicke had already more or less predicted on theoretical grounds that there had to be a separate area for auditory language comprehension. He must have realized that the commonalities in these cases outweighed their differences, although he retained some doubt (see also the previously cited remarks of Wernicke from his 1876 paper). It has often been said that the strength of Wernicke's work lies in his theories and not in the case descriptions or pathological findings at the post-mortem examinations [4, 46]. In this respect, there are parallels to Broca's work and in particular his first two cases of motor aphasia.

2.5 Wernicke and the Anatomy of Language Areas

So what can we conclude from Wernicke's monograph about the anatomical localization of language areas? Wernicke considered that the entire perisylvian area and the insula were involved in language processing. Although he assumed that memory images were located at homologous areas in both hemispheres, he adopted the view that the left hemisphere was dominance for speech. 'But only the left sound center is effectively connected with the motor speech center, and thus probably only the left sound center has established well-worn connections with the conceptual regions' [1]. That both hemispheres are functionally interconnected was not known at the time. It is of interest to note that Wernicke considered the superior temporal and the inferior frontal gyrus to be one continuous gyrus that wraps around the Sylvian fissure.

> The entire region of the first convolution, which circles around the fossa Sylvii serves in conjunction with the insular cortex as a language centre. The first frontal convolution, which is a motor area, is the center of representations of movement; the first temporal convolution, a sensory area, is the center for sound images. The fibrae propriae which come together in the insular cortex form the mediating psychic reflex arcs. The first temporal convolution should thus be considered the central termination of the acoustic nerve, and the first frontal convolution (which includes Broca's area) the central end of the nerves of the muscles of speech. [1]

There are no anatomical drawings in Wernicke's monograph, and in his later works, language areas are also depicted fairly schematically (see Fig. 2.7) [12]. These drawings are clearly intended as theoretical and not as anatomical models. Wernicke

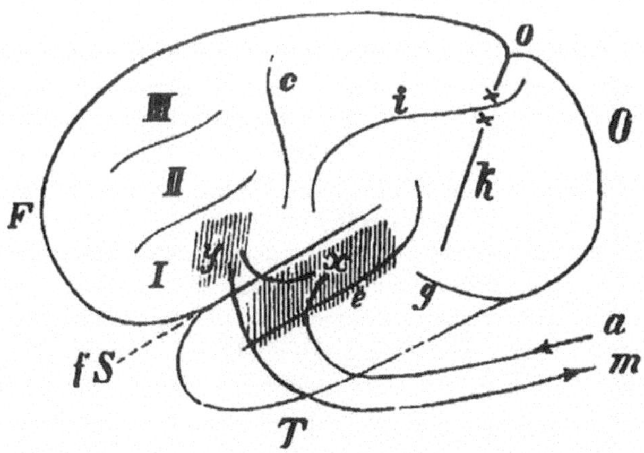

Fig. 2.7 Figure from Wernicke's *Lehrbuch der Gehirnkrankheiten* (1881), indicating cortical areas that are involved in language functions. Wernicke explicitly stated in his treatise that he considered the entire inferior frontal gyrus as the motor centre for speech, although he depicted only part of it in his scheme. In a similar manner, Wernicke describes in his book that the auditory language centre *x* occupies the left superior temporal gyrus and part (German: Randzone) of the second temporal gyrus. Again, this area seems larger than is actually depicted in the figure. (Figure and quotation taken from Wernicke, 1881 [12]). Note: frontal convolutions were numbered in relation to the Sylvian fissure (following Leuret's convention)

indicates language 'centres' *a*, *b*, etc., but these are not really the 'centre' of a very specified anatomical area, but rather a crude indication of the cortical territory that is involved. Interestingly, Wernicke explicitly mentions in the text that accompanies this figure that the seat of the auditory image centre is partly located also in the *second* temporal gyrus.

Let us now look back at the models in Fig. 2.3. What about the areas *c* and *d* that Wernicke depicted here, should these be labelled 'language' areas or not? Wernicke left them out of his model that describes the processing of auditory information. Lesions of *c* or *d* seem to him not directly related to aphasia [1]. In German, the word for 'language' is 'Sprache', which literally means 'speech'. But Wernicke meant 'language' in the broader sense that we use it for, thereby focusing on verbal comprehension. On the one hand, we have seen that the areas *c* and *d* do play a critical role in language processing, as they select the motor images during speech. At the same time, they play an important role in the understanding [Begriff] of objects, forming a concept at a more fundamental level. Although Wernicke strictly distinguishes between *content* and *form* of language (between disorders of intelligence and language) he is not always explicit on how this translates to his models for language and gnosis. In line with the fact that it is not always possible to classify a disorder uniquely into either an aphasia or an agnosia, it is difficult to label brain areas uniquely as having only one specific function. This is one of the important consequences of the models that Wernicke presents in his monograph and in his later works. The connectionist architecture lets different brain regions collectively perform a certain task and, conversely, lets individual areas participate in more than one function.

Wernicke's connectionist approach (which is hardly ever found in neurological textbooks) also explains why he never gave a precise anatomical localization for the sensory language area. In his view, there are no very focal or strictly localized language areas; he favoured a model where information is distributed over larger parts of the brain, organized around a number of distinct language centres. Wernicke gave indications as to the location of these epicentres, but was not very specific. Why, for instance, is the auditory image centre *a* in Fig. 2.3 drawn at two different locations (temporal pole and middle part of the superior temporal gyrus)? Is it to underline his idea that language areas are not strictly localized? Is it out of a kind of nonchalance, in a way that the schemes are incongruously drawn on the *right* hemisphere instead of the left? Or do these areas *a* represent two different auditory images, one for sound of the *word* (i.e. the phonological representation) and one for the sound of the *object* itself?

Similar to Broca's case, when Wernicke published his monograph on aphasia, there were earlier descriptions of auditory comprehension disorders and temporal lesions, but these either lacked autopsy findings or failed to have a lasting impact on the neurological community. Whitaker and Etlinger cite a virtually unknown work of Meynert from 1866 and claim that on this basis he should have had the historical credit instead of Wernicke [5], but others disagree [47]. These discussions predominantly involve the classical contributions of Wernicke, notably his sensory language area. But Wernicke's contribution is much greater than is generally recognized, in particular his idea that knowledge of a given concept is distributed throughout the brain and cannot be located in one or a few areas. Ironically, localism received great support from his theories, in particular in the first half of the twentieth century.

References

1. Wernicke C. Der aphasische symptomencomplex: eine psychologische studie auf anatomischer basis [reprint from Nabu Public Domain reprints]. Breslau: Cohn and Weigert; 1874.
2. Bastian HC. On the various forms of loss of speech in cerebral disease. Br Foreign Med Chir Rev. 1869;43:209–36.
3. Johann BF, Schmidt B. A pioneer in the history of aphasia. Arch Neurol. 1977;34:306–7.
4. Tesak J, Code C. Milestones in the history of aphasia: theories and protagonists. Hove: Psychology Press; 2008.
5. Whitaker HA, Etlinger SC. Theodor Meynert's contribution to classical 19th century aphasia studies. Brain Lang. 1993;45:560–71.
6. Schmahmann JD, Pandya DN. Fiber pathways of the brain. Oxford: OUP; 2006.
7. Meynert T. Ein Fall von Sprachstörung, anatomisch begründet. Medizinische Jahrbücher der Zeitschrift der KK Gesellschaft der Ärzte in Wien. 1866:152–189.
8. Seitelberger F. Theodor Meynert (1833-1892), pioneer and visionary of brain research. J Hist Neurosci. 1997;6:264–74.
9. Geschwind N. Selected papers on language and the brain. Berlin: Springer; 1974.
10. Mesulam MM. Large-scale neurocognitive networks and distributed processing for attention, language, and memory. Ann Neurol. 1990;28:597–613.
11. Gage N, Hickok G. Multiregional cell assemblies, temporal binding and the representation of conceptual knowledge in cortex: a modern theory by a "classical" neurologist, Carl Wernicke. Cortex. 2005;41:823–32.
12. Wernicke C. Lehrbuch der Gehirnkrankheiten für Ärzte und Studierende, vol. 1. Kassel: Fisher; 1881.
13. Wernicke C. Das Urwindungssystem des Menschlichen Gehirns. Arch Psychiatr Nervenkr. 1876;6:298–326.
14. Wernicke C, Hahn E, Sachs H, et al. Photographischer atlas de gehirns. Schnitte durch das menschliche Gehirn in photographischen Originalen. Breslau: Schletter'schen Buchhandlung (Franck & Weigert); 1897.
15. Henderson VW. Early concepts of conduction aphasia. In: Kohn SE, editor. Conduction aphasia. Hillsdale: Lawrence Erlbaum; 1992. p. 23–38.
16. Kussmaul A. Die Stoerungen der Sprache. Versuch einer Pathologie der Sprache. Leipzig: FCW Vogel; 1877.
17. Wernicke C. Nervenheilkunde: Die neueren Arbeiten ueber Aphasie. Forschr Med. 1886;4:371–7.
18. de Bleser R. Wernicke's (1903) case of pure agraphia: an enigma for classical models of written language processing. In: Code C, Wallesch CW, Joanette Y, Roch A, editors. Classic cases in neuropsychology. Hove: Psychology Press; 1996. p. 13–30.
19. Mesulam MM. Principles of behavioral and cognitive neurology. Oxford: Oxford University Press; 2000.
20. Damasio AR, Damasio H. Aphasia and the neural basis of language. In: Mesulam MM, editor. Principles of behavioral and cognitive neurology. New York: Oxford University Press; 2000. p. 294–315.
21. Roux FE, Draper L, Köpke B, Démonet JF. Who actually read Exner? Returning to the source of the frontal "writing centre" hypothesis. Cortex. 2010;46:1204–10.
22. Charcot JM. Des différentes formes de l'aphasie. Progr Med (Paris). 1883;11:441–4.
23. Morian. Zwei Fälle von Kopfverletzungen mit Herdsymptomen. Langebeck's Archiv für Chirurgie. 1885;21:898–914.
24. Grashey H, De Bleser R. On aphasia and its relations to perception. Cogn Neuropsychol. 1989;6:515–46.
25. Bartels C, Wellesch C-W. Nineteenth-century accounts of the nature of the lexicon and semantics: riddles posed by the case of Johann Voit. In: Code C, Wallesch C-W, Joanette Y, Roch A, editors. Classic cases in neuropsychology. Hove: Psychology Press; 1996. p. 53–68.

26. Dejerine J. Sur un cas de cécité verbale avec agraphie, suivi d'autopsie. Mém Soc Biol. 1891;3:197–201.
27. Dejerine J. Contribution a l'étude anatomo-pathologique et clinique des différentes variétés de cécité verbale. Mém Soc Biol. 1892;4:61–90.
28. Epelbaum S, Pinel P, Gaillard R, et al. Pure alexia as a disconnection syndrome: new diffusion imaging evidence for an old concept. Cortex. 2008;44:962–74.
29. Montant M, Behrmann M. Pure alexia. Neurocase. 2000;6:265–94.
30. Dehaene S. Reading in the brain: the new science of how we read. New York: Viking Adult; 2009.
31. Geschwind N, Devinsky O, Schachter SC. Norman Geschwind: selected publications on language, epilepsy, and behavior. Boston: Butterworth-Heineman; 1997.
32. Bernal B, Ardila A. The role of the arcuate fasciculus in conduction aphasia. Brain. 2009;132:2309–16.
33. Geschwind N. Disconnexion syndromes in animals and man. Part I. Brain. 1965;88:237–94.
34. Poeppel D, Hickok G. Towards a new functional anatomy of language. Cognition. 2004;92:1–12.
35. Shuren JE, Schefft BK, Yeh HS, et al. Repetition and the arcuate fasciculus. J Neurol. 1995;242:596–8.
36. Berthier ML, Lambon Ralph MA, Pujol J, Green C. Arcuate fasciculus variability and repetition: the left sometimes can be right. Cortex. 2012;48:133–43.
37. Buchsbaum BR, Baldo J, Okada K, et al. Conduction aphasia, sensory-motor integration, and phonological short-term memory—an aggregate analysis of lesion and fMRI data. Brain Lang. 2011;119:119–28.
38. Hickok G, Buchsbaum B, Humphries C, Muftuler T. Auditory-motor interaction revealed by fMRI: speech, music, and working memory in area Spt. J Cogn Neurosci. 2003;15:673–82.
39. Bizzi A, Nava S, Ferre F, et al. Aphasia induced by gliomas growing in the ventrolateral frontal region: assessment with diffusion MR tractography, functional MR imaging and neuropsychology. Cortex. 2012;48:255–72.
40. Catani M, Thiebaut de Schotten M. A diffusion tensor imaging tractography atlas for virtual in vivo dissections Cortex. 2008;44:1105–32.
41. Makris N, Kennedy DN, McInerney S, et al. Segmentation of subcomponents within the superior longitudinal fascicle in humans: a quantitative, in vivo, DT-MRI study. Cereb Cortex. 2005;15:854–69.
42. Hickok G, Erhard P, Kassubek J, et al. A functional magnetic resonance imaging study of the role of left posterior superior temporal gyrus in speech production: implications for the explanation of conduction aphasia. Neurosci Lett. 2000;287:156–60.
43. Luria AR. Basic problems of neurolinguistics. 1976
44. Geschwind N. Wernicke's contribution to the study of aphasia. Cortex. 1967;3:449–63.
45. Head H. Aphasia and kindred disorders of speech. Cambridge: Cambridge University Press; 1926.
46. Marx OM. Aphasia studies and language theory in the 19th century. Bull Hist Med. 1966;40:328–49.
47. Eling P. Meynert on Wernicke's aphasia. Cortex. 2006;42:811–6.
48. Grashey H. Uber aphasie und ihre beziehungen zur Wahrnehmung. Archiv Psychiatr Nervenkr. 1885;16:654–88.

3 Aphasia or Agnosia?

Wernicke proposed that knowledge of the outside world was conceptualized and that the necessary information that makes up these concepts is stored in many different and interconnected areas of the brain. He clearly made a distinction between knowledge of a word itself (i.e. the sound or pronunciation of it) and knowledge of its meaning or 'concept'. Since the era of the diagram makers (see also Chap. 4), there has been a discussion of how concepts are anatomically represented in the brain. Several of the earlier researchers truly considered a 'centre' for conceptual knowledge, similar to the anatomical centres for motor or speech functions. Some, for example Mills, even believed that this concept centre was identical to the naming centre [1]. Most others, like Wernicke or Lichtheim, considered the concept centre more of a theoretical construct with a largely heuristic purpose, at least not something with a strict anatomical definition. To them, the true meaning of words and sentences emerged as the result of the complex interplay (association) between many different areas.

With this in mind, it can be better understood that when a patient has difficulties with naming, it does not automatically imply that there is a language disorder, or that the lesion is necessarily located in language areas. There can also be a dysfunction in several other systems that precludes correct naming, in particular in areas that sustain conceptual information. In fact, the localization of an anomic disturbance has been a point of controversy for a long time [2]. As put forward by Geschwind (1967), there were two opposing views in the literature: 'one which insisted that anomia resulted from a lesion of the left temporoparietal regions while the other claimed that it was the result of a diffuse disorder of the brain' [3]. Here is how Benson and Ardila formulated this (1996):

> Word-finding deficits (anomia) are noted and complained about by patients with structural damage to any cerebral area in either hemisphere. Virtually every aphasic patient suffers some degree of naming disturbance; however, the characteristics of the word-finding problem can vary considerably in the different aphasia syndromes. (…) It is important to bear in mind that anomia is a term with a double meaning in aphasiology. In one usage, the term is synonymous with naming disorder; in this broad sense all aphasic patients are anomic.

G.-J. Rutten, *The Broca-Wernicke Doctrine*, DOI 10.1007/978-3-319-54633-9_3

When used in the broad sense of a word-finding disorder (eg, decreased performance on a confrontation naming task), anomia is not of localizing value. In attempts to be more specific, some aphasiologists limit use of the term to those patients whose word-finding difficulty leads to circumlocutions and/or verbal paraphasias as observed in some patients with fluent aphasia. In this more tightly defined sense, anomia becomes synonymous with anomic aphasia, nominal aphasia, or amnestic aphasia. It is important to keep in mind that naming difficulties, often called anomia, are present in all aphasics but that the term anomia is also used to refer to a particular aphasia syndrome (anomic aphasia).

Naming difficulties can result from a deficit at different stages of the naming process: perception (decoding), storage, selection, retrieval, or actual production of the word (encoding). Furthermore, acquired naming difficulties can be restricted to specific semantic categories and even to a particular modality of representation. Naming disorders can result from the patient's inability to perceive or to identify the target object and as such can be considered a perceptual or agnostic deficits [4].

As we have seen in the previous chapter, Wernicke was well aware of the intimate relationship between aphasia and agnosia, and these disorders were already the subject of research and discussion in his time.[a] One of Wernicke's assistants, Heinrich Lissauer (1861–1891), published an early case of visual agnosia in 1890 [6]. Lissauer is also considered the first to have provided a detailed account of this type of agnosia as well as an important classification that is still used today [7].[b] Wernicke was involved during clinical observation and treatment of the patient and had already given a talk about this case at a meeting in Breslau [8, 9]. Another important paper that was published at that time (1889) was on a closely related disorder. It was written by Carl Freund (1862–1932), and it described a patient with a modality-specific naming disorder that he referred to as optic aphasia (the term optic anomia is probably more appropriate) [10]. Both papers not only share a similar topic (impaired naming) but also a common historical basis: at the time of publication, both Lissauer and Freund were working in Breslau under the direction of Wernicke [10]. They are discussed here at some length not only for historical reasons but also to illustrate how difficult, if not impossible, it is to disentangle language and non-language disturbances both from a practical and a more theoretical point of view.

3.1 Lissauer

Lissauer starts his seminal paper by referring to related works of Wilbrand (1887) [11] and Freund (1889) [12], but states that 'in particular a case such as the one described below cannot be surpassed by any clinical cases reported to date in terms of the severity of its symptoms and their clear-cut manifestation' [9]. He gives an extensive description of the medical history and the physical status of the patient.

[a] Lissauer used the term 'mindblindness' (German: Seelenblindheit). The term 'agnosia' was coined by Freud in his book *On Aphasia* (1891), but several descriptions of the disorder predate Freud [5].

[b] Between apperceptive and associative visual agnosia, I will explain this later in this chapter in more detail.

The latter is in fact a very comprehensive neurological examination. Today, the largest part of such an examination would be performed by neuropsychologists and ophthalmologists rather than neurologists. Over a span of 12 pages, Lissauer lists observations and test results in 13 categories that cover several different cognitive domains. These categories are summed up here to illustrate the thoroughness of Lissauer's report: 'refraction; visual fields; visual acuity; colour perception; visual estimating; stereoscopic vision; visual memory; memory for past visual experience; drawing; reading and writing; form perception; topographical orientation; reaction times to visual stimuli'. The examination revealed a complete right-sided hemianopia, but no other focal neurological deficits. Speech was fluent with intact comprehension and an unimpaired vocabulary. We should keep in mind that the tests that were applied by these historical authors were not standardized and that population norms were lacking. Still, their level of observation and methods of physical examination were often impressive, even by our current standards.

The patient was an 80-year-old former shopkeeper who reported that he had hit his head against a wooden fence during a severe storm. He retired to bed for 2–3 days because he did not feel well and complained that he was not able to see as well as before. Here is a part of the description given by Lissauer:

> In the morning, when he wanted to have a wash, the patient searched his room for the washstand, which was in its usual place. He also searched for his boots which were, as usual, under the bed, but he looked for these behind the stove and in the kitchen. He frequently mistook articles of clothing, for instance mistaking his jacket for his trousers. He thought that a number of pictures in his room were boxes and tried to search in them for things he had lost. When eating he mixed up pieces of cutlery. He used his spoon wrongly, by dipping the handle into the soup. Once he tried to put his hand into the food and once into a cup of coffee.
>
> Since his accident the patient no longer read. He gave his letters to his daughter saying that his vision was not clear enough. He continuously complained about the deterioration of his eyesight. He insisted that this visual problem had started suddenly on 3rd August following his accident and fall, though in view of his weak memory one cannot rely too much on this statement [9].

When objects were visually presented to the patient he was unable to recognize many of them, although there were marked fluctuations in his performance. When stimuli were perceived auditorily or via touch, his responses were normal. The patient seemed for the most part unaware of his inappropriate responses, although, as Lissauer writes, 'His answers were never given with the complete assurance with which a normal subject would make a statement about the name or characteristic of a familiar object' [9]. Clearly there was a naming disorder, but without dysphasic characteristics. Lissauer therefore considered the visual anomia in his patient a disorder of recognition and not of language: a visual agnosia.

> The way in which our patient tended to express himself could have produced the suspicion that there were some transcortical speech disturbances. In fact this was not the case. Our patient never mixed up words in a paraphasic fashion. For example, when he spoke of spectacles he meant just that: an instrument made of glass which he had put on to read or write a hundred times. However, when he said that a fork was a pair of glasses it was not that he used the wrong word for the correct concept but that the concept itself was wrong. For anyone who worked with the patient there was not the least doubt about this [9].

Lissauer divided the process of recognition into two stages, based upon his connectionist view of cortical areas with subcortical connections. In the first stage, causing apperceptive agnosia, recognition of an object may fail because of impaired visuospatial perception (strictly speaking this is not a real disorder of recognition, but of perception only). The lesion is located in the visual cortex. Lissauer gives credit to Munk for the term 'visual agnosia' [13]. In the second stage, recognition is not possible despite intact perceptual skills. This is caused by a disconnection whereby the visual cortex itself remains largely intact, but white matter connections are lesioned in such a manner that information is unable to reach distant areas that are needed for higher-order recognition [8, 14]. The resulting disorder is called associative agnosia. It is to be noted that Lissauer clearly states that he does not expect to find patients with a pure form of one of these agnosias. He considers a pure associative agnosia 'a contradiction in terms'. His patient had impairments in apperception, but his associative agnosia dominated.

> It would seem to be obvious that apperception as a special mental process should be thought of as separate from the understanding of its meaning with its manifold associations. I came to this conclusion for two reasons. First, according to the theory of localization with which I agree completely, those processes which occur in only one modality and are therefore localized should be separated from those which involve a variety of associations and so are the product of the whole cortex. A process of the first kind involves only apperception and it must be followed by a process of the second kind before the act of recognition can be completed. Secondly my observations force me to make this distinction. There is no doubt that our patient perceived many things without comprehending them; that is, he purely apperceived some objects but did not recognize them. Therefore the first stage occurred without the second stage. Thus indeed under certain pathological conditions the first stage may occur in isolation [9].

Lissauer's observations and theories are strongly rooted in the work of the connectionists. Remember that Lissauer was Wernicke's pupil.

> What is different in my presentation from those of other authors is my attempt to put special emphasis on the transcortical tracts and their importance in the manifestation of visual agnosia. This approach to the problem was suggested by the theories and systems which Wernicke and Lichtheim have postulated for the organization of speech functions with such important consequences [9].

The associations that subserve recognition are manifold, says Lissauer, and he takes 'a simple example which involves all sensory modalities', that of a violin.

> With the violin's image there are connected a number of recollections which concern its name, its sound, its image. The sound of the instrument, the sensations and tactile experience which go with the handling. In addition there may be the optical image of the violinist in his characteristic pose. It is only when these associations between the percept of the instrument and the above-mentioned recollections occur promptly in consciousness that one is enabled to interpret the object as a musical instrument and differentiate it from other instruments and generally to categorize it. If, however, this association is delayed or disrupted through some pathological process then even if the image of the violin is perceived, however precisely, there are no associations with prior experiences and recognition is therefore not possible [6].

This is the connectionist view that was characteristic of the Meynert and Wernicke schools, and Lissuaer's example strongly resembles that of Wernicke's bell that we saw in the previous chapter. Other assistants of Wernicke also based their theories on disruption of subcortical network connections. One of them is Liepmann, who in the period 1895–1899 developed a theory of apraxia [8, 15]. Another, less well-known assistant was Freund, who was mentioned above. Freund probably reported on the first disconnection syndrome of a cognitive function (other than conduction aphasia which had already been reported by Wernicke). He certainly was the first to explain disorders by a model that involved areas in both hemispheres. His discussion of interhemispheric connections via the corpus callosum predates that of Dejerine's famous papers on alexia [16].

3.2 Freund

Freund's patient was 57-year-old Carl Schluckwerder who was referred to the hospital because of 'physical weakness, weight loss, headache, clumsiness, emotional lability and speech difficulty' [10]. Eventual post-mortem examination revealed a tumour in the posterior parts of the brain and the splenium of the corpus callosum. Freund distinguishes two stages in the illness of his patient and attributes the 'unusual speech' (optic aphasia) during the first stage to a disorder of cerebral visual function. Initially there was an accompanying right-sided hemianopia. In a later stage, tumour progression led to more cognitive impairments. 'As well as a right-sided hemianopia, there was a defect of the major part of the left visual half-field [i.e. the tumour has now progressed to the right hemisphere]. The number of objects not recognized steadily increased. The patient's orientation became worse. Finally, he showed clear symptoms of total visual agnosia [asymbolie]'. The language disturbances make this case different from that of Lissauer, and led Freund to conclude that there must be some new form of aphasia.

> A review of the various forms of aphasia distinguished by Wernicke leads to the conclusion that the present case does not belong to any of them. Our patient is thus not genuinely aphasic, however much his actual manner of speech and the occurrence of agraphia and alexia initially give the impression of an aphasic disturbance. In attempting to explain this complex disorder it seems to me of fundamental importance that the patient immediately and correctly names familiar objects if he handles them while his eyes are shut, that is to say, if the touch imaginations [Tastvorstellungen] are activated. However, during visual inspection, when objects are merely shown to him, he does not find the appropriate name.
>
> Consequently, the suspicion arises as to whether our patient's unusual way of speaking might also be related to a disturbance of visual function. Here I refer to the connection between concrete ideas and words discussed by Wernicke. According to Wernicke, normal comprehension of speech, as well as voluntary, spontaneous speech, requires not only that the idea of the word [der Begriff des Wortes] is intact but also that of the object concerned. The idea of an object is made up of a number of components, that is of the representations in memory of the fundamental properties of an object which are stored in the different sensory regions of the brain. These representations in memory are closely linked to another and represent a functional unity, the 'idea' of an object. Activation of each individual stored

> representation is transmitted to others belonging to the same idea. This association-chain [Associationskette] becomes functional in spontaneous speech in so far as the 'idea of the word' of the object concerned is aroused. According to this view, an impairment in relation to the visual memory pictures should have a noticeable effect on spontaneous speech [10].

Freund reviews other cases in the literature (presented by Wilbrand and Stenger) and concludes that this must be a specific language disorder, which he names 'optic aphasia'. He more specifically suggests a distinction between Wernicke's sensory–acoustic aphasia and the new sensory–visual aphasia. The latter condition can coexist with agnosia, hemianopia and/or sensory–acoustic aphasia. To explain these various conditions better, Freund refers to a diagram (see Fig. 3.1) from which he postulates nine different variations of optic aphasia. Freund hastens to say that this is a purely theoretical construct: in practice, single conduction pathways will never be destroyed in isolation. The Schluckwerder case is explained by a lesion of two subcortical tracts (I and O2S, see Fig. 3.1). Because of the right-sided hemianopia due to the lesion of the left optic pathway (I), perception and recognition of objects could only have been performed by the right hemisphere. The lesion of O2S then permits visual information to reach the speech centre. However, connections from other sensory modalities to the speech centre have remained intact, which explain the intact verbalization of an object once it is presented in another sensory modality. Freund states that 'the anatomic course of the connection O2S is as yet largely

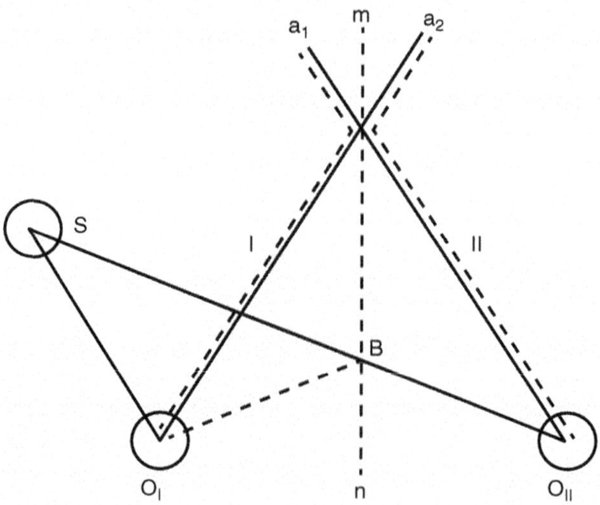

Fig. 3.1 Freund's diagram (1889) from which he derived various different (theoretical) forms of optic aphasia. S represents the speech centre [German: Sprachcentrum] 'incorporating not only the word sound memory centre [Klangbildcentrum] but also the memory centre for word articulation' [Centrum für Sprachbewegungsvorstellungen]. O1 and O2 are, respectively, the left and right visual cortex in which the visual representations are stored. I and II are the left and right optic radiations. mn: longitudinal fissure. a1 and a2 represent the (hemi)retinas. Point B is the posterior part of the corpus callosum. Figure taken from the paper by Beaton (1991) that translated and commented on Freund's original paper 'On optic aphasia and visual agnosia' [10]

unknown'. He (correctly) assumes that these fibres first run in the right hemisphere, then in the splenium of the corpus callosum and finally in the white matter of the left hemisphere to temporal and frontal speech centres. Postmortem examination indeed showed a swollen and damaged splenium. Later research confirmed Freund's speculation of the functional and anatomical role of the callosum.

To this day, discussion continues about the existence of optic aphasia as a more or less isolated phenomenon; the most common explanation is an anatomical visuo-verbal disconnection [17, 18]. Several authors have doubted the existence of optic aphasia [19–21]. The borders between various types of agnosia and aphasia remain fuzzy and are not strictly defined, which is perfectly understandable in a complex associative system where any brain region can be involved in many different functions. Generally speaking, the term visual agnosia describes a patient who fails to name a visually presented object but has apparently normal spontaneous language functions and no obvious visual disorder [22]. Optic aphasia points to a more selective language impairment.

3.3 A Systematic Approach to the Anomic Patient

Current scientific opinion is that information that enters the brain is processed in a hierarchical manner, whereby the content of information gradually increases in complexity (although this process is not necessarily linear or unidirectional) [23]. Many papers and textbooks have tried to disentangle and categorize these processes. In his paper, 'Disorders of visual processing', De Renzi systematically describes how to test patients with naming difficulties in order to detect the 'level' of their processing deficit [24]. I will summarize this process briefly.

As a prerequisite, patients should be alert and able to communicate. Next, one needs to ensure that visual stimuli are perceived normally and that basic visual skills are intact. Visual details must be able to be integrated into 'a detailed structural description of the stimulus', apt to be matched with its 'representation stored in visual memory' [24]. Patients with deficits at this lower cognitive level, for instance, have problems with matching figures or objects that have minor differences or that are viewed from different perspectives. They are then said to suffer from apperceptive agnosia. De Renzi continues with a next step where the examiner needs to decide whether the patient is able to match his or her perceptual construct with the representation of the same object that is stored in visual memory. There are various specific experiments available to test the integrity of the internal representations and to detect deficits at this processing stage. Here De Renzi makes an interesting (but rather subjective) observation: 'If the search is positive, a feeling of familiarity ensues'. This feeling is well known; it typically arises (and is particularly annoying) when you just know that a word or some other piece of information is known to you, but for some reason you cannot yet express or verbalize it. This *just knowing* and tip-of-the-tongue feeling relates to the De Renzi's *feeling of familiarity*.[c] In the final stage of processing,

[c] There is actually quite a lot of research done on these so-called metacognitive judgments. A quick search, for instance on www.pubmed.com, gives several starting points [25, 26].

once a representation has been formed and we have had our feeling of familiarity, identification is completed: a meaning is attributed to it, and a name is retrieved from memory. A deficit at this last stage is considered to be a language disorder.

It should always be kept in mind that terminology can be confusing or misleading. What clinicians do in everyday practice is try to categorize the patient's neurological symptoms into 'known' functional disorders, such as Broca's aphasia or visual agnosia. This has obvious advantages: it facilitates communication (e.g. between doctors or therapists) and also gives the impression that one has grasped and understood all of the patient's cognitive impairments. However, the use of such a taxonomy can make us blind to impairments that do not fit predescribed disorders. This was eloquently formulated by Penfield and Roberts in their book *Speech and Brain Mechanisms* (1959):

> Theory is indispensable, but terms such as those of agnosia, particularly when subdivided into visual verbal, visual literal, etc., do nothing but confuse us. There is not a single case in the literature of visual verbal agnosia without other defects, together with the ability to recognize some word at some time if the examination is detailed enough. We must record what the patient sees and does under this and that circumstance, and not use such terms as visual verbal agnosia and auditory agnosia unless these things actually exist, which, as far as we are concerned, has never been proved [27].

References

1. Nielsen JM. Agnosia, apraxia, aphasia: their value in cerebral localization (second edition, completely revised). New York: Hafner Publishing Company, Inc.; 1962.
2. Geschwind N. Selected papers on language and the brain. Berlin: Springer; 1974.
3. Geschwind N. The varieties of naming disorders. Cortex. 1967;3:97–112.
4. Benson DF, Ardila A. Aphasia: a clinical perspective. New York: Oxford University Press; 1996.
5. Freud S. Zur Auffassung der Aphasien: Eine kritische Studie. Leipzig: F. Deuticke; 1891.
6. Lissauer H. Ein Fall von Seelenblindheit nebst einem Beitrage zur Theorie derselben. Eur Arch Psychiatry Clin Neurosci. 1890;21:222–70.
7. Riddoch MJ, Humphreys GW. Visual agnosia. Neurol Clin. 2003;21:501–20.
8. Catani M, ffytche DH. The rises and falls of disconnection syndromes. Brain. 2005;128:2224–39.
9. Lissauer H, Jackson J. A case of visual agnosia with a contribution to theory. Cogn Neuropsychol. 1988;5:157–92.
10. Freund CS, Beaton A, Davidoff J, Erstfeld U. On optic aphasia and visual agnosia. Cogn Neuropsychol. 1991;8:21–38.
11. Wilbrand H. Die Seelenblindheit als Herderscheinung und ihre Beziehungen zur homonymen Hemianopsie zur Alexie und Agraphie. JF Bergmann; 1887.
12. Freund CS. Über optische Aphasie und Seelenblindheit. Arch Psychiatr Nervenkr. 1889;20:276–97.
13. Munk H. Über die Functionen der Grosshirnrinde. Berlin: Verlag von August Hirschwald; 1890.
14. Teuber HL. Alteration of perception and memory in man. In: Weiskrantz L, editor. Analysis of behavioral change. New York: Harper & Row; 1968. p. 268–375.
15. Liepmann H. Das Krankheitsbild der Apraxie ("motorischen Asymbolie") auf Grund eines Falles von einseitiger Apraxie (Fortsetzung.). Monatsschr Psychiatr Neurol. 1900;8:15–44. 102

16. Dejerine J. Contribution a l'étude anatomo-pathologique et clinique des différentes variétés de cécité verbale. Mém Soc Biol. 1892;4:61–90.
17. Beauvois MF. Optic aphasia: a process of interaction between vision and language. Philos Trans R Soc Lond Ser B Biol Sci. 1982;298:35–47.
18. Schnider A, Benson DF, Scharre DW. Visual agnosia and optic aphasia: are they anatomically distinct? Cortex. 1994;30:445–57.
19. Geschwind N. Disconnexion syndromes in animals and man. Part I. Brain. 1965;88:237–94.
20. Critchley M. The problem of visual agnosia. J Neurol Sci. 1964;69:274–90.
21. Vogel P. Sigmund Freud zur Auffassung der Aphasien–eine kritische Studie. 2001
22. Damasio AR, Geschwind N. Anatomical localization in clinical neuropsychology. In: Vinken PJ, Bruyn GW, Klawans HL, editors. Handbook of clinical neurology. Amsterdam: Elsevier Science Publishers BV; 1985. p. 7–22.
23. Mesulam MM. From sensation to cognition. Brain. 1998;121:1013–52.
24. De Renzi E. Disorders of visual recognition. Semin Neurol. 2000;20:479–85.
25. Hart JT. Memory and the feeling-of-knowing experience. J Educ Psychol. 1965;56:208–16.
26. Maril A, Simons JS, Weaver JJ, Schacter DL. Graded recall success: an event-related fMRI comparison of tip of the tongue and feeling of knowing. NeuroImage. 2005;24:1130–8.
27. Penfield WP, Roberts L. Speech and brain mechanisms. Princeton University Press: Princeton; 1959.

The Diagram Makers and Their Critics

4

In the late nineteenth century, the theory of localization was strongly supported by experimental evidence from both electrocortical stimulation and cytoarchitectonics (for details, see Chaps. 5 and 6, respectively). Fritsch and Hitzig had demonstrated in 1870 that cortical stimulation at specific sites of the frontal lobe of a dog elicited contralateral muscular reactions, whereas stimulation of other regions did not result in any noticeable response [1]. Brodmann had parcellated the cortex of the human brain on the basis of histological differences in cell population and neuronal architecture and came up with 43 different areas in his now famous cytoarchitectonical brain maps [2].

4.1 Lichtheim

Following Broca and Wernicke, several authors published on language models and/or language-related anatomical regions. Ludwig Lichtheim (1845–1928) expanded Wernicke's diagram and explicitly added a 'concept centre' (Fig. 4.1a). This subsequently enabled him to explain why some patients had intact repetition despite a comprehension disorder or non-fluent speech. Lichtheim adhered to the same principles as Wernicke, namely, that speech is a learned behaviour that depends on centres for motor and sensory images and a reflex arc. He also assumed—as did Wernicke—that pure lesions (such as those specified with the numbers 1–7 in his model) would only seldom occur, as damage from pathological lesions was generally non-selective and more extensive. Lichtheim added anatomically distinct centres for reading and writing as a new layer to his model (compare Figs. 4.1b and 2.3c) [3]. He did not support his theoretical assumptions with significant experimental evidence and seemed aware of the hypothetical status of his new classification. Still, he associated his centre with anatomical regions:

> The motor-image center is localized in the '[part ...] of the lowermost left frontal convolution lying against the Sylvian Fossa' and the sound image center 'in the temporal convolution lying on the opposite site.' The connection between the two centers goes through the

G.-J. Rutten, *The Broca-Wernicke Doctrine*, DOI 10.1007/978-3-319-54633-9_4

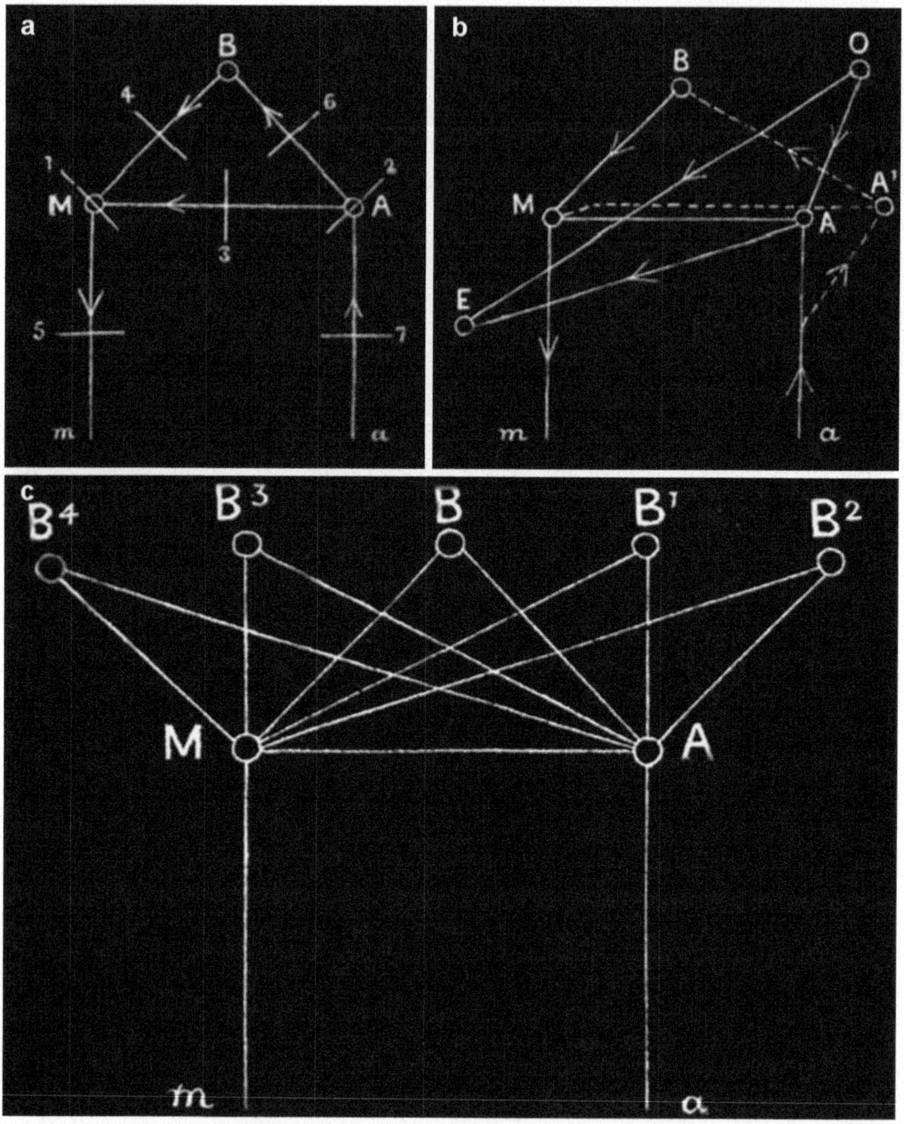

Fig. 4.1 (*Top left*, **a**) The famous Wernicke–Lichtheim or 'house model' in which Lichtheim described four new possible aphasic syndromes. Numbers *1–7* refer to theoretical lesion sites that are explained in the paper. *A* centre of auditory images, *B* concept centre, *M* centre of motor images, *a* acoustic pathway, *m* motor pathway. (*Top right*, **b**) The model has been expanded with centres to enable reading and writing. *O* centre for optical images, *E* centre from which the organs for writing are innervated. (cf. Wernicke's diagram in Fig. 2.3c). In addition to that, Lichtheim here indicated that a new set of pathways has shown up (*dotted lines*) after destruction of the original pathway AB, suggesting rewiring and plasticity of the brain. (Bottom, **c**) The concept centre 'B' is anatomically distributed over a wide region of the sensorial space and is not located in one particular region of the brain (Figures taken from Compston, 2006 [34])

insula or directly adjacent regions. The center for conceptions represents the exception: 'My view tends to assume [...] that the concept formation is not linked to a location in the brain but is a common function of the entirety of the sensory areas instead'. [4]

Although the word 'centre' suggests otherwise, Lichtheim did not believe that concepts could be localized to a particular brain area, as he explicitly stated in his paper:

Though in the diagram B is represented as a sort of centre for the elaboration of concepts, this has been done for simplicity's sake; with most writers, I do not consider the function to be localized in one spot of the brain, but rather to result from the combined action of the whole sensorial sphere. Hence the point B should be distributed over many spots; and the commissures MB and AB would not form two distinct and separate paths, but consists of converging radiations from various parts of the cortex to the point A and M [he refers to Fig. 4.1c; Lichtheim later called this the 'semantic field']. This admission does not do away with the possibility of the interruptions in the commissures BM, B1M, B2M, &c; but leads us to expect that any simultaneous break in them must occur close to their entrance into the lower centers M and A. [5]

Because of this new module 'B', the two language centres were not only connected anatomically but also via multiple conceptual representations that were spread throughout the cortex. The famous 'house model' was created (in line with previous and similar ideas of Wernicke that were discussed in Chap. 2). Lichtheim was able to add four new categories to the three forms of aphasia that had already been dealt with in Wernicke's model. Roth and Heilman describe these transcortical aphasias as follows (2000):

When connections from Wernicke's area to the concept center are disrupted, comprehension is impaired, because a semantic analysis of words cannot be performed. However, repetition is spared, because auditory information can access Wernicke's area and be transmitted to Broca's area for production of speech. This disorder is called 'transcortical sensory aphasia'. In contrast, when connections between the concept center and Broca's area are disrupted, internally generated speech (spontaneous speech or naming) will be halting and effortful, as in Broca's aphasia. However, because Wernicke's arc is intact, repetition is normal. This type of aphasia is called 'transcortical motor aphasia'. [6]

4.2 Kussmaul

In 1877, 8 years before Lichtheim, Adolph Kussmaul (1822–1902) also presented a patient with a transcortical sensory aphasia (i.e. with a comprehension disorder but intact repetition). His explanatory model resembled that of Lichtheim but differed in the fact that the 'concept centre' was only accessible via the phonological lexicon (i.e. Wernicke's area). Because of this, his model was unable to explain transcortical motor aphasia. But Kussmaul's ideas on language went beyond that of single word processing; he also considered the lexical and the sentence level and introduced the term 'agrammatism' to describe impairments in grammatical formulation. The diversity of symptoms in aphasia had also been recognized by other authors.

Bateman (1870), for example, pointed to the fact that aphasias can differ in their degree of severity, and deficits can be modality specific [4, 7]. Furthermore, he noted that speech with an emotional content ('automatic speech') can be preserved in aphasic patients and that in multilingual patients languages can be selectively affected. Both Kussmaul and Bateman emphasized the heterogeneity of aphasic phenomena and deliberately refrained from anatomical localizations as they believed that language covered an 'enormous association area' in the brain (Fig. 4.2). Kussmaul considered it naive to think that there is a set of language in any particular convolution of the brain and wrote that: 'Wernicke made the mistake of plotting the centre in specific areas of the brain. The localization of elementary functions of language is not mature enough for this' [8]. This criticism is not entirely justified, however. We have seen in Chap. 2 that Wernicke proclaimed an associationist view for cognitive functions and considered the entire perisylvian area to be involved in language processing.

The Wernicke–Lichtheim model had its flaws and, for instance, could not explain anomic aphasia. Many others adapted or augmented this model, sometimes with reference to specific brain areas. Variations were proposed among others by Bastian (1887), Charcot (1889), Dejerine (1891, 1892), von Monakow (1905), Henschen (1922) and Kleist, to name some of the more famous researchers

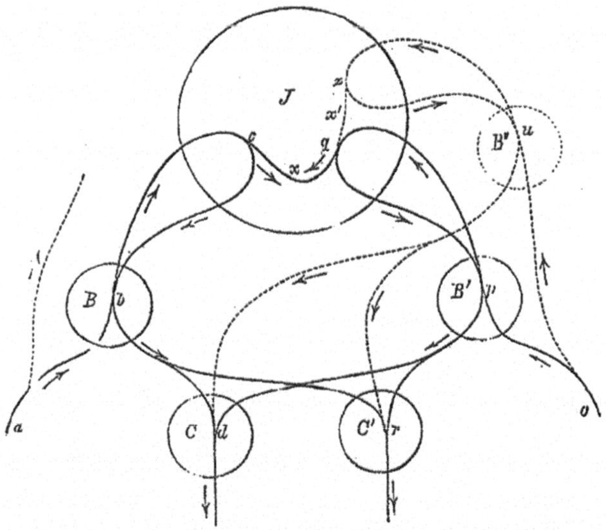

Fig. 4.2 Kussmaul's language model (1877) for word processing. Although the model resembles that of Wernicke, Kussmaul explicitly omitted any reference to brain anatomy. He considered 'language' far more complex than word processing alone. *d* route for speech, *r* route for writing, *B* centre for sound images (cf Wernicke's area), *B′* centre for optical images, *B″* centre for optical images for (deaf–mute) lip-readers only, *C* centre for coordination of sound movements into spoken words (cf Broca's area), *C′* centre for writing, *J* concept centre (Figure taken from Tesak and Code, 2008 [4])

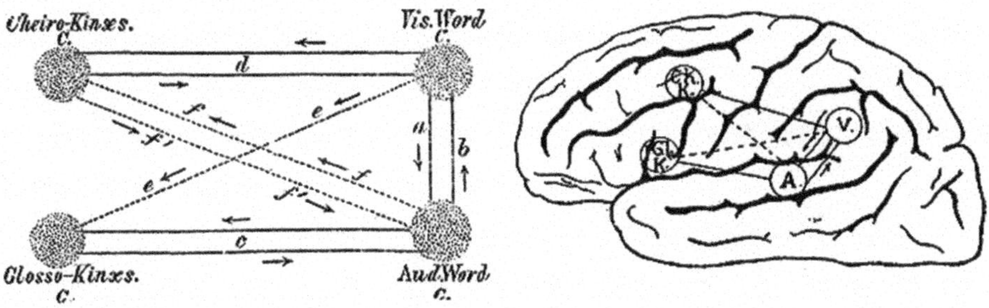

Fig. 4.3 Bastian's language diagram (1887) for word processing and corresponding anatomical sites and pathways. Note the absence of a concept centre, which was rejected by Bastian. The model has visual (V) and auditory (A) word centres and two sensorimotor centres that control kinaesthetic information from the tongue and hand muscles (hence, their respective names: glosso-kinaesthetic (GL.K.) and cheiro-kinaeshetic (CK.K.) centre (Figures taken from Head, 1926 [9])

(see Figs. 4.2, 4.3, 4.4, 4.5, 4.6 and 4.7) [4, 9–11]. It must be said that some of these diagram makers acknowledged the limitations of their models (and frankly admitted that they were best used for teaching or heuristic modelling), but many others seemed to believe that clinical language disorders could be deduced from their 'box-and-arrow' models.[a] This debate, in fact, is still continuing today. One of the main questions that remains to be answered is to what extent individual language maps (either healthy or diseased) differ from group language maps. The fact that even normal brains can significantly differ in size and shape (e.g. the pattern of gyri and sulci) obviously limits generalization of results into detailed anatomical-based models; see Chap. 5 for a detailed account. Additional difficulty is that brains are constantly changing their functional configuration because we never stop learning, or adapting otherwise to our environment. Under pathological conditions, functions can reorganize to such an extent that the resulting anatomo-functional configuration becomes significantly different from that of healthy subjects [12]. This pathology-driven process of reorganization seems particularly effective in slowly growing brain tumours (Fig. 4.9).

4.3 Hughlings Jackson

An alternative view to the static and neuroanatomical models was proposed by John Hughlings Jackson (1835–1911), who became one of the most influential neurologists in England. He made several important contributions to neurology and is also seen by many as a great pioneer in clinical neuroscience [13]. His ideas on aphasia and language representation in the brain were quite different from the localizationist

[a] An example of a more recent box-and-arrow model is shown in Fig. 4.8.

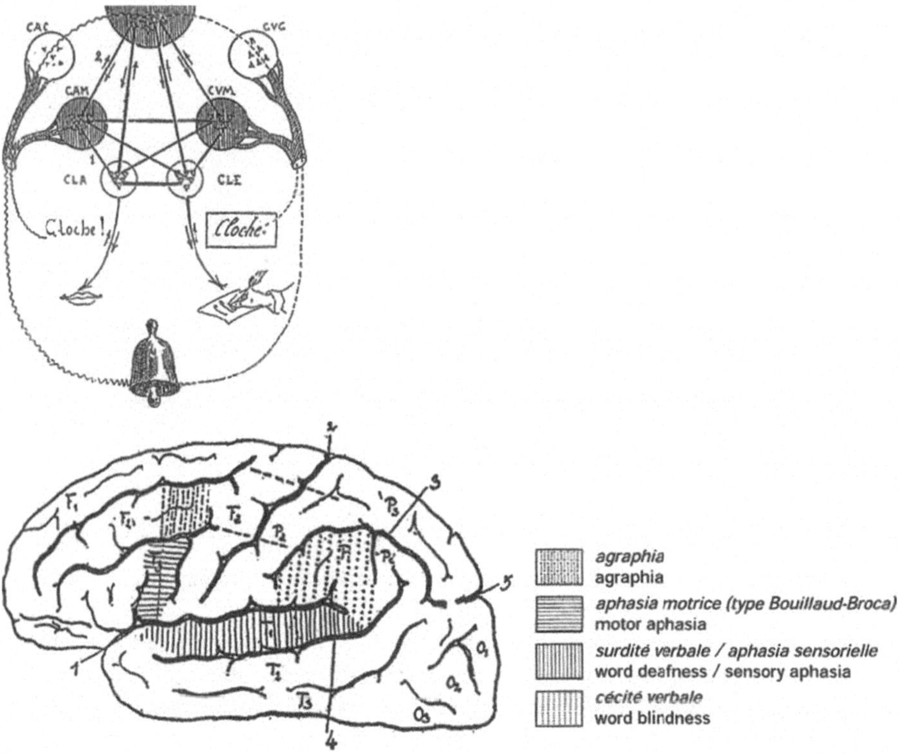

Fig. 4.4 Charcot's localization of language centres and aphasias (1889). (*Top*) Charcot's famous 'bell' diagram which is essentially the same as the model of Kussmaul. Again, there are four memory centres which are connected to an association centre. (*Bottom*) Localisation of aphasias. Some authors, for instance, Bogen and Bogen (1976), have argued that Charcot included T2 in his language regions [35]. *IC* association centre, *CAC* general auditory centre, *CAM* hearing centre for words, *CLA* centre for articulated speech, *CVC* general visual centre, *CVM* visual centre for words, *CLE* centre for writing, *1* Sylvian fissure, *2* Rolandic fissure, *F1/F2/F3* first/second/third frontal gyrus, *T1/T2/T3* first/second/third temporal gyrus, *O1/O2/O3* first/second/third occipital gyrus, *Ps* superior parietal lobe, *Pi* inferior parietal lobe (Figures and legends taken from Tesak & Code, 2008 [4])

views that prevailed in Germany and France at that time. Although he agreed that Broca's aphasia resulted from damage to the left frontal lobe, he did not believe there was a specific faculty of articulate language.

Hughlings Jackson was influenced by the works of Spencer and Darwin, whose *On the Origin of Species by Means of Natural Selection* was published in 1859. Hughlings Jackson's works on aphasia (published in the period 1864–1894) were rooted in these new biological and evolutionary principles, and he investigated language (and brain functions in general) from a more dynamic and psychological point of view than was done before. He considered functions to be organized

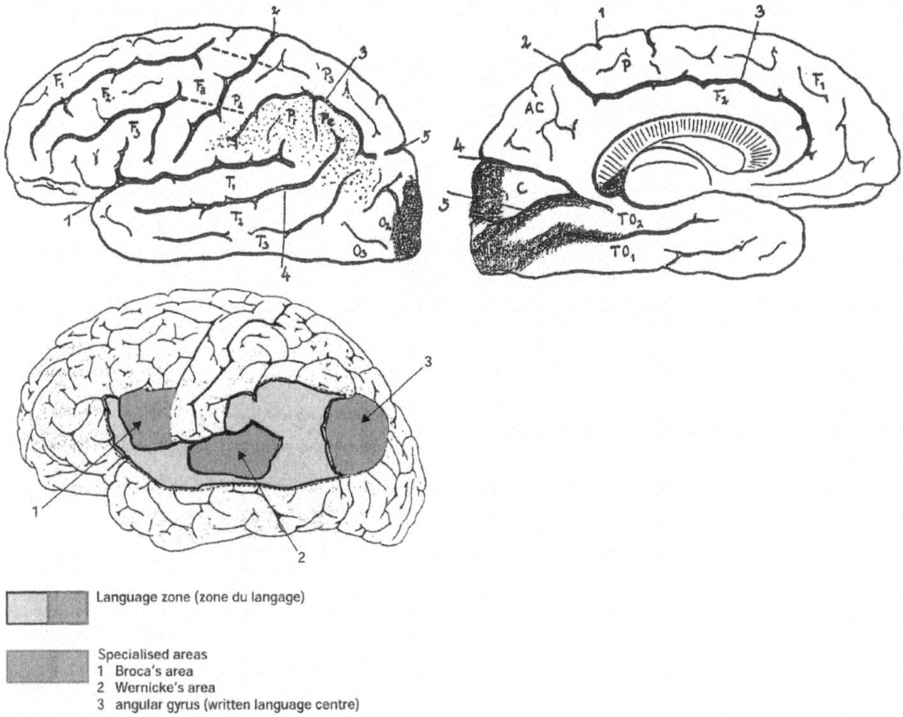

Fig. 4.5 Dejerine's work became important because of his descriptions of isolated disorders of reading (word blindness or alexia) and writing (agraphia). In 1891 he described a case of alexia with agraphia due to a lesion in the left angular gyrus [36]. In 1892 he described a 63-year-old man with pure alexia (i.e. without agraphia) due to a stroke. The patient died 5 years later due to a second stroke that affected the left angular gyrus (and led to paraphasia and agraphia). This case is shown in the figure and discussed here further. Post-mortem examination revealed that the old lesion (which had caused the reading disorder) occupied 'the [left] occipital lobe, and particular the circumvolutions of the occipital pole, starting at the base of the cuneus, as well as those of the lingual and fusiform lobules'. Dejerine argued that the extensive destruction of white matter in the left occipital lobe had destroyed the connecting fibres from the right occipital lobe to the language areas necessary for reading in the left posterior (and inferior) temporal lobe. This area is sometimes called the visual word form area, although its existence is disputed [10]. *Top figures* show the old and newer lesion (*dark and stippled areas,* respectively). For in-depth description, see works by Geschwind [37] and Dehaene [38]. (*Bottom figure*) Four years later, Dejerine (together with Mirallie) described another type of alexia ('third alexia') which can occur with Broca's aphasia in frontal lesions [39]. His explanation for this alexia was somewhat different and originated in his concept of a 'language zone'. Within this zone there are specialized cortical centres that are functionally integrated. According to Tesak and Code (2008): 'Cortical lesions in the language zone lead to a disorder of "inner speech" and create supramodal disorders such as alexia in motor aphasia' [4]. Although Dejerine's model of pure alexia has strong roots in connectionism, he also moved away from Wernicke's views when he assigned a specialized role to the angular gyrus in the visual representation of words. For Wernicke, higher functions were the product of connections, not cortical areas (*Top figures* taken from Geschwind, 1962 [37]; *bottom figure* taken from Tesak and Code, 2008 [4])

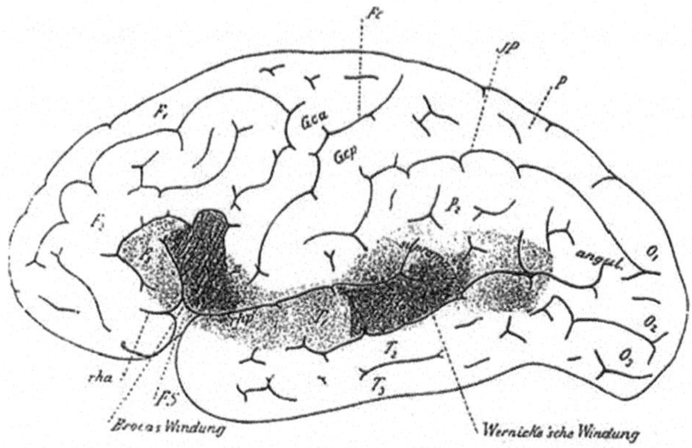

Fig. 4.6 The borders of the language regions in von Monakow's model (1905) are deliberately blurred. He was convinced that language regions could not be drawn with lines, as they probably abated gradually in all directions into the neighbouring gyri, and advocated a wide perisylvian language region [4, 11]. von Monakow was the first to describe the phenomenon of diaschisis, whereby a lesion could affect remote areas via its long-range connections. Part of the functional deficits could therefore be explained by distant effects, and as such he argued against strict localism. The figure shows language areas (*dark*) and 'relative fields' (*light*) (Figure taken from Tesak and Code, 2008 [4])

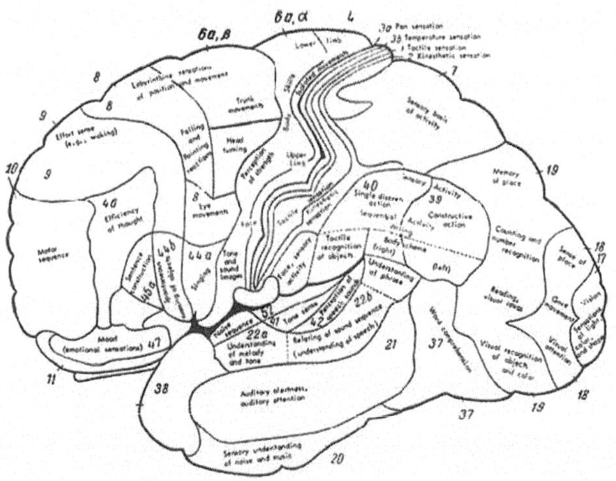

Fig. 4.7 Kleist (1934) was an 'extreme' localist. His detailed maps were based upon symptom–lesion studies in hundreds of brain-damaged soldiers from the World War I and more or less followed the cytoarchitectonic maps of Brodmann. He located sensory aphasias to Brodmann areas 42 (perception of speech sounds), 22a (understanding melody and tone) and 22b (understanding speech; understanding phrases). Motor aphasias were located to Brodmann areas 44a (singing), 44b (spoken naming) and 45a (syntactic speech). His maps were so detailed that they recall the maps of the phrenologists. Ironically, Kleist had been one of Wernicke's assistants (between 1903 and 1905) (Figure taken from Nieuwenhuys, Voogd en van Huijzen, 2007 [40])

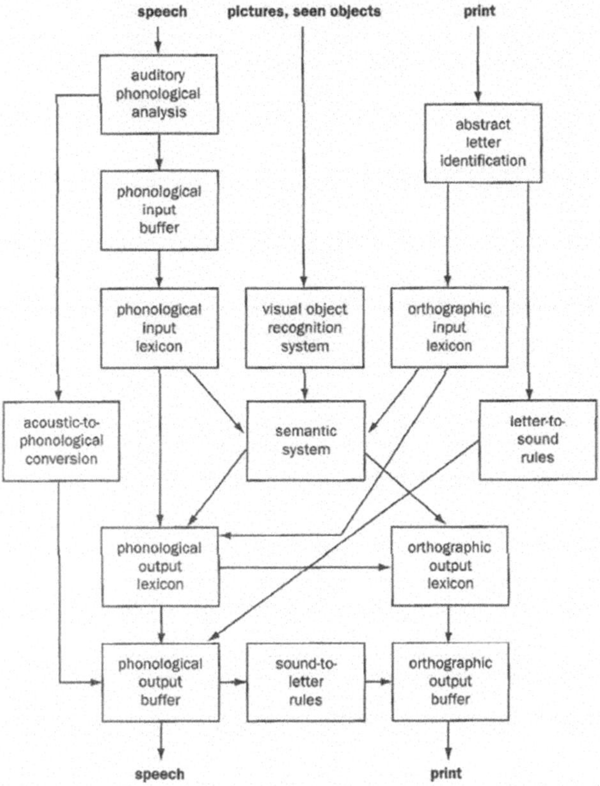

Fig. 4.8 A model is a human construct that aims to predict behaviour of real-world systems. It does so by simplification of various processes. An example of a box-and-arrow model for language is shown in the figure. This PALPA (Psycholingual Assessment of Language Performance in Aphasia) model is based on the assumption that the language system is organized in separate modules that can be selectively impaired by brain damage (Kay 1992) [41]. The more complex models are, the more difficult it is to localize functions, as these are 'distributed' across many different *boxes* and *arrows*. The visual system of the monkey, for example, can be deconstructed into more than 30 areas and over 300 connections [42]. Strict anatomical localization of function has now become impossible

hierarchically in the nervous system, with a gradient from the oldest functions to the most recently developed (both on an individual/ontogenetic level and on a species/phylogenetic level) and from the simplest to the most complex. Higher cognitive functions such as language were therefore controlled by the younger neocortical parts of the brain [13]. Hughlings Jackson stressed the importance of 'positive symptoms' and did not only focus on the 'negative symptoms' (i.e. neurological deficits). Positive symptoms were new behavioural phenomena from lower brain regions that occurred when damaged higher brain regions lost their inhibitory control over these regions. Lesions can thus influence the functioning of remote brain areas, hence his famous statement 'to locate the damage which destroys speech and

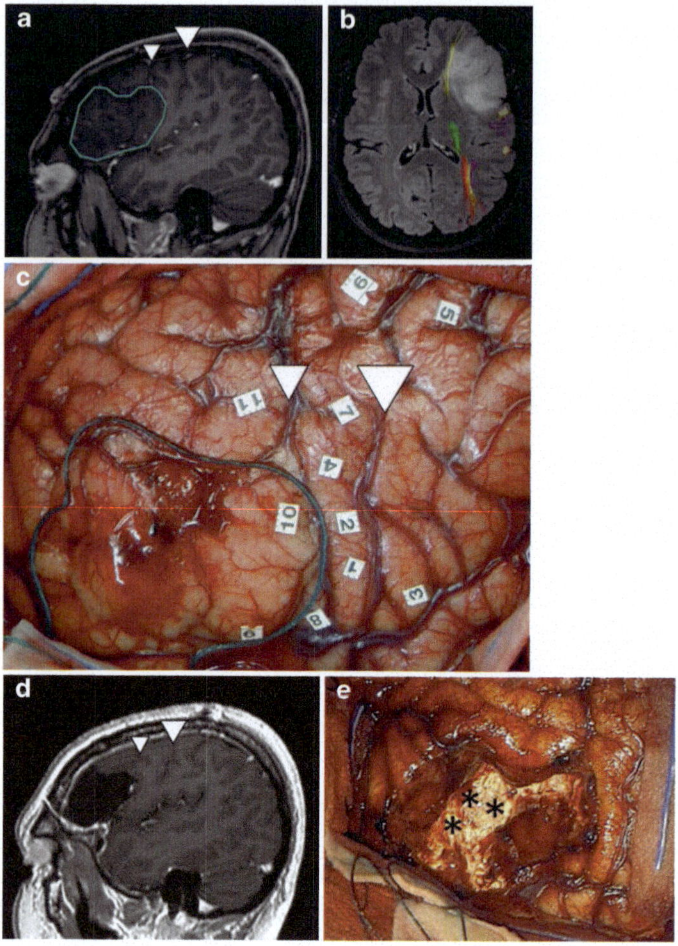

Fig. 4.9 (**a**, **b**) MR images of a 29-year-old patient with a low-grade glioma in the left inferior frontal gyrus (this is a type of brain tumour that typically takes years to become symptomatic). Clinical debut was a seizure; there were no (neurological) deficits. The tumour had invaded classic Broca's area. The posterior border was the precentral sulcus (*small arrow*), and cranial border was the inferior frontal sulcus. The *large arrow* points to the central sulcus. Information from MR tractography in figure (**b**) is shown with colours. The *yellow* tract is the inferior fronto-occipital fasciculus (IFOF), which represents the medial functional border of the resection. (**c**) Intraoperative photograph before tumour resection. Contour of the tumour has been marked with a small cord. *Small and large arrows* indicate precentral and central sulcus, respectively. *Numbered markers* indicate sensorimotor and language areas that were found with electrocortical stimulation. Markers *1, 2, 4, 7,* and *8* indicate the primary motor cortex. A speech arrest was found at markers *8* and *11*. Stimulation at markers *9* and *10* did not yield consistent language errors, and these were not considered critical language areas. (**d**) Postoperative MR image demonstrated macroscopical complete resection. (**e**) Tumour resection was performed up to sulcal borders. Asterisk indicates sites where subcortical stimulation resulted in (semantic) language impairments, likely due to stimulation of the IFOF. The patient had transient speech disorders that became clinically manifest on the second day after surgery. Three months after surgery, these had resolved. Such a case demonstrates that the functional and anatomical localization of Broca's area do not necessarily coincide and strongly suggests that reorganization of function took place prior to—and possibly also after—surgery

to locate speech are two different things' [14]. Hughlings Jackson made a distinction between propositional speech (whereby original ideas are encoded into novel utterances) and non-propositional speech (automatically generated speech of 'old' and overlearned utterances such as cursing, counting or praying) [4]. His ideas were neither recognized nor accepted for many years [15]. The same holds for Sigmund Freud (1856–1939) who, strongly influenced by the works of Hughlings Jackson, wrote a book on aphasia (1891) that explicitly criticized his contemporary 'diagram makers'. This (first) book *On aphasia: a critical study* [*Zur Auffassung der Aphasien: eine kritische Studie*] has generally been overlooked both in the psychoanalytical and neurological community [16]. It only sold 257 copies in 10 years and was, for instance, excluded from the Freud Standard Edition as not sufficiently psychological. It was not translated into English until 1953.

4.4 Freud

Some have criticized Freud's contribution to language, saying that he did not offer anything new and just reinterpreted and used the ideas of his predecessors [17]. However, the importance of this book, in which he not only criticized his contemporaries for their oversimplified views on language organization in the brain but also presented his own alternative views, is now widely recognized. Damasio pointed out that some of Freud's insights resonate in modern language theories [18]. Freud himself seemed very content with his book. In a letter (1891, to William Flies) he wrote:

> In a few weeks I shall afford myself the pleasure of sending you a small book on aphasia for which I myself have a great deal of warm feeling. In it I am very impudent, cross swords with your friend Wernicke, with Lichtheim and Grashey, and even scratch the high and mighty idol Meynert.[b]

In the opening sentence of his book, Freud boldly stated his goals and provided the reader with a list of his contemporaries all of whom—we know now—became famous for their contributions to the theories of aphasia and language organization in the brain:

> If, without having new observations of my own, I attempt to treat a topic to which the best minds in German and foreign neuropathology—such as Wernicke, Kussmaul, Lichtheim and Grashey, Hughlings Jackson, Bastian and Ross, Charcot, and others—have devoted their efforts, I had best immediately indicate the few aspects of the problem which I hope to advance through my discussion. [19]

[b] Quotation taken from Greenberg (1995) [19]

Greenberg wrote a fascinating book that places Freud on the crossroads of neurology and psychoanalysis.[c] According to Greenberg:

> Freud's purposes in 1891 were to refute prevailing theories and to propose one of his own. (...) The theories Freud wanted to refute are (1) that there is a distinction between aphasia caused by destruction of nerve centers and aphasia caused by destruction of neural pathways and (2) that the functions of the nervous system are restricted to anatomically definable areas, or localized. (...) Freud argued for a 'functional explanation' beyond the localizing one. He believed that the damage caused by a brain lesion could affect areas of the brain at remove from the lesion site. Therefore the many complex subvarieties of aphasic disturbances could not be accurately predicted by the site of, for example, a brain tumor. [19]

In 1883 Freud worked in Meynert's laboratory and psychiatric clinic. In 1885–1886 he visited Charcot, who at that time was not only an authority on aphasia but also on hysteria (he used hypnosis to treat hysterical patients) [20]. Freud also had good acquaintance with the English cultural and scientific community. He made his first trip to England when he was 19 years old, and there, according to Greenberg, 'his Anglophilia was confirmed' [19]. For his book Freud was particularly influenced by Hughlings Jackson who proposed that an individual psychological state could not immediately be attributed to a 'direct' cause in the brain. This gave Freud a neurological basis for his observations in hysteria, where patients presented with a paralysis or speech disturbances that could not be explained by organic lesions. Some authors have suggested that one of the reasons for writing the aphasia study was to understand the speech phenomena in these patients. Another probable reason is that Freud wanted to investigate in what manner the body is represented in the brain, whereby he suggested that these representations have no distinct localization and are more functional than topographical in nature [21].

Freud started his book with a lengthy and critical review of the models of Wernicke, Lichtheim and Grashey. He meticulously identified several alleged errors and did not refrain from bold statements such as 'Wernicke's conduction aphasia does not exist' (for the reason that he did not believe that repetition disorders were possible with intact spontaneous speech and comprehension). He also made a case against transcortical motor aphasia. Another important point is his discussion of paraphasic symptoms, which for him are not related to specific focal brain lesions. For Wernicke, paraphasias (he himself used 'Verwechslungen der Wörter') were the main characteristics of conduction aphasia with a lesion of the pathway between frontal and temporal areas (tract *ab* in Wernicke's figures in his monograph). However, Freud objected that self-monitoring remains possible even without this tract, namely, via the conceptual centres. Freud pointed to several other inconsistencies in Wernicke's description of the function of tract *ab*, a point on which Wernicke indeed remained somewhat vague about in his monograph (see Chap. 2) [22]. Freud proposed that in case of disruption

[c] At that time these specialities had not been 'born' and were only gradually evolving; Freud called himself a 'lecturer on neuropathology' on the title page of his book [19].

of tract *ab*, repetition would be kept intact only for 'known' words. Repetition of unknown words (for instance, words in an unknown foreign language) would no longer be possible because these words were never conceptualized. But he was sceptical of the existence of such a syndrome. Freud considered paraphasia more a property of the language system and a 'positive symptom' (in Hughlings Jackson's terminology) than the immediate consequence of brain damage:

> The paraphasia observed in patients does not differ in any respect from that mixing up and garbling of words which the healthy person can observe in himself when tired or distracted, or under the influence of disturbing affects.' For this reason, Freud concludes, 'paraphasia might well be regarded as a purely functional symptom, a sign of reduced efficiency of the apparatus of speech association'. By inserting an observation at the level of psychology, he has been able to expand the concept of a phenomenon otherwise limited in the neurological perspective and to fit the expanded concept like a supportive brick into the theoretical construct he is building, which argues for functional rather than purely anatomic and site-specific definition. Yet—and here we see the rhetorician in a characteristic maneuver—'this does not exclude its appearing in the most consummate form as an organic focal symptom [19]'.

Freud repeatedly referred to the work of Starr, who provided evidence from post-mortem investigations that supported Freud's statements.

> However, one meritorious author, Allen Starr, took the trouble of investigating the anatomic foundations of paraphasia. He came to the conclusion that paraphasia can be produced by lesions in very different regions. He found it impossible to discover a consistent difference in pathology of cases of sensory aphasia with or without paraphasia [19].

Freud considered that 'all aphasias are due to interruptions of association, i.e. conduction' [19]. He introduced a large 'language field' (German: Sprachfeld) in the left perisylvian cortex and proposed that disruption of fibres causes a disorganization within the language field and subsequently a specific type of aphasia. In this manner different language centres were created, although from Freud's point of view these had only pathological and no physiological significance. A later publication (1893) provides us with an illustration of Freud's language field in the brain (Fig. 4.10) [23]. Freud must have struggled with his attitude towards the diagram makers as his illustrations comprise more or less similar centres and pathways. It is not apparent to me on what evidence Freud based this anatomical drawing, but the four centres are strongly reminiscent of the language centres of Charcot (cf. Fig. 4.4). Freud greatly admired Charcot, after whom he named his first son (Jean Martin). Like Wernicke, Freud used words and their related concepts as the basis for language. He considered words and objects as complex association structures (Fig. 4.11). Between words and objects, there existed a 'symbolic' relationship. From this, Freud classified language disorders from a psychological viewpoint, and he distinguished between verbal aphasia, asymbolic aphasia and agnostic aphasia. His classification, and his book in general, was little noticed by most of Freud's contemporaries, and his influence on aphasiology has remained very limited [9, 15, 19, 24].

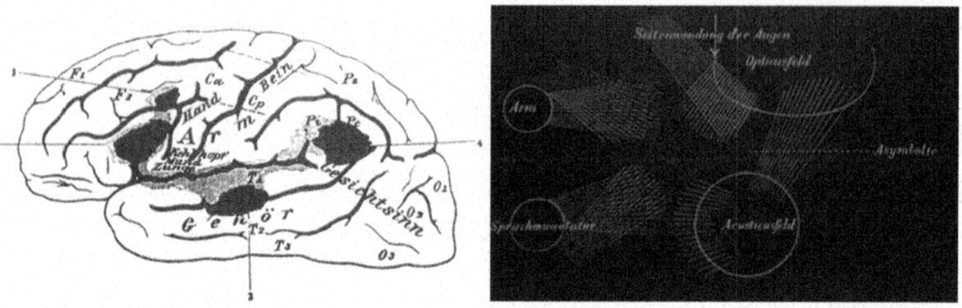

Fig. 4.10 (*Left*) Freud's representation (1893–1894) of his language field in the brain (hatched areas). A large part of this field lies deep within the Sylvian fissure. *Dark areas* represent the language centres. (1) Area where a lesion causes agraphia (the area is bordering the centre for hand motor control). (2) Broca's area where a lesion causes motor aphasia (adjoining the centres for muscles controlling speech and larynx). (3) Wernicke's area where a lesion causes word deafness (directly adjoining the field where the acoustic nerve or part of it terminates). (4) The area where a lesion causes alexia (directly adjoining the visual centres in the cortex). (*Right*) This is Freud's 'anatomic scheme of the language association field'; his abstracted explanation of the language centres. 'The cortical fields for acoustical, optical and arm muscles are schematized by *circles*; the association pathways that reach from them to the interior of the language field are represented by pencils of rays. Wherever the latter are crossed by pencils that have been cut off from their origins, a 'centre' for the relevant association element is created. For the acoustic field, the double-sided connections are not indicated, partly in order not to complicate the diagram and partly because of lack of clarity that exists about the precise relationship between the auditory field and the acoustic language centre. Dividing the connections with the optical field, also, spatially into two pencils permits taking into consideration the fact that eye movements are enlisted in a special way in the association process of reading' (*Left figures* and legend taken from Greenberg, 1997 [19] reproduced from Kästle (1987) [23]. *Right figure* and legend taken from Greenberg, 1997 [19] reproduced from Freud (1891) [43])

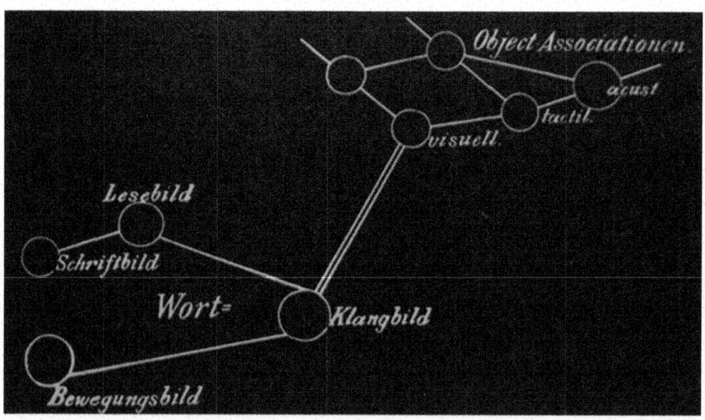

Fig. 4.11 Freud's *Psychological diagram* (1891) of a word presentation [Vorstellung]. The word presentation is shown as a closed complex of presentations, whereas the object presentation is shown as an open one. The word presentation is not linked to the object presentation by *all* its constituent elements, but only by its sound image [Klangbild]. Among the object associations, it is the visual ones that stand for the object, in the same kind of way as the sound image stands for the word. *Lesebild* reading image, *Schriftbild* writing image, *Bewegungsbild* motor image (Figure and legend taken from Greenberg, 1997 [19] that reproduced it from reference [43])

4.5 Marie, Head and the Decline of Localism

A number of developments eventually led to the decline of localism and its almost complete disappearance in the first half of the twentieth century. Besides new scientific points of view, changes in the political and cultural climate probably played an important role.

> The explosion of the exact sciences before the Great War created a mechanical image of the world that excited great confidence in scientific progress. The mechanized slaughter of the Great War shook that confidence and produced a postwar cultural and literal backlash that had an antimechanical, even antiscientific tinge (Forman, 1984). The industrial carnage in the trenches of World War I intensified antimechanistic views in Europe that carried over to localization theory. Medical science was as tainted as physical science; if science could not be trusted to provide answers to the problems of existence in the modern world, then society must find some alternative in order to fill those needs. A view of life, the mind, and the nervous system as an integrated whole produced an approach to scientific neurology that stood in marked contrast to reductionist localization theory. [25]

Geschwind (1964) added to that the effects of the German defeat:

> Head (1926) had been shrewd enough to point out that much of the great German growth of neurology had been related to their victory in the Franco-Prussian war. He was not shrewd enough to apply this valuable historical lesson to his own time and to realize that perhaps the decline of the vigor and influence of German neurology was strongly related to the defeat of Germany in World War I and the shift of the center of gravity of intellectual life to the English speaking world, rather than necessarily to any defects in the ideas of German scholars. [26]

Gradually, the models and theories of 'association psychology' were replaced by theoretical developments from other schools. The many brain-injured soldiers from the war resulted in a renewed interest in rehabilitation and therapies for aphasia. Theories of brain plasticity and reorganization did not fit well with the static and classical language approach [11]. There was also a growing interest in language processing on a sentence level, and these processes and their grammatical impairments were difficult to implement in the Wernicke–Lichtheim model (that was constructed on the level of word processing). Steinthal and Pick, among others, tempted to incorporate linguistics into aphasiology [27]. One of the most well-known advocates for a more holistic approach was Goldstein, who was initially influenced by Wernicke (his teacher) and later by Freud's book on aphasia [22]. Goldstein referred to the old theories, with their anatomically based centres and impairments of discrete linguistic faculties, as 'atomistic'. Instead he believed that:

> every individual speech-performance is understandable only from the aspect of its relation to the function of the total organism in its endeavor to realize itself as much as possible in the given situation. [28]

Although to a modern reader such a statement is probably 'obtuse and awkward', as Henderson formulated it, this fitted well within the school of Gestalt psychology [22]. For those who adopted a more holistic approach, language deficits did not merely reflect a loss of function of a particular brain region, but provided important information about the remaining functionality of undamaged regions. Goldstein (1948) examined thousands of brain-damaged patients and noted that the classic models were unable to explain the many complex behavioural and neurological pictures that he observed [28]. He was convinced that 'we are by no means justified in inferring directly from a correlation between a localized defect and a defect in performance a relationship between the concerned area and a definite performance corresponding to the defect' [28]. Such a statement would only be true when the brain is completely static and unresponsive to the effects of damage, which is obviously not the case. Still, direct and causal inferences between the site of the brain damage and the indispensability of that region are still—erroneously—often made. Goldstein formulated new theories whereby he considered the nervous system as a network that always acts and reacts as a whole. He emphasized, in the spirit of Hughlings Jackson, that both normal and abnormal reactions (he refers to the latter as 'symptoms') 'are only expressions of the organism's attempt to deal with certain demands of the environment' [29]. Symptoms are therefore not merely the result of focal damage, but the expression of a new functional equilibrium that has been reached by the diseased organism.

It is tempting to classify any of the approaches to aphasia and language modelling either as anatomically based localism or psychologically based holism. However, almost all early researchers used ideas from both approaches and also acknowledged that in a pure form either localism or holism is of no practical use (or even theoretical use, for that matter). Regarding Goldstein only as a holist is just as wrong as characterizing Wernicke solely as a localist [30]. Geschwind noted that Goldstein, 'despite the holistic views which he expressed in his philosophical discussions, actually explicitly stated his support of the classical localizations throughout his career' [31]. Goldstein readily acknowledged that brain areas were to some extent functionally specialized, as his work with brain-damaged patients had clearly taught him. But, he remarked, with the lesion-deficit approach 'only very gross localizatory distinctions are possible' [28]. Goldstein thus did not strictly oppose localism, but merely pointed out that it had severe limitations and was of limited use in the individual patient.

In any case, the more holistic theories prevailed in the period after the World War I. In many historical accounts, Marie and Head are given crucial roles in the downturn of what Head had derogatorily had called the 'diagram makers'. Marie had done extensive work in aphasia. According to Tesak and Code (2008), he produced at least 14 articles on aphasia in the period 1906–1907, reporting cases whereby damage to the inferior frontal gyrus had not led to aphasia and vice versa, cases where Broca's aphasia was related to lesions outside this area [4]. Marie had also re-examined Broca's patients Leborgne and Lelong and claimed that these cases did not support Broca's theory.[d] He concluded from his own anatomo-clinical studies that (1) there is only one single form of aphasia (sensory); (2) lesions that cause aphasia are located in the territory of Wernicke's area (gyrus supramarginalis, gyrus angularis and feet of the first two temporal gyri); (3) aphasic patients always show a decline in general cognitive functioning, because language and thinking cannot be separated; (4) a motor Broca's aphasia does not exist, and the foot of the third

[d] See Chap. 1 for a detailed account.

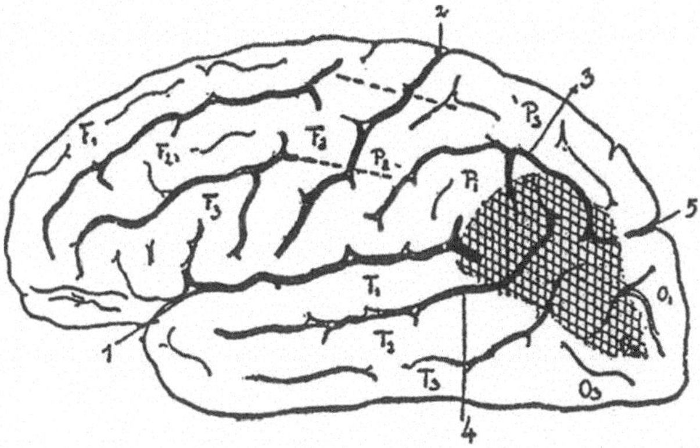

Fig. 4.12 Marie's (1906) depiction of Wernicke's area. As Marie did not believe in the existence of Broca's area as a language area, he left it out of the picture (for Marie, a failure to comprehend language was a definite sign of aphasia, a problem with was speech not). As noted by Bogen and Bogen (1976), Marie slightly modified the language territory in later years, whereby he included also the inferior part of the Rolandic cortex [35, 44] (Figure taken from Tesak and Code, 2008 [4] who reprinted it from Marie, 1906 [45])

frontal (gyrus) has nothing to do with language functions; and (5) motor aphasia is a combination of sensory aphasia with anarthria (Fig. 4.12).[e]

In 1908 there was a heated debate between Marie and his rival Dejerine, whereby Marie passionately opposed the doctrine of localization of function. The meeting ended inconclusive regarding a consensus on language localization in aphasic patients, but 'by the end of the debate most neurologists seemed to agree that there was a need to question the traditional teachings on aphasia' [32, 33].

Many consider the influential work of Head to have strongly contributed to the downfall of the localization theories. In his comprehensive overview, *Aphasia and Kindred Disorders of Speech*, Head (1926) dismissed and even denigrated the information processing models that had been developed by investigators like Wernicke or Lichtheim. Head wrote that 'the writers of this period were compelled to lop and twist their cases to fit the procrustean bed of their hypothetical conceptions' [9].

After a period of silence in the first half of the twentieth century, in which a less anatomical and more holistic approach prevailed in aphasiology, Geschwind would reintroduce the concept of cortical disconnection in the 1960s (see Chap. 7) and renew the interest in language pathology. Like Head, he had thoroughly re-examined the historical literature. Benson and Ardila (both renowned aphasiologists and trained under Geschwind and Luria, respectively) concluded that: 'While not excluding considerable validity to the holistic concept, Geschwind warned that failure to consider the anatomical basis of language was actually misleading. His combination of pertinent clinical observations, superb scholarship, and vigorous presentation of a rational scientific-philosophic approach won the day' [15].

[e] Note that Marie was convinced that the left inferior frontal region (i.e. Broca's area) has nothing to do with language.

References

1. Fritsch GT, Hitzig E. Über die elektrische Erregbarkeit des Grosshirns. Arch Anat Phys. 1870;37:300–32.
2. Brodmann K. Vergleichende lokalisationslehre der Grosshirnrinde in ihren prinzipien dargestellt auf grund des Zellenbaues. Leipzig: J.A. Barth; 1909.
3. Lichtheim L. On aphasia. Brain. 1885;7:433–84.
4. Tesak J, Code C. Milestones in the history of aphasia: theories and protagonists. Hove: Psychology Press; 2008.
5. Davidoff J, editor. Brain and behaviour: critical concepts in psychology. London: Routledge; 2000.
6. Roth HL, Heilman KM. Aphasia: a historical perspective. In: Nadeau SE, Gonzalez-Rothi LJ, Crosson BA, editors. Aphasia and language: theory to practice. New York: Guilford Press; 2000. p. 3–29.
7. Bateman F. On aphasia or loss of speech, and the localization of the faculty of articulate language. London: John Churchill & Sons; 1870.
8. Kussmaul A. Die Störungen der Sprache. Versuch einer Pathologie der Sprache. Leipzig: FCW Vogel; 1877.
9. Head H. Aphasia and kindred disorders of speech. Cambridge: Cambridge University Press; 1926.
10. Price CJ, Devlin JT. The myth of the visual word form area. NeuroImage. 2003;19:473–81.
11. von Monakow C. Gehirnpathologie. Vienna: Alfred Hölder; 1905.
12. Duffau H. Lessons from brain mapping in surgery for low-grade glioma: insights into associations between tumour and brain plasticity. Lancet Neurol. 2005;4:476–86.
13. Franz EA, Gillett G. John Hughlings Jackson's evolutionary neurology: a unifying framework for cognitive neuroscience. Brain. 2011;134:3114–20.
14. Jackson JH. On the nature of the duality of the brain. 1874:41.
15. Benson DF, Ardila A. Aphasia: a clinical perspective. New York: Oxford University Press; 1996.
16. Vogel P. Sigmund Freud zur Auffassung der Aphasien–eine kritische Studie. Frankfurt: Fischer Taschenbuchverlag; 2001.
17. Marx OM. Aphasia studies and language theory in the 19th century. Bull Hist Med. 1966;40:328–49.
18. Damasio AR. Aphasia. N Engl J Med. 1992;326:531–9.
19. Greenberg VD. Freud and his aphasia book: language and the sources of psychoanalysis. Ithaca: Cornell University Press; 1997.
20. Bogousslavsky J. Sigmund Freud's evolution from neurology to psychiatry: evidence from his La Salpetriere library. Neurology. 2011;77:1391–4.
21. Rizzuto AM. The origins of Freud's concept of object representation ('Objektvorstellung') in his monograph 'on aphasia': its theoretical and technical importance. Int J Psychoanal. 1990;71:241–8.
22. Henderson VW. Early concepts of conduction aphasia. In: Kohn SE, editor. Conduction aphasia. Hillsdale: Lawrence Erlbaum; 1992. p. 23–38.
23. Kästle OU. Einige bisher unbekannte Texte von Sigmund Freud aus den Jahren 1983/94 und ihrstellenwert in seiner wissenschaftlichen Entwicklung. Psyche. 1987;41:508–23.
24. Jelliffe SE. Sigmund Freud as a neurologist. J Nerv Ment Dis. 1937;85:696–711.
25. York GK. Localization of language function in the twentieth century. J Hist Neurosci. 2009;18:283–90.
26. Geschwind N. The paradoxical position of Kurt Goldstein in the history of aphasia. Cortex. 1964;1:130.
27. Pick A. Aphasie und Linguistik. Germanisch-Romanische Monatsschrift. 1920;8:65–72.
28. Goldstein K. Language and language disturbances. Aphasic symptoms complexes and their significance for medicine and theory of language. New York: Grune and Stratton; 1948.

29. Goldstein K. The organism: a holistic approach to biology derived from pathological data in man. New York: American Book Company; 1934.
30. Ludwig D. Language and human nature: Kurt Goldsteins' neurolinguistic foundation of a holistic philosophy. J Hist Behav Sci. 2012;48:40–54.
31. Geschwind N. The organization of language and the brain. Science. 1970;170:940–4.
32. Brais B. The third left frontal convolution plays no role in language: Pierre Marie and the Paris debate on aphasia (1906-1908). Neurology. 1992;42:690–5.
33. Paciaroni M, Bogousslavsky J. Jules Joseph Dejerine versus Pierre Marie. Front Neurol Neurosci. 2011;29:162–9.
34. Compston A. From the archives. On aphasia. Brain. 2006;129:1347–50.
35. Bogen JE, Bogen GM. Wernicke's region—where is it? Ann N Y Acad Sci. 1976;280:834–43.
36. Dejerine J. Sur un cas de cécité verbale avec agraphie, suivi d'autopsie. Mém Soc Biol. 1891;3:197–201.
37. Geschwind N, Devinsky O, Schachter SC. Norman Geschwind: selected publications on language, epilepsy, and behavior. Boston: Butterworth-Heineman; 1997.
38. Dehaene S. Reading in the brain: the new science of how we read. New York: Viking Adult; 2009.
39. Henderson VW. Jules Dejerine and the third alexia. Arch Neurol. 1984;41:430–2.
40. Nieuwenhuys R, Voogd J, Huijzen CV. The human central nervous system: A synopsis and atlas. Springer; 2007.
41. Kay J, Lesser R, Coltheart RM. Psycholinguistic assessment of language performance in aphasia. Lawernce Erlbaum: Hove; 1992.
42. Felleman DJ, Van Essen DC. Distributed hierarchical processing in the primate cerebral cortex. Cereb Cortex. 1991;1:1–47.
43. Freud S. Zur Auffassung der Aphasien: Eine kritische studie. Leipzig: F. Deuticke; 1891.
44. Marie P, Foix C. Les aphasies de guerre. Masson; 1917
45. Marie P, Moutier F. Nouveau cas d'aphasie de Broca sans lésion de la troisième frontale gauche. Bull Mem Soc Med Hop Paris. 1906;23:1180–3.

Naming and Numbering the Convolutions

5

Early anatomists failed to see order amidst the apparent chaos of the brain's convolutions. It was only in the second half of the nineteenth century that studies of the topographical pattern of gyri and sulci gradually gained weight and importance. Alexander Ecker (1816–1887) was among the first who mapped the convolutions and sought for anatomical order among individual brains (Fig. 5.1). In the Introduction of his book, *On the Convolutions of the Human Brain* (1869), he wrote:

> For men were wont to regard the convolutions as a series of folds without order or arrangement, and draughtsmen represented them much as they would a dish full of macaroni. It was only by degrees that certain sulci and gyri came to be recognized as more constant than others; but as long as attention was confined only to the fully-developed human brain, real progress was not possible. Comparative Anatomy and the History of Development—those beacons of Human Anatomy—have been also the first to shed light upon this dark corner; for it was the labours of Huschke and, in particular, of Gratiolet, directed towards the brain of apes, that have established the conformity, in structural style, of the brain of apes with that of man, and have thereby for the first time paved the way towards a comprehension of the latter. [1]

Ecker and his contemporaries considered embryology (i.e. the study of the developing brain) and comparative anatomy (i.e. the study of the brains of different mammals and primates) indispensable:

> to learn some day or other to recognise a law for the formation of the convolutions - that is to say, to learn to recognise and comprehend the formation of the convolutions as a necessary consequence of certain mechanical antecedents in the growth of the brain and the skull. [1]

G.-J. Rutten, *The Broca-Wernicke Doctrine*, DOI 10.1007/978-3-319-54633-9_5

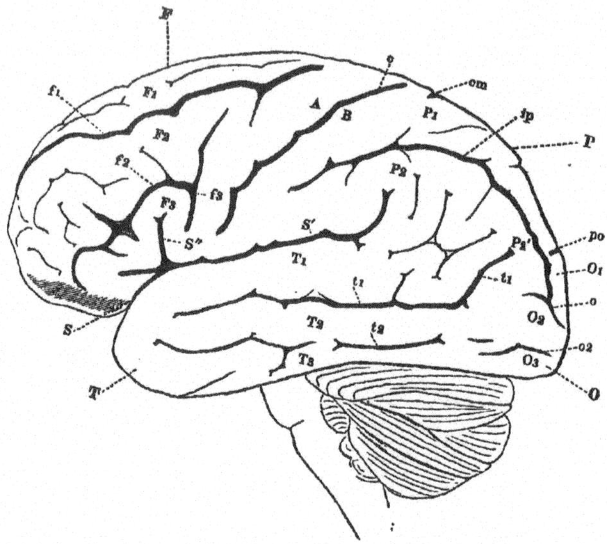

Fig. 5.1 Schematic brain from Ecker's *On the Convolutions of the Human Brain* (1873). Ecker says about his illustrations that they 'should be considered less in the light of pictures than of maps by the aid of which the traveller will be in a position to better shape his course in the district which he is exploring' (Figure taken from Ecker, 1873 [1])

5.1 Ecker, Leuret and Gratiolet: Order Out of Chaos

Ecker especially wrote his book for the physician who studied the brain, 'so that he may be capable of registering with accuracy the all-important observations upon the pathological changes in the cortex of the cerebrum' [1]. He included a short but systematic guideline for practical identification of the convolutions in the Appendix, which is still very useful today.[a] Ecker advises his readers always to make a sketch of the portion of the cortical surface where the convolutions are difficult to identify or have an abnormal arrangement. He specifically refers to the diopter of Lucae for this purpose (Fig. 5.2) and to the—now famous—wax models of Ziegler which Ecker had helped to design.[b]

Credit, for naming of the gyri and sulci, should probably go to Louis Pierre Gratiolet (1815–1865) and his teacher François Leuret (1797–1851). Their two-volume book (1839; 1857) was a unique source for several generations; it was the first large work to offer an organized description of the convolutions (Fig. 5.3) [3, 4]. The second volume was written solely by Gratiolet because of the illness and death of Leuret. Their book also paid homage to Rolando by coining the term *fissure of Rolando* [3]. Luigi Rolando (1773–1831) had expressed the new concept of the

[a] It starts with what we would now refer to as the anterior Sylvian point, the anterior part of the Sylvian fissure where the horizontal and anterior ascending rami arise [2]. From there on, the operculum and the central sulcus are identified, and these are subsequently used as landmarks for further topographical description.

[b] See for examples of the latter the website of the Hunterian Museum in Glasgow; http://www.huntsearch.gla.ac.uk.

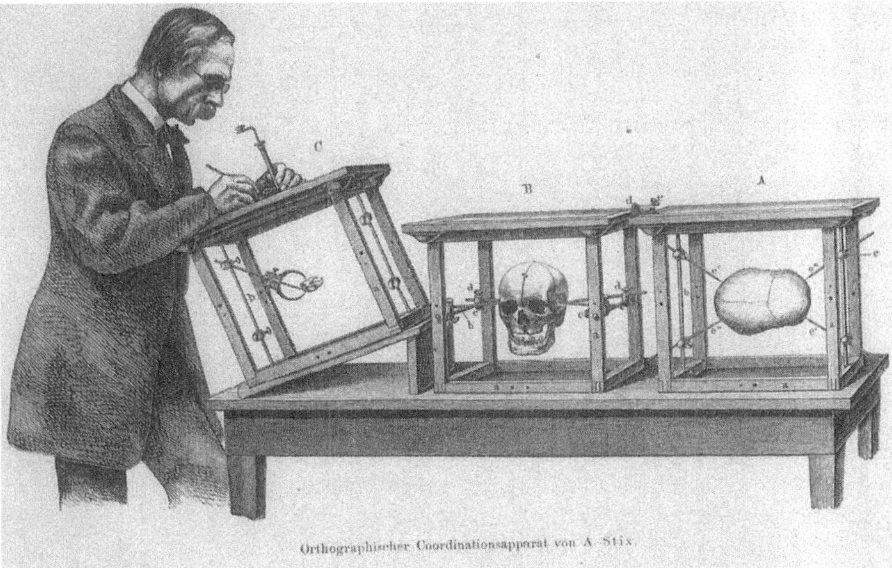

Fig. 5.2 Anthropologists sought for methods to standardize their observations. They generally accepted that drawing, rather than photography, was the most accurate means of representing skulls, because the expert could then control the representation. One of the most famous examples is the device invented by Lucae. The figure shows a small skull that is immobilized and an observer who uses a diopter in his left hand. This setup assures that his gaze always has a perpendicular perspective on the object. The object is drawn on a glass plate, and the ink on the glass is later transferred to paper. The result will be a geometric projection of the object (Figure taken from Zimmerman, 2001 [9])

regularity of convolutions 10 years earlier (Fig. 5.4) [5]. Gratiolet also completed a monograph of his own, *Mémoire sur les plis cérébraux de l'homme et des primatès (Cerebral folds of man and the primates)* [6]. However, he was initially reluctant to assign functions to the structures he named:

> In a general manner, I agree with M. Flourens that the intelligence is one, that the brain is one, that it acts above all as a whole; but this does not exclude the idea that certain faculties of the mind stand in special relation, although not exclusively, with certain cerebral regions.[c]

In 1861, Gratiolet and Broca engaged in a famous dispute in the Anthropological Society in Paris. Broca, at that time, had established himself as the head of French anthropology. He had been the founder of the Society (in 1859) and already enjoyed a good scientific reputation. Gratiolet's work, in stark contrast, was greatly undervalued during his life and he lived in great poverty and hardship [7]. The topic

[c] Quotation taken from Pearce (2006) [7]

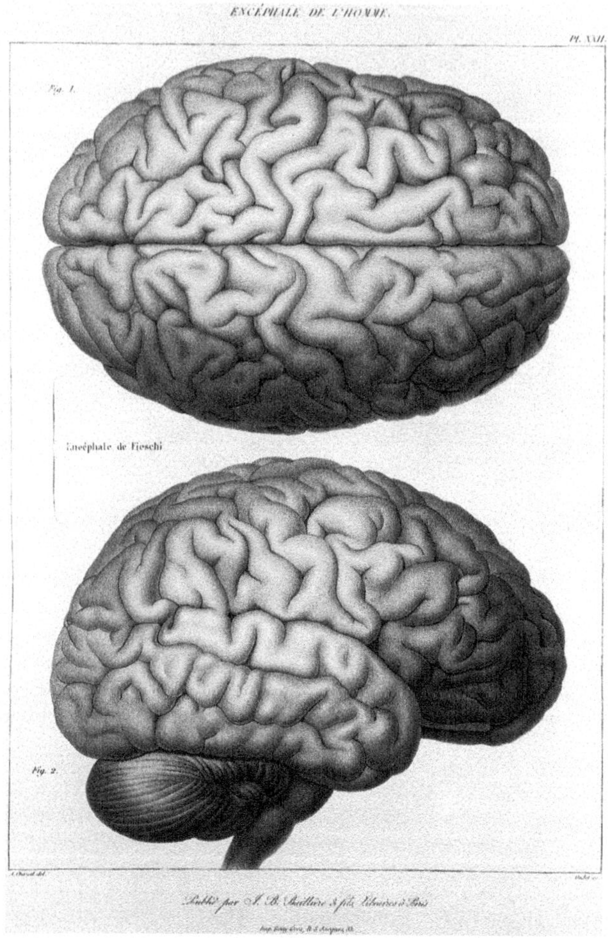

Fig. 5.3 Human brain (drawings from Gratiolet and Leuret (1839–1857) [4])

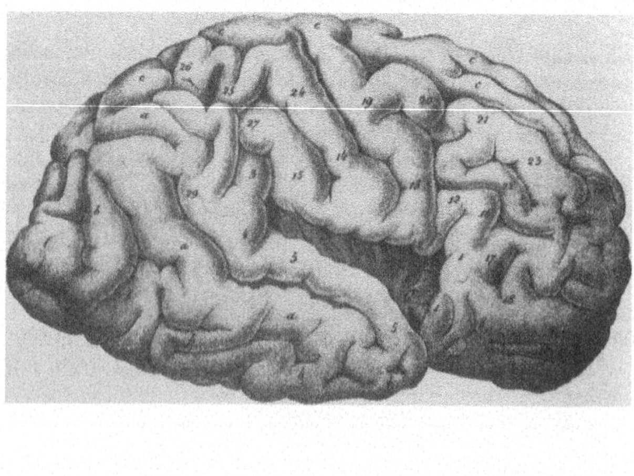

Fig. 5.4 Rolando's illustration of the human cerebral cortex (1830) [5]

of the debate was the relation between intelligence and brain size, a common topic of discussion at that time. Broca was a strong proponent of a direct and causal relationship and sought evidence for why some individuals or groups were more successful than others:

> In general, the brain is larger in men than in women, in eminent men than in men of mediocre talent, in superior races than in inferior races. Other things equal, there is a remarkable relationship between the development of intelligence and the volume of the brain.[d]

It is here where anthropology, phrenology and evolutionary theories meet. 'Races' were compared and ranked according to scientific measurements, including those of the skull and the brain [9]. A good friend of Broca, Alphonse Bertillon (1853–1914), was a Parisian police officer who believed that criminals could be recognized on the basis of physical characteristics. He developed techniques and instruments to measure various features that would not change in adult life, e.g. eye colour, the shape of the ears and the distance between the eyes. One of the most famous proponents of 'anthropometrics' was Cesare Lombroso (1835–1909), an Italian physician and psychiatrist. His work on criminality, which is now discredited, laid the foundations for modern criminology. In his book *Criminal Man*, he argued that some people were born criminals and that they were throwbacks (*atavistics*) to a primitive stage of evolution. Lombroso believed that this primitiveness could be read from their bodies and their habits. His theories were met with fierce criticism, as others argued that criminals were not genetically predisposed but rather a product of social inequality and poverty. In later years Lombroso acknowledged that criminality is the result of both individual and social factors. Despite his views he was one of the first who advocated the rehabilitation and humane treatment of prisoners, in particular because he considered them not responsible for their own behaviour.

Broca greatly contributed to anthropology, not only because he devised numerous measuring devices but also because he systematically questioned the generally made assertions about 'inferiority' of races and thereby pointed out several fallacies [3]. But, as Schiller (1992) puts it in his biography of Broca:

> We must not expect a Broca, a Lincoln, indeed any enlightened minds, to have believed in racial equality. The attitude was humanitarian, at best. Plain common sense, and even the most careful observation by the means then available, clearly showed that other races were unable to meet white standards measured by white values, in science, technical achievement, or art. [3]

Broca and Gratiolet share a common history. Coincidentally, they were both born in Sainte-Foy-la-Grande, a small village some 70 km east of Bordeaux.[e] Both had also been students of Leuret. Their debate lasted for 5 months and centred around

---

[d] Quotation taken from Gould (1992) [8]

[e] In Google street view, you can walk from Boulevard Gratiolet to the avenue Paul Broca [7].

the massive head and large brain of Georges Cuvier (1769–1832), one of the scientific giants of that era. Cuvier's brain weighted 1830 g, 400 above the average. Steven J Gould wrote a fascinating and amusing essay on the debate title 'Wide hats and narrow minds'; here is an excerpt:

> Thus, when Cuvier died, his colleagues, in the interest of science and curiosity, decided to open the great skull. (...) They began with the internal organs and, 'finding nothing very remarkable', switched their attention to Cuvier's skull. 'Thus', wrote the physician in charge, 'we were about to contemplate the instrument of his powerful intelligence.' And their expectations were rewarded. (...) Broca pushed his advantage and rested a good part of his case on Cuvier's brain. But Gratiolet probed and found a weak spot. In their awe and enthusiasm, Cuvier's doctors had neglected to save either his brain or his skull at all. The figure of 1830 g for a brain could not be checked; perhaps it was simply wrong. Gratiolet sought an existing surrogate and had a flash of inspiration: 'All brains are not weighted by doctors', he stated, 'but all heads are measured by hatters and I have managed to acquire, from this new source, information which, I dare to hope, will not appear to you as devoid of interest.' In short, Gratiolet presented something almost bathetic in comparison with the great man's brain: he had found Cuvier's hat. And thus, for two meetings, some of France's greatest minds pondered seriously the meaning of a worn bit of felt. [8]

Although Gratiolet did not relate brain function to size, he eventually did relate it to structure [7]. Others, like Leuret, related intelligence to the number of convolutions and the complexity of their pattern, as this had obviously increased over time in species and primates [7].

5.2 Microscopic Cartography

Over the course of the nineteenth century, the microscope was improved to such an extent that it became possible to visualize individual brain cells and characterize their structure and architecture. Investigators such as Jan Purkinje (1787–1869), Otto Deiters (1834–1863), Wilhelm His (1831–1904) and Theodor Meynert (1833–1892) described different types of nerve cells with their axons and dendritic branches [10]. What was also observed was that these nerve cells formed 'nets' and were somehow connected to one another (Fig. 5.5). This posed the question whether the nervous system was one huge reticulum or made up of individually and anatomically distinct units. Were nerve cells directly connected to each other (as believed by the *reticulists*) or was there a gap between cells (as believed by the *neuronists*)? This 'neuron doctrine' was the topic of a long (and often bitter) scientific struggle, personified by two great names that both of whom received the Nobel Prize for their work in 1906: Camillo Golgi (1843–1926) and Santiago Ramón Cajal (1852–1934). The work of Cajal eventually provided most of the evidence for what Heinrich Waldeyer (1836–1921) would name 'the neuron'. Cajal proved that axons communicate with other cells across a gap (later

Fig. 5.5 von Kölliker's 1867 portrayal of spinal nerve branches forming nets (Figure taken from Finger, 2001 [10])

named *synapse* by Sherrington), although the synaptic space itself was only made visible with electron microscopy well into the twentieth century. For an in-depth and fascinating view, I recommend the book of Rapport, which tells the story of Golgi and Cajal from both a neuroscientific and a romantic historical perspective [11].

The fact that neurons seemed independent units fueled early twentieth-century concepts that different brain areas carry specific functions and operate independently of each other. Even more so did the discovery that the cerebral cortex is made up of different cellular layers (Fig. 5.6) and the fact that there are large differences in layer architecture between different cerebral regions. Several anatomists each came up with their own cytoarchitectonic maps, the most famous ones being Brodmann, Campbell, Vogt and Vogt, Smith and von Economo and Koskinas (see Figs. 5.7 and 5.8) [12]. For some reason, it is predominantly the work of Korbinian Brodmann (1868–1918) that is remembered from this period. His maps of the cerebral cortex have been widely reproduced in the neurological and neuroscientific literature and almost become archetypical figures. 'Brodmann areas' (BA) are still frequently used in modern neuroscience to indicate brain areas and have in particular been closely linked, or even become synonymous with specific functions: for instance, BA 4 for primary motor functions, BA 17 for visual functions and BAs 44 and 45 as synonym for Broca's area and 'motor language'

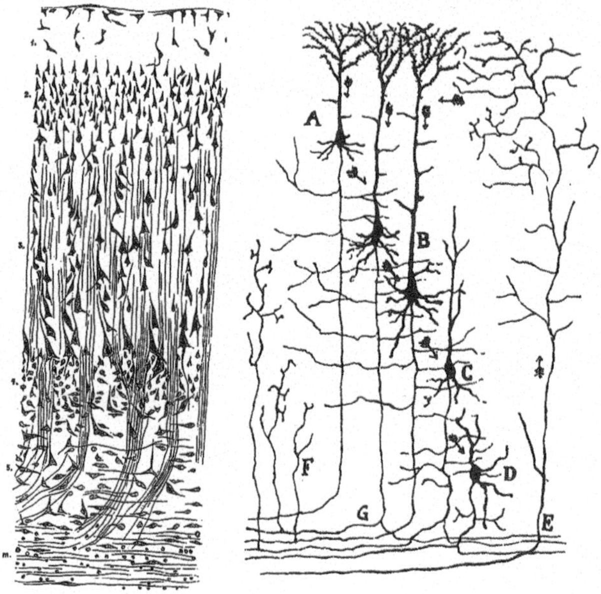

Fig. 5.6 Meynert's 1885 representational drawing of the then known five numbered layers of the cerebral cortex (*left*), compared with Cajal's detailed drawing of the same layer published 7 years later (*right*). Cajal later described the now accepted sixth layer (Text and figures taken from Rapport, 2005 [11], in which references are given to the original works of Meynert [69] and Cajal [70])

functions. Despite all this, Brodmann's famous book of 1909, entitled *Vergleichende Lokalisationslehre der Grosshirnrinde in ihren Prinzipien dargestellt auf Grund des Zellensbaues*, was not reprinted until 1985 and was only translated in English in 1994[f]. The editor and translator, Garey, himself a neuroscientist, wrote in his Translator's introduction:

> Few people have ever seen a copy of the 1909 monograph, and even fewer have actually read it! There has never been an English translation available, and the original book has been almost unavailable for years, the few antiquarian copies still around commanding high prices. As I, too, used Brodmann's findings and maps in my neurobiological work, and have the good fortune to have access to a copy of the book, I decided to read the complete text and soon discovered that this was much more than just a report of laboratory findings of a turn-of-the-century neurologist. It was an account of neurobiological thinking at that time, covering aspects of comparative neuroanatomy, neurophysiology, and neuropathology, as well as giving a fascinating insight into the complex relationships between European neurologists during the momentous times when the neuron theory was still new. [13]

[f]K. Brodmann. The Principles of Comparative Localisation in the Cerebral Cortex Based on Cytoarchitectonics.

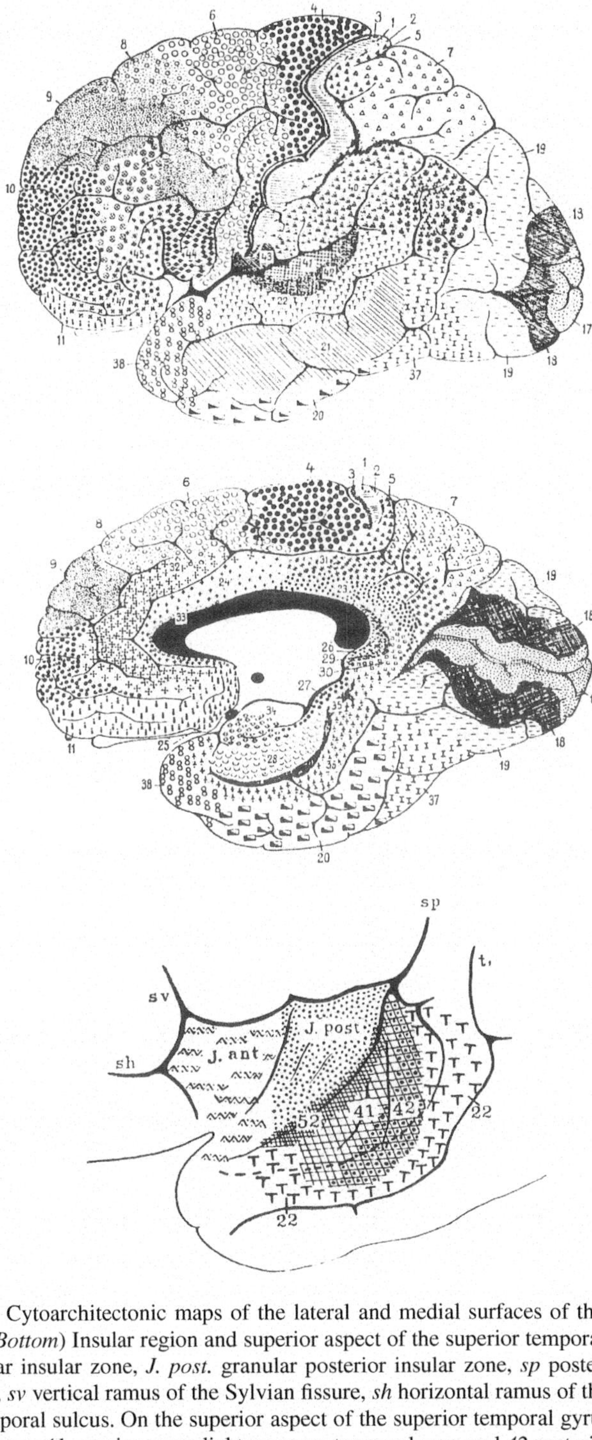

Fig. 5.7 (*Top*) Cytoarchitectonic maps of the lateral and medial surfaces of the human cerebral hemispheres. (*Bottom*) Insular region and superior aspect of the superior temporal region exposed. *J. ant.* agranular insular zone, *J. post.* granular posterior insular zone, *sp* posterior ramus of the Sylvian fissure, *sv* vertical ramus of the Sylvian fissure, *sh* horizontal ramus of the Sylvian fissure, *tl* superior temporal sulcus. On the superior aspect of the superior temporal gyrus are three areas: *52* parainsular area, *41* anterior or medial transverse temporal area, and *42* posterior or lateral transverse temporal area (Figures taken from Brodmann's *Localisation in the Cerebral Cortex* [13])

PLATE I

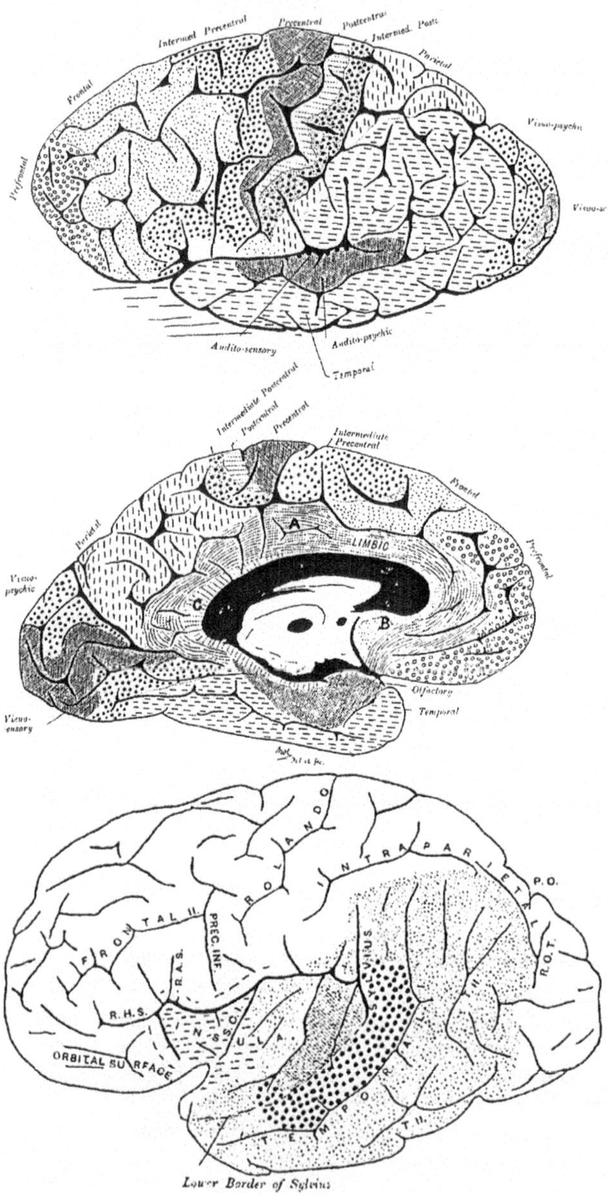
Precentral
Postcentral
Parietal
Frontal
Prefrontal
Audito-sensory
Audito-psychic
Temporal
Intermediate Postcentral
Postcentral
Precentral
Intermediate Precentral
Frontal
Parietal
Prefrontal
Visuo-psychic
A
LIMBIC
C
B
Olfactory
Temporal
ROLANDO
INTRAPARIETAL
P.O.
FRONTAL II
PREC. INF.
R.A.S.
R.H.S.
INSULA
ORBITAL SURFACE
R.O.T.
TEMPORAL
T. II.
Lower Border of Sylvius

5.2.1 Brodmann

Brodmann initially worked on the instigation of Oskar Vogt and Cécile Vogt-Mugnier in their neurological centre (*Neurologische Zentralstation*) in Berlin. Between 1901 and 1909 Brodmann systematically described the cytoarchitectonic architecture of the mammalian cortex. Brains were generously donated by the nearby zoological garden, whereafter they were fixed in formalin, embedded and sectioned in paraffin and then stained with cresyl violet. In this manner Brodmann meticulously analysed 63 different species, ranging from rat and cat to seal, tiger and chimpanzee. This was in addition to a number of human brains. It remains unclear on how many human specimens Brodmann's work is based. The exact answer is not known (to me), but is certainly more than one, as Brodmann repeatedly mentions regional variations in microscopic and macroscopic (e.g. sulcal) borders between individuals. Brodmann and his contemporary cartographers all provide a comparative neuroanatomical approach because they believed in 'the important biological principle that the genesis of mammalian cortex is not only conceived according to a common plan, but it completes its further development according to standard rules' [13]. As Zilles and Amunts (2010) write in their 'Centenary of Brodmann's map':

> Based on this integrative concept (histology with phylogeny), Brodmann indicated through his numbering system homologies between the cortical areas of different mammals. (…) Implicitly, Brodmann demonstrated that the architectonic parcellation of the human cortex can be understood only by comparison with different mammalian brains. (…) Each cortical area of his human map is labelled by a number between 1 and 52, but areas with the numbers 12–16 and 48–51 are not shown in his map. Brodmann explained these 'gaps' with the fact that some areas are not identifiable in the human cortex but are well developed in other mammalian species. This holds true particularly for the olfactory, limbic and insular cortices. [12]

Fig. 5.8 Campbell's schematic drawing of the human brain (1905). (*Top*) 'Human brain, M. aet. 41. Orthogonal tracings of the lateral and mesial surfaces (the former somewhat tilted to show the convexity) of the left cerebral hemisphere, with a representation of the extent of the various areas defined therein from an examination of the cortical nerve fibres and nerve cells.' 'In a surface diagram it is impossible to give a true idea of many of these fields, because cortex concealed within fissures cannot be indicated, and unfortunately the figures are especially misleading in regard to some of the most important areas; thus the floor, not the lip, of the fissure of Rolando is the boundary between the precentral and postcentral fields, and accordingly the concealed portion of these areas is almost equivalent to that exposed: the same applies to the calcarine or visuo-sensory field, while that marked "audito-sensory" is almost completely hidden in the Sylvian fissure.' (*Bottom*) 'Drawing of the left cerebral hemisphere (human) with Sylvian fissure opened out. Showing (1) the audito-sensory area (*shaded*) confined to the two transverse temporal gyri and not extending on to the insula; (2) the audito-psychic area (*large dots*) on the free surface of the posterior three-fifths of the first temporal gyrus; (3) the extent, on the lateral surface of the hemisphere, of the common temporal cortex (*small dots*).' *SCI* sulcus centralis insulae, *RHS* ramus horizontalis Sylvii, *RAS* ramus ascendens Sylvii, *Prec. Inf.* sulcus precentralis inferior, *PO* parieto-occipital fissure, *ROT* ramus occipitalis transversus, *TH* sulcus temporalis secundus (Text and figures taken from Campbell, 1905 [19])

Brodmann thus divided the human cerebral cortex into 43 areas. According to Zilles and Amunts:

> Only those regional differentiations in the cortical structure had been taken into account, which are apparent in the laminar organization of a cross-sectioned gyrus, in the positioning, size, packing density and distribution of cells, that is, in the cytoarchitectonic differences. Histological differences sensu strictu, that is, details of single cells, appearance of fibrils and tigroid substance as well as details of the structure of the cell nucleus, etc., are not used topographically. [12]

It is important to note that Brodmann also did not include white matter architecture. 'Myeloarchitecture' was studied by the Vogts who used myelin-stained histological sections for this purpose. With their findings they further subdivided the 43 areas of Brodmann into approximately 200 areas, adhering to the major cortical areas as described by Brodmann. Although Brodmann areas are today still frequently used in scientific papers, they do not account for regional differences in fibre connectivity, a fact that is not often realized. This is important because modern neurocognitive theories focus more and more on the dynamic interaction between areas and not only on their localized functions. Brodmann was convinced that cytoarchitectonically different areas must subserve different functions. However, he did not conclude from this that functions were necessarily located in only one area. He seemed well aware of the limitations of his work and in particular of the 'functional' conclusions that could be drawn from it:

> I remain hopeful that the results of histological localisation will not be without influence on the histopathology of the cerebral cortex. I am however not so optimistic as to believe that areal topography, as I have described in this treatise, will at present lead to cortical localization of individual psychiatric disorders or even individual psychological symptoms. [13]

Still, Brodmann discusses the possible functional consequences of his work in the last chapter of his book, *Physiology of the cortex as an organ*:

> Although my studies of localisation are based on purely anatomical considerations and were initially conceived to resolve only anatomical problems, from the outset my ultimate goal was the advancement of a theory of function and its pathological deviations. Now the question arises as to what we can deduce from our histotopographical findings in terms of physiology of the cerebral cortex. [13]

Brodmann also reviewed contemporary opinions and quoted Meynert, Exner and Wundt that all considered nerve cells to play a subordinate role in functional differentiation. According to them, functional specialization mainly resulted from differences in excitation patterns between brain areas and not from differences within the grey matter itself. One of the great merits of Brodmann's work is that he tried to refute localist concepts on histological and ontological grounds. In later work he added the results from lesion and stimulation studies in animals to support his views [14]. In proposing new hypotheses, Brodmann firmly disagreed with theories that reduce functional 'concepts' to specific cells, cellular layers or even brain areas:

> The first thing to say is that just as untenable as the idea of a 'concept cell' or an 'association layer' is the assumption of specific 'higher order psychic centres'. Indeed recently

> theories have abounded which, like phrenology, attempt to localize complex mental activity such as memory, will, fantasy, intelligence or spatial qualities such as appreciation of shape and position to circumscribed cortical zones. Older authors such as Goltz, Rieger, Wundt, and recently, particularly outspokenly, Semon, have already quite rightly expressed their opposition to such a 'naive view' and pleaded simple psychological facts against it. [13]

It was Brodmann's strong opinion that the higher mental faculties, in particular, can only take place:

> through an infinitely complex and involved interaction and cooperation of numerous elementary activities, with the simultaneous functioning of just as many cortical zones, and probably of the whole cortex, and perhaps also including even subcortical centers. Thus we are dealing with a psychological process extending widely over the whole cortical surface and not a localised function within a specific region. (…) One must therefore also assume a certain regional preference for higher activities, sometimes more in occipital and temporal areas, sometimes more in frontal. Such activities are, however, always the result (and not merely the sum) of the function of a large number of suborgans distributed more or less widely over the cortical surface; they can never be the product of a morphologically or physiologically independent 'centre'. [13]

Brodmann thus did not pursue a strict localist view on function, and it was clear to him that language functions could not be linked to single brain areas.

> It would be particularly tempting (…), considering the controversy recently engaged by Pierre Marie about aphasia [see chapter 1 of this book], to also engage in a discussion of the specific localisation of speech. However, it seems to us that the time is hardly ripe for this for most of the necessary physiological preparatory work is lacking. What is more, it is in no way to be seen as definite that the cortical localisation of speech coincides with that of aphasia. In relation to aphasia, however, one can already immediately conclude two things from the psychophysiological considerations described above. First, an aphasia, regardless of whether it belongs to the motor or sensory subcategory, can never be linked to a single structural centre, and always includes a complex of such areas, forming a larger region. Secondly, the 'aphasia centre' covers a much greater expanse than one was formerly accustomed to believe. [13]

Still, Brodmann mentioned 'Broca's area' when he gave anatomical descriptions of areas 44 and 45. However, he did not do it very consistently: first he linked it to area 44, later to both areas. What is of more importance is that Brodmann identified the pars opercularis as the seat of area 44 and the pars triangularis as the seat of area 45. This connection of microscopically and macroscopically defined brain areas has become common practice and is still frequently taken for granted in neurocognitive studies. It is not justified, however, as there is no invariant relationship between cytoarchitectonic areas and gyral/sulcal topography, a fact that has nowadays been convincingly demonstrated [12, 14]. Macro-anatomical landmarks are not reliable in identifying cytoarchitectonic regions [15]. Brodmann was already well aware that there was no perfect match. In his descriptions of cytoarchitectonic areas, he explicitly mentions that these are *approximately* bounded by the various sulci. He also repeatedly commented on the intersubject variability of gyri and sulci, to the extent that he almost seemed to warn his readers. Here is his literal account of the areas of the inferior frontal gyrus:

> **Area 44**—the opercular area—is a well-differentiated and sharply circumscribed structural region that on the whole corresponds quite well to the opercular part of the inferior frontal gyrus—Broca's area. Its boundaries are, posteriorly, approximately the inferior precentral gyrus, superiorly the inferior frontal sulcus and anteriorly the ascending ramus of the Sylvian fissure. Inferiorly or medially it encroaches on the frontal operculum and borders on the insular cortex. The area then stretches around the diagonal sulcus, and there are again minor structural differences between the cortex in front of and behind this sulcus to justify the separation of an anterior opercular area from a posterior opercular area by the diagonal sulcus. As there is much variability and inconsistency of these sulci one will find rather mixed topographical relationships of these structural areas in individual cases.
>
> **Area 45**—the triangular area—is cytoarchitectonically closely related to the previous area [area 44] that corresponds approximately to the triangular part of the inferior frontal gyrus. Consequently its caudal border lies in the ascending ramus of the Sylvian fissure, its dorsal border in the inferior frontal sulcus and its rostral border near the radiate sulcus of Eberstaller, although it may extend in places beyond this last sulcus as forward as the frontomarginal sulcus of Wernicke, and this area may also encroach partially on the orbital part; on the inferior surface of the inferior frontal gyrus it borders the insular cortex.
>
> Concerning the exact morphological borders of the last two areas [44 and 45], that are so extremely important on account of their relationship to the motor speech area, I should like once again to expressly point out the great individual variations of the sulci in this region. As emerges from Rezius' great monograph 'Das Menschenhirn', the diagonal sulcus is not infrequently fused with the inferior precentral sulcus or communicates with the ascending ramus, is often very strongly developed, but sometimes is entirely absent. The radiate sulcus and the ascending ramus vary widely in shape and structure so that naturally the relations of areas 44 and 45 to these sulci must be subject to major individual variations.(…)
>
> **Area 47**—the orbital area—shares certain architectonic affinities with areas 44 and 45 such that it can be combined with them to form a subfrontal region. It lies essentially around the posterior branches of the orbital sulcus, generally well differentiated from area 11, but with constant morphological borders. Laterally it crosses the orbital part of the inferior frontal gyrus. [13]

Note that Brodmann thus grouped together areas 44, 45 and 47 on the basis of cytoarchitectonic similarities and clustered them into a subfrontal region. He assumes that this region is a more suitable candidate for Broca's area than area 44 alone, although he does not provide any 'functional' evidence in his book. He also refrains from speculation about brain areas that might be related to 'sensory aphasia':

> I have already made brief reference elsewhere to the fact that in particular, according to all that can be concluded from anatomical localisational data, the seat of motor aphasia must extend much further anteriorly than appears from Broca's classic theory, and that at least the anterior sections of the inferior frontal gyrus, and perhaps even part of the actual orbital surface, must be included in it (thus, apart from area 44, also areas 45 and 47). [13]

A major drawback of Brodmann's work is that his observed intersubject variability was not reflected in his maps. This simplification has undoubtedly led to

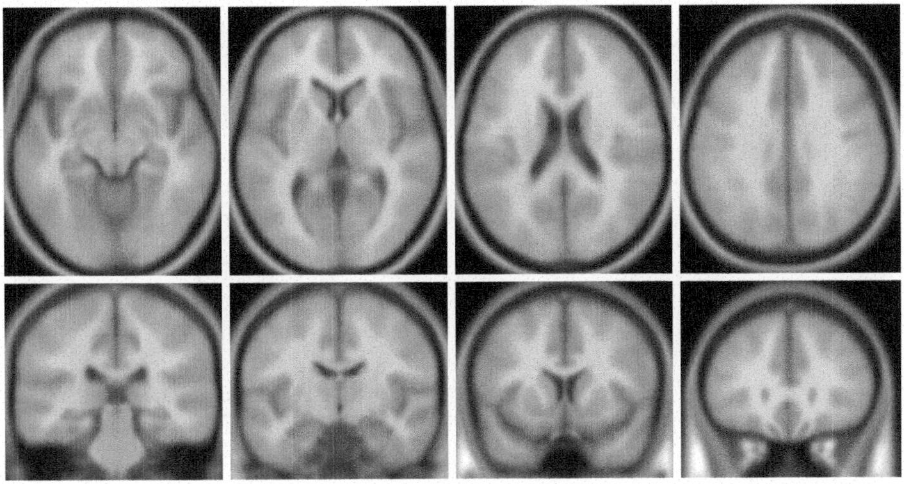

Fig. 5.9 MRI scans of the brains of 150 normal subjects were co-registered and transformed into a common stereotactic space, creating an 'average' brain. Note that details near the cortex are more blurred than deeper structures (e.g. basal ganglia) due to higher interindividual variability of cortical than subcortical structures

misinterpretation of his work and further supported the dogma of a strict correspondence of brain structure and function. Along a similar line of reasoning, it can be said that Wernicke's language schemes have been interpreted too literally; his idea that a large part of the temporal lobe was involved in language processing was generally overlooked in the historiography, presumably because Wernicke indicated language areas in his schemes only with small circles. The main reason for all this could well be that the maps and schemes of Brodmann and Wernicke are so visually appealing and convincing in their simplicity (as maybe any good model should be) that the textual nuances that were made by their authors got lost over time. It is also not easy to visualize uncertainty and variability across different individuals in a single schematic drawing. Nowadays, imaging studies use standardized and averaged brains for display of group results (see for an example Fig. 5.9). The disadvantage of these computerized brains is easily seen: details are lost and the cortical surface has a 'smoothed' appearance because intersubject variability has been largely averaged out. The brain is in fact—again—reduced to a fairly simple drawing, with only the ventricles and major sulci still identifiable. Statistical MRI atlases were developed to overcome the idiosyncrasies of using the brain of a single subject as a template. A more recent example of such a single brain atlas is the stereotactic atlas of Talairach and Tournoux (1988) (Fig. 5.10) [16]. For many years this coordinate system served as a guiding tool for neurosurgical procedures and also as the standard for reporting imaging results. Brodmann areas were added to later electronic versions of the atlas [17]. However, the original atlas was based on the postmortem section of a single subject: a 60-year-old French woman with a smaller than

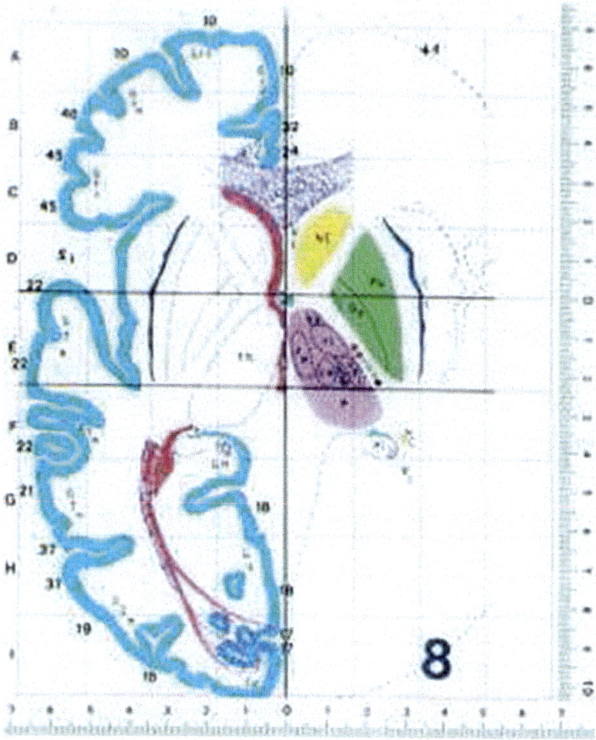

Fig. 5.10 Original digitized axial plate from the Talairach and Tournoux atlas (1988) (Figure taken from Nowinski and Belov, 2003 [71])

average brain size. Most brains would need a significant deformation if they were to match such a 'standard' brain.

So which map is best? Or, as put by Jones in a critical review in *Brain* (2008):

> Who can say whether the map of the human cerebral cortex by von Economo and Koskinas with its 107 areas is any more 'correct' than that of Campbell with 14, of Brodmann with 44, of the Vogts with more than 200 (…). [18]

Indeed, who can say? (By the way, note that Jones refers here to incorrect numbers of areas from the historical authors.) It may be apt to assume that the work of most authors has been largely forgotten because their maps have simply been surpassed in quality by those of Brodmann. However, it is not simply a question of 'right' and 'wrong'. We have already seen the problems with any composite map due to intersubject variability. Another difficulty is that the historical maps were all the result of subjective visual inspection of histological sections. Investigators searched for more or less abrupt microscopic changes (for instance, in the presence of large pyramidal cells or the distinctiveness of the laminar and columnar organization of the cortex) in order to assess whether or not one area significantly differed from another [14]. This basically requires multivariate analysis and is better done

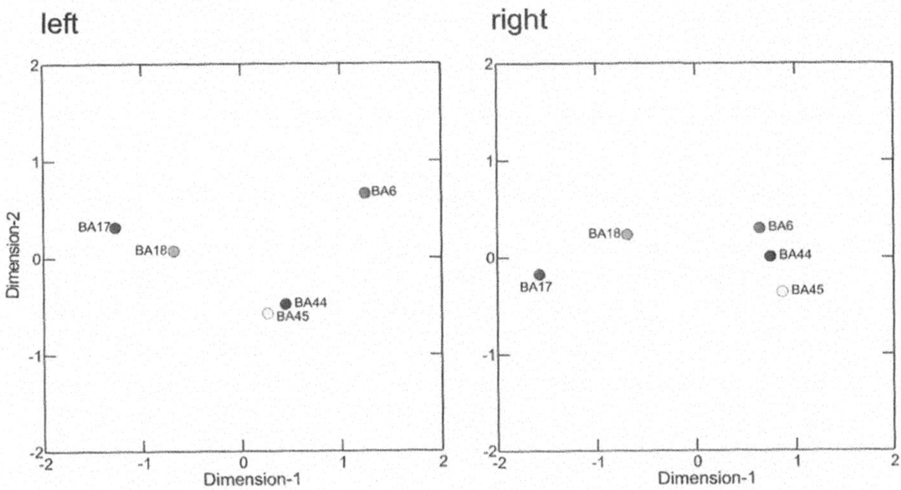

Fig. 5.11 Modern cytoarchitectonic studies use computerized detection methods to evaluate areal borders. These algorithms search for statistical changes in parameters that reflect laminar characteristics, such as the local volume density of cell bodies. Because several parameters are usually evaluated, multivariate statistics are required. The figure shows a graphic representation of interareal differences: the greater the dissimilarity in cytoarchitecture, the greater the distance between the areas in the graph. Left hemispheric Brodmann areas 44 and 45 are very similar, but differ from BA 6 and visual areas BA 18 and BA 19. Differences were less pronounced between BAs 45, 45 and 6 in the right hemisphere, whereby BA 44 seems to take an intermediate position between BA 45 and BA 6 with respect to its cytoarchitecture. Data were based on the analysis of ten human post-mortem brains (Figure taken from Grodzinsky, 2006 [14])

with a computer (see Fig. 5.11). And, of course, maps are nearly impossible to compare because they are all influenced by the various presumptions and axioms that were made by their creators. Campbell's maps, for instance, were not only based on findings from cytoarchitecture but also on myeloarchitecture and clinico-pathological deficits. It is therefore not surprising that the number and territories of his areas substantially differ from those of Brodmann.

5.2.2 Campbell

Let us compare, as an illustrative example, Brodmann's maps with those of Campbell and in particular compare the opinions of both men on Broca's area and its functions. Wernicke's area is not really discussed by them; Campbell only briefly mentions the concept of 'word deafness' and possible anatomical substrates in a review of the literature.

Alfred Walter Campbell (1868—1937) published his monograph *Histological studies on the localization of cerebral function* in 1905 [19]. He described 17 different cortical areas, a number that is significantly less than that of Brodmann. Campbell's intermediate precentral area includes both the inferior frontal gyrus as well as Brodmann's area 6 (see Fig. 5.8). Considering the localization of the motor speech area, Campbell states that it:

> is probably not so restricted as previously supposed, and (…) the forward extension of the 'intermediate precentral' cortex in the inferior frontal gyrus may have the same function as the cortex of Broca's area. In support of this assumption, it is pointed out, in the first place, that histologically the cortex of all this part of the 'intermediate precentral' area is alike, that is to say, the area of Broca is not distinguishable by any localised specialisation of structure; and, secondly, that it is a common matter of clinical experience that a superficial lesion confined to the cortex of Broca's area, is not wholly effective in the production of motor aphasia: in other words: if the disability is to be permanent the lesion must be deep and penetrating. The explanation given for the occurrence of complete and permanent motor aphasia after a deep-seated lesion in the pars basilaris [this probably equals the pars opercularis, GR] is that all connections between the 'intermediate precentral' cortex and the direct labial, lingual and laryngeal centres occupying the lower end of the precentral area proper—and by the way remaining intact—are severed. Such a lesion therefore produces an effect equivalent to destruction of the whole of the 'intermediate precentral' cortex coating the inferior frontal gyrus. [19]

Campbell dedicated an entire chapter in his book to his intermediate precentral area, of which this excerpt is only one of twelve concluding points. The chapter provides an in-depth discussion not only of the architecture of the cortical layers but also of association fibres and possible functions. Campbell claims that this area is different from the 'precentral area' because of histological and functional differences. For the latter he refers to the electrocortical stimulation studies of Sherrington and Grunbaum [20]. He also argues that because motor aphasia is not accompanied by an actual paralysis of muscles that are required for articulation, therefore motor language centres must lie apart from primary motor centres. Campbell (1905) defined motor aphasia as 'an annihilation of the power to call into action and execute the complex associated series of oral, lingual and laryngeal movements resulting in articulate speech'. He also quotes Hughlings Jackson's classification of movement in the central nervous system into three regions and proposes a modification whereby he considers the 'intermediate precentral area' to be of 'the highest or third level':

> I am of opinion that this particular stretch of cortex is specially designed for the execution of complex movements of an associated kind of skilled movements, of movements in which consciousness or volition takes an active part, as opposed to automatic movements, and my remarks will now be devoted to the development of this thesis. [19]

Campbell thus objects to the views of 'most clinicians' (his words) who consider destruction of the cortex of Broca's area alone as the essential factor in motor aphasia, and claims that a surface lesion is inadequate to produce a complete and permanent aphasia. He describes the case of middle-aged man who suffered from a complete motor aphasia, but who completely recovered after a few months. Twelve years later the man died suddenly from a perforated ulcer, never having had any language problems anymore:

> At the autopsy an old-standing patch of softening was found in the left hemisphere, and this we attributed to embolism, because there were coexisting signs of old mitral endocarditis; the distribution of the lesion was curious, the lower two-thirds of the pars basilaris were completely destroyed, but on making a series of horizontal sections we found that the destruction did not extend inwards beyond the plane of the surface of the insula. It thus left the white substance anterior to the lenticular nucleus and internal capsule—destroyed in the other cases—intact, some of the fronto-parietal operculum further back, and a portion of the temporal operculum, and also some of the cortex of the insula was obliterated, but the orbital operculum and the pars triangularis were quite untouched. Now I am quite certain that anyone inspecting the specimen would at once say that the individual must have suffered from permanent motor aphasia; placed beside the two hemispheres from cases of complete motor aphasia already alluded to, the area of destruction, as seen from the surface, is quite correctly placed, and indeed more extensive. [19]

Campbell's work did not fit well with the classic dogma of cortical localization of function, and that may be a reason why his work is largely forgotten. In 2005, ffytche and Catani wrote about Campbell's monograph that:

> One hundred years on, Campbell's integrative approach, combining anatomical, pathological and physiological insights, resonates far more with contemporary cognitive neuroscience than Brodmann's comparative anatomy. Indeed, Campbell's use of hodology—the white matter connections of each brain region—as a guide to cortical function foreshadows an approach to the study of functional anatomy that has only recently become possible in the living brain—that of diffusion tensor tractography. [21]

Until today, there remains a vivid discussion about the brain regions that constitute Broca's area. Brodmann objected against Campbell's 'intermediate precentral area' and wrote that area 6 'is undoubtedly to be separated as a special region' as areas 44, 45 and 47 posses 'a distinct granular layer, a feature that Campbell overlooked' [13]. Indeed, modern studies confirm Broca's opinion that a granular layer is lacking in BA 6, and this area—that is located on the ventral part of the precentral gyrus—is nowadays considered premotor cortex. However, a granular layer is also less present in BA 44, which is commonly described as dysgranular [22, 23]. In this respect, BA 44 is considered part of the motor territory, whereas BA 45, that is granular, has a similar architecture to other brain regions of the prefrontal cortex. More recent tractography and functional neuroimaging studies indicate that primary motor cortex and premotor cortex are more intimately linked to language functions than classically assumed [24–26].

5.3 Language Areas Defined in Terms of Gyri and Sulci

Any good anatomist, of course, notices that there are substantial differences between individual brains. Anatomical variability, in fact, is one of the reasons why classic models, such as those of Broca and Wernicke, can only be a coarse reflection of findings in an average population. This makes it difficult, and frequently even impossible, to match clinical syndromes to similar brain structures across individuals. As Ecker wrote in 1873:

> The difficulties which stand in the way of a solution to this problem [the structure-function correlation, GR] are numberless, not the least of these being one which is inherent in the very study of the convolutions, viz. the difficulty of recognizing a constant unity of form which underlies the multiplicity of individual variations. [1]

Physiological variability was thus a well-recognized 'problem' at the time.[g] Despite this, Ecker and others looked for similarities rather than differences and were in search of an underlying invariant anatomical scheme. Ecker systematically described primary, secondary and tertiary convolutions and drew an archetypical scheme of the brain that to me gives a very 'modern' impression (actually it is almost 150 years old; see Fig. 5.1). Ecker defined a number of primary convolutions that are always 'pretty regularly disposed'; these were the convolutions that were separated by the Sylvian fissure, central sulcus and parieto-occipital fissure. In addition to that, he stated that secondary and tertiary convolutions exhibited numerous variations.[h] Ecker gave several examples in his book and occasionally had to conclude that some gyral tracks are very difficult to unravel. He suggested that, in order to attain a complete understanding of such a gyrus, 'it is absolutely necessary to have recourse to brains poor in convolutions for reference, and, above all things, the brain of a fetus in one of the later months of embryonic existence' [1]. The foetal brain:

> is reduced to its simplest expression, and one is able to recognize, almost, as it were, in a diagrammatic sketch, all the essentially typical factors, without having the view rendered hazy, and without being diverted from that which is essential by secondary sulci and gyri, which answer, in a certain measure, to a more florid style of decoration. [1]

Study of macroscopic brain variability among individuals took a large step forward with the development of CT and MRI. This enabled non-invasive examination of the brain without opening of the skull, whereby the convolutions of healthy and young subjects could be virtually dissected and investigated. Powerful examples are studies in monozygotic and dizygotic twins which indicate that brain size is almost completely genetically determined, but that the variation in gyral patterns seems mostly related to environmental factors [28]. MRI confirmed what was already suspected from post-mortem and CT studies: there is an enormous variability of the gyral and sulcal patterns in normal subjects. Everybody's brain basically has its own topographical 'fingerprint'.

---

[g] Left–right differences were also likely part of the discussion, as Vick d'Azyr, Gratiolet, Leurat and several others had described differences in maturation, weight and topological structure between both hemispheres.

[h] This was well in advance of modern findings; see for example Ono 1990 [27].

5.3.1 Broca's Area

Studies of the anatomical variability of language areas have for the larger part focused on Broca's area. But there is an important caveat in all these studies, namely, that there is no uniform agreement on definitions. Most scientists and clinicians, when they speak of Broca's area, refer to an anatomical area and not per se to an area that represents specific (language) functions. Others seem to adhere more to historical 'definitions'. There is another difficulty, namely, that the presumed functions of Broca's area have been redefined many times since Broca's original description and are still not universally agreed upon. Despite these several possibilities, most authors use an anatomical perspective and locate Broca's area within the posterior part of the left inferior frontal gyrus and more specifically to the pars triangularis and pars opercularis. Some authors have included part of the precentral gyrus or the anterior extension of the inferior frontal gyrus into the orbital surface [29, 30].

By definition, the inferior frontal gyrus is separated ventrally by the Sylvian fissure and dorsally by the inferior frontal sulcus. The gyrus can roughly be divided into three parts by the ascending and horizontal ramus of the Sylvian fissure. Most authors use these rami to demarcate the pars triangularis (bordered left and right by the rami) and the pars opercularis (bordered left by the ascending ramus and right by the inferior part of the precentral sulcus) (see Fig. 5.12) [27, 31]. However, there are frequently two other sulci present that can make identification of rami difficult: within the pars opercularis, there is in approximately 50% of cases a diagonal sulcus; within the pars triangularis, there is even more frequently the presence of a triangular sulcus (Fig. 5.12). The morphology of rami and sulci of the inferior frontal gyrus is not very consistent, and there are many variations that prohibit a standard representation.

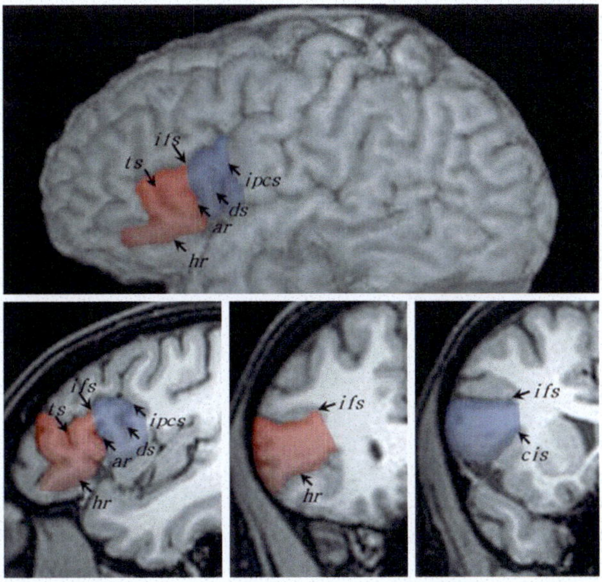

Fig. 5.12 Rendering of the left hemisphere (*top image*) in an individual subject based on MR images. The pars opercularis (*blue*) and pars triangularis (*red*) are delineated based on sulcal contours. Note that some borders are artificially drawn in absence of anatomical landmarks. *ar* ascending ramus, *cis* circular insular sulcus, *ds* diagonal sulcus, *hr* anterior horizontal ramus, *ifs* inferior frontal sulcus, *ipcs* inferior precentral sulcus, *ts* triangular sulcus (Figure taken from Keller, 2007 [17])

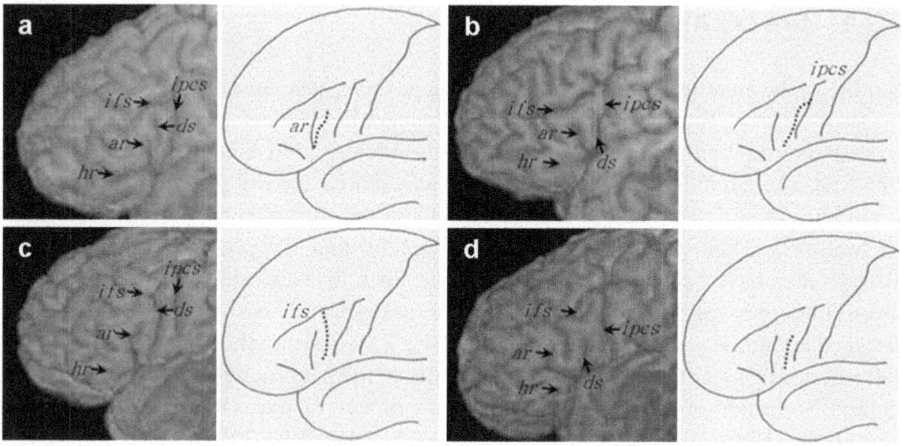

Fig. 5.13 Four variations of the diagonal sulcus, shown for every subject on an MRI-based surface rendering and a corresponding schematic illustration. (**a**) Connection with the anterior ascending ramus of the Sylvian fissure, (**b**) connection with the inferior precentral sulcus, (**c**) connection with the inferior frontal sulcus, (**d**) no connection with surrounding sulci. See for legend Fig. 5.12 (Figure taken from Keller, 2007 [17])

Keller (2007) studied brain morphology with MRI in 50 healthy subjects and noted the following sulcal variations (see also Figs. 5.12 and 5.13):

> (1) **inferior frontal sulcus**: continuous (one segment) or discontinuous (two or more segments); connection with the inferior precentral sulcus: long connection (where the inferior precentral sulcus anastomoses with a continuous, uninterrupted inferior frontal sulcus), short connection (where the inferior precentral sulcus anastomoses with a discontinuous, interrupted inferior frontal sulcus), superficial connection (where a connection is apparent from the surface of the brain, but a submerged bridge of cortex separates the sulci) or no connection.
> (2) **inferior precentral sulcus** (ventral most region): single or dual; connection or no connection with Sylvian fissure.
> (3) **anterior ascending ramus of the Sylvian fissure**: present or absent.
> (4) **anterior horizontal ramus of the Sylvian fissure**: present or absent; common or separate origin from the anterior ascending ramus.
> (5) **diagonal sulcus**: present or absent; connection to inferior precentral sulcus, inferior frontal sulcus or anterior ascending ramus, or no connection to these sulci (see also Fig. 5.12). [17]

In his review of the literature, Keller concludes:

> There is however great inter-individual variability in the shape, length, and number of these sulcal contours which gives rise to great variability in size, surface area and volume of the pars opercularis and pars triangularis. [17]

An exact anatomical definition of the pars opercularis and pars triangularis is thus difficult and sometimes even impossible. There is an additional problem with the use of anatomical landmarks; because more than half of the human cortex lies buried within sulci, not all anatomical details are visible from the surface [32]. The pars opercularis, for example, can be partially or even completely submerged [33]. In the latter situation, the ramus ascendens and inferior precentral sulcus are not visible on

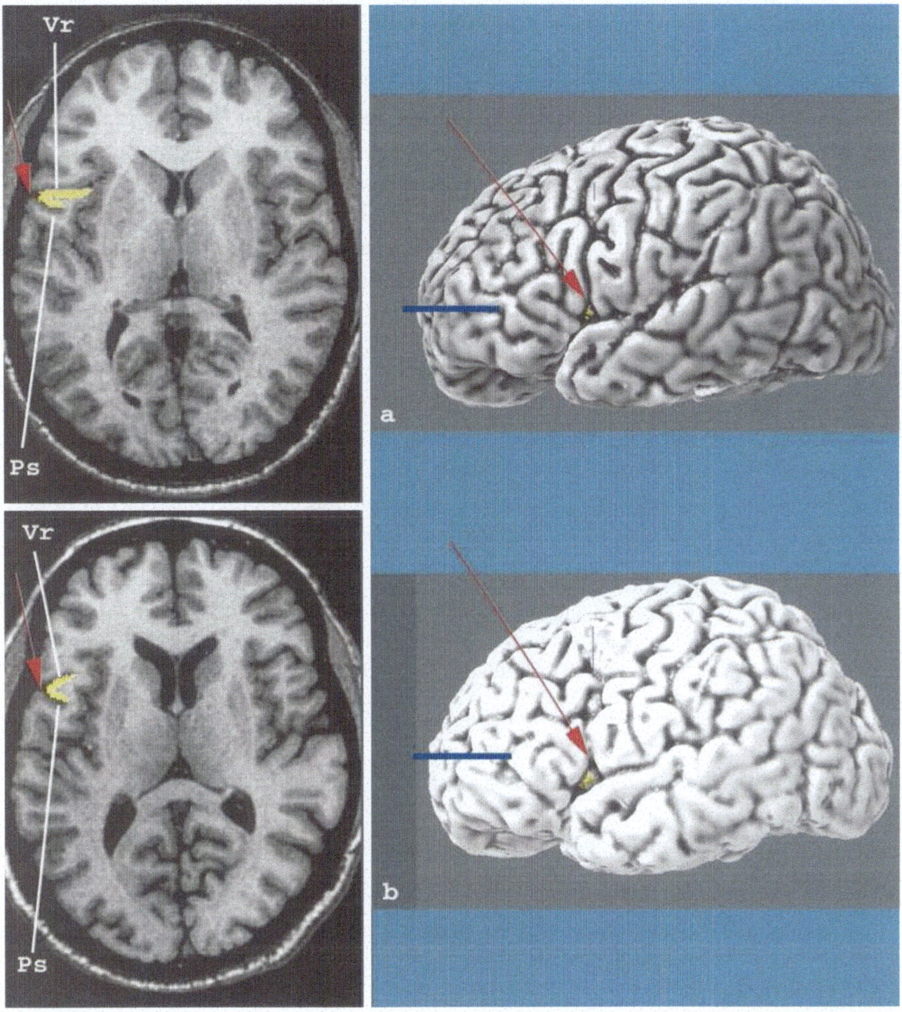

Fig. 5.14 Two brains taken from the study of Tomaiuolo (1999), whereby the pars opercularis (*yellow area*) is hidden beneath the cortical surface. In these cases the ramus ascendens and the inferior precentral sulcus cannot be distinguished on the surface of the brain (*red arrow*). They are, however, easily identified on axial MRI images. *Vr* vertical (or ascending) ramus, *Ps* inferior precentral sulcus (Figures taken from Tomaiuolo, 1999 [33])

the surface of the brain, and this will certainly lead to confusion and errors when classifying different parts of the inferior frontal gyrus from superficial inspection. On axial MRI images, these sulci, and the submerged convolution, are easily identified (see Fig. 5.14). Tomaiuolo (1999) pointed out that these anatomical variations have relevance to the literature on electrocortical stimulation. During an operation the surgeon probes the cortical surface with an electrode, but he or she is normally unable to stimulate within sulci. Results of these mapping procedures initially offered strong support for localization theory, as in particular Broca's area was in most patients consistently located in the posterior part of the inferior frontal gyrus.

Despite the problems with exact anatomical definitions, several authors have calculated asymmetry indices for presumed frontal language area. The fact that a substantial part of the pars opercularis may not be visible at the cortical surface probably contributed to a bias in study results in the pre-imaging era. Some authors described a rightward asymmetry of the region of the inferior frontal gyrus that included the pars triangularis and opercularis [34]. Later, with MRI, most studies tended towards a leftward asymmetry, although the issue has not been settled [30].

Broca's area is by tradition mostly defined in anatomical terms and not in functional terms. With Wernicke's area, it seems the other way around. There was never any agreement on its anatomical territory (see Chap. 4). Bogen and Bogen wrote an influential paper about the problem of locating Wernicke's area on anatomical grounds; they took a historical perspective and listed various examples and figures. Even when they narrowed down Wernicke's function to verbal comprehension only (as was originally done by Wernicke himself) definitions vary widely. Wernicke's area varies from a small area in the posterior part of the superior temporal gyrus to a large area that includes large parts of the temporal and parietal lobe. Here is Bogen and Bogen (1976) on anatomical definition of language areas:

> Broca's area is defined anatomically. It is the foot, that is, the posterior third of the inferior frontal gyrus. The question about Broca's area is not 'where is it?'—there is no question about where it is. The question about Broca's area is 'what good is it?' At this time we only consider Wernicke's region, of which the question is not 'what good is it?' because it is defined in terms of what it's good for—it's the area where the lesion will cause language comprehension deficit. The question with Wernicke's region is 'where is it?'. [35]

Bogen and Bogen conclude that a map that shows a probability distribution would probably better represent Wernicke's area, but realize that a vast amount of information is necessary to construct such a map. They end their commentary by saying that:

> we should welcome an approach that can make use of a vast amount of data which has heretofore been conveniently ignored by the simplified schemes with which we have struggled in the past. [35]

5.3.2 The Planum Temporale

In 1968, Geschwind and Levitsky reported in *Science* that the planum temporale, the area that lies directly posterior to Heschl's gyrus, is larger in the left hemisphere than in the right hemisphere [36]. They studied 100 post-mortem brains, whereby in 65% the planum temporale was found to be larger on the left side. In 11% it was larger on the right side. On average, the left planum temporale was one-third longer than the right one. These findings have generally been interpreted as the first important neuroanatomical evidence for left hemisphere specialization of language, and ignited a further interest in anatomical asymmetries. The fact that the brain is asymmetrical was not new: already at the end of the nineteenth century, it was reported that there were differences in the length and the curvature of the Sylvian fissure in both hemispheres [37–39]. This asymmetry is present in newborns but becomes progressively pronounced through adolescence to adulthood [40]. In most people there is a sharp upward angulation of the posterior part of the right Sylvian fissure into the inferior parietal lobule. On the left side, the Sylvian fissure courses more posteriorly and along a more horizontal trajectory than on the right [40]. This latter pattern is usually seen in right-handed subjects with left hemisphere language dominance. The consequence of this asymmetry, as formulated by Rubens (1976), is that:

> on the right, there is a smaller parietal operculum, a shorter planum temporale, a higher sylvian point, and compensatory expansion of the inferior parietal region posterior to the lateral fissure. [41]

The planum temporale has attracted so much attention as it is considered by many a key element of Wernicke's region. Pfeifer (1920) was the first to describe the asymmetry of the posterior temporal region in terms of anatomical structures and boundaries [39]. Most people nowadays would agree that the planum temporale covers the part of the superior temporal gyrus that is located posteriorly to the transverse (Heschl's) gyrus. In left hemispheres there is usually one strongly developed transverse (Heschl's) gyrus, whereas in right hemispheres there are frequently two [42, 43]. However, this situation is certainly not exemplary for every brain and precise identification of gyri and sulci can be difficult (see for details Fig. 5.15 and the critical appraisal of the borders of the planum temporale in the paper of Shapleske [39]).

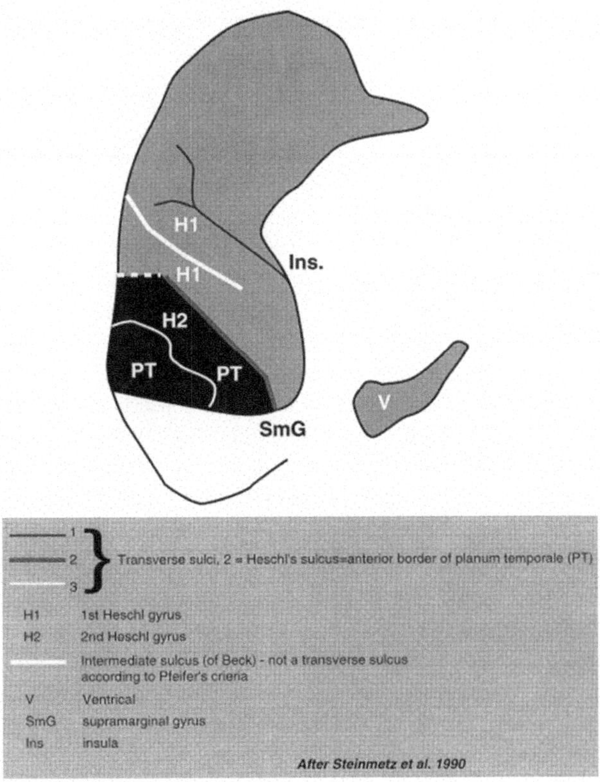

Fig. 5.15 Schematic axial view (i.e. from above) of a left superior temporal gyrus. Pfeifer (1920) defined the planum temporale (PT) as the area behind the transverse (i.e. Heschl's) gyrus [42]. There can be more than one transverse gyrus present. In this example, there are two, demarcated by three, transverse sulci. According to Pfeiffer, any transverse sulcus has to originate from the retroinsular region. Sometimes there is an additional sulcus (the intermediate sulcus of Beck) that separates a Heschl's gyrus into an anterior and posterior part (Figure taken from Shapleske, 1999 [39])

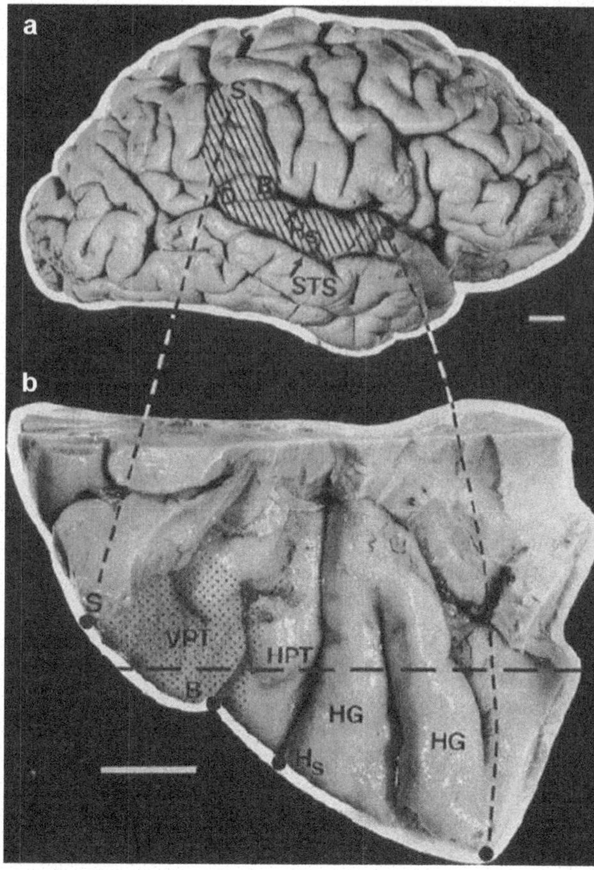

Fig. 5.16 Photographs of the lateral surface of the right hemisphere (**a**) and a superior view on a block (**b**) that was removed and contains Heschl's gyrus/planum temporale. The horizontal and vertical parts of the planum temporale (respectively, HPT and VPT) were removed in the same block. The photograph is in the plane of HPT; the VPT is foreshortened. *HG* Heschl's gyrus, *HS* Heschl's sulcus, *B* point of bifurcation, *S* end of the posterior ascending ramus of the Sylvian fissure, *D* end of the posterior descending ramus of the Sylvian fissure (Figure taken from Witelson 1995 [45])

Definition of the posterior border of the planum temporale is also not without controversy: first, because of anatomical ambiguities in some hemispheres and, second, because of differences in the definitions that are used by researchers [44, 45]. In most hemispheres, the Sylvian fissures bifurcate into two branches, respectively, the posterior ascending ramus and posterior descending ramus (see Figs. 5.16 and 5.17 for more details) [46]. Some authors have used this bifurcation point as the posterior border of the planum temporale, whereas others have included the banks of either the ascending or descending posterior ramus. Witelson lists the definitions most commonly used in the literature (see Fig. 5.17) [45]. The problem is that the endpoint that is chosen significantly influences the results on (laterality) measurements of the planum temporale. If the posterior ascending ramus is not included, as was done by Geschwind and Levitsky in their landmark studies, the planum temporale is usually larger in the left hemisphere. This asymmetry can be dramatic, even to such an extent that the left planum is ten times larger than the right [47]. Asymmetry is also accompanied with an increase in grey matter volume and myelinization of axons within this region [48]. However, if the posterior ascending ramus is included in the definition of the planum temporale, as in more recent studies, size differences between hemispheres

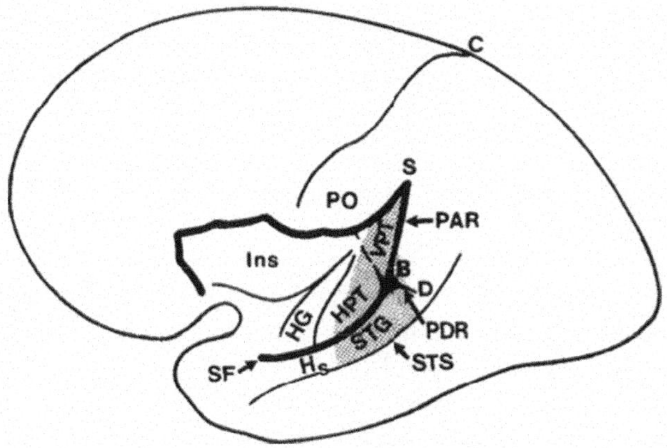

Fig. 5.17 Schematic representation of gross morphological features of the planum temporale (PT) and its surroundings. A left hemisphere with typical morphology is shown, with the upper and lower walls of the Sylvian fossa pulled apart (heavy lines represent the lateral edges), exposing the superior surface of the superior temporal gyrus. The Sylvian fissure (SF) bifurcates at point B, into the posterior descending ramus (PDR) ending at D and the posterior ascending ramus (PAR) ending at S. The floor of the Sylvian fissure exposes Heschl's gyrus (HG). The horizontal part of the PT (HPT), located in the supratemporal plane, is by most people typically considered to be the PT. The vertical part of the PT (VPT) is located on the inner (posterior) wall of the upward curve of the PAR. However, variable definitions of the PT can be found in the literature, ending the PT at points B [36, 43], D [73, 74] or S [45, 74]. In hemispheres where the upward swing of SF is very sharp and anterior in its origin, PT has often been considered to be absent [34]. *Hs* Heschl's transverse sulcus, *STS* superior temporal sulcus, *PO* parietal operculum, *Ins* insula (Figure and text (modified) taken from Witelson, 1995 [45])

seem to disappear [39, 49]. The planum temporale now consists of a horizontal as well as a vertical part. It is of further note that the planum parietale (i.e. the bank along the posterior ascending ramus that is part of the supramarginal gyrus) is usually larger in the right hemisphere than in the left. This is related to differences in the upswing of the ascending rami on the right and left side, as is also the case for the planum temporale measurements. Interestingly, there seems to be no significant relationship between the size of the left planum temporale and right planum parietale, suggesting functional independence of these two regions [50]. The planum parietale is a relatively 'new' area that has not been studied much; initial results indicate that it may be involved in voice processing [51]. A famous example of aberrant sulcal anatomy of the posterior Sylvian region was found in the brain of Einstein [52]. In his case, the posterior end of the Sylvian fissure had a relatively anterior position, and the posterior ascending ramus was continuous with the postcentral sulcus. Consequently, the parietal operculum was absent (this area normally lies between the postcentral sulcus and the posterior ascending ramus), but the inferior parietal region was significantly larger than in normal subjects:

> A further consequence of this morphology is that the full supramarginal gyrus lies behind the Sylvian fissure, undivided by a major sulcus as is usually the case. (…) the compactness of Einstein's supramarginal gyrus within the inferior parietal lobe may reflect an extraordinarily large expanse of highly integrated cortex within a functional network. [52]

> The inferior parietal lobe is well developed in the human brain; it is a secondary association area that provides cross-modal associations among visual, somaesthetic, and auditory stimuli [53]. Visuospatial cognition, mathematical thought [54], and imagery of movement [55] are strongly dependent on this region. Einstein's exceptional intellect in these cognitive domains and his self-described mode of scientific thinking [56] may be related to the atypical anatomy in his inferior parietal lobules. Increased expansion of the inferior parietal region was also noted in other physicists and mathematicians. [52]

Macroscopic differences within the posterior Sylvian region are paralleled by differences in cytoarchitecture. Classically, the planum temporale is seen as secondary or associative auditory cortex that lies posteriorly to the primary auditory cortex of Heschl's gyrus. The planum temporale itself is made up of several different cytoarchitectonic areas, suggesting functional differentiation [57]. Galaburda, Sanides and Geschwind (1978) were the first to describe that the area Tpt (in the posterior part of the superior temporal gyrus and temporoparietal junction, i.e. the posterior part of the planum temporale) was larger on the left side [58]. This area corresponds to area TA1 of von Economo and Koskinas (1925) and the posterior part of area 22 of Brodmann. They concluded that the asymmetries of the planum temporale 'probably reflect asymmetries in an auditory cytoarchitectonic area and therefore may represent, at least in part, the anatomic substrate for language lateralization' [58]. In a later paper, they also concluded that Tpt is not per se related to auditory information:

> Area Tpt represents a transitional type of cortex between the specialized isocortices of the auditory region and the more generalized isocortex (integrated cortex) of the inferior parietal lobule.[i] [59]

In the classical cytoarchitectonic view, the primary auditory cortex (A1) is located on Heschl's gyrus and corresponds to BAs 41 and 42 [60]. The primary auditory cortex is buried within the Sylvian fissure (Fig. 5.7) and surrounded posterolaterally by the secondary auditory cortex (BA 22). This region is involved in higher-order auditory processing, including the perception of speech. Each primary auditory area has access to information from both ears, and both primary and secondary auditory cortices have major connections to the thalamus (via the mediate geniculate body). Although there appears to be stronger influence of the contralateral ear, primary auditory information is not lateralized to one hemisphere. Complete cortical deafness is therefore only expected with bilateral damage to primary and secondary auditory areas and generally requires two separate lesions [60]. Some authors, however, have restricted the primary auditory cortex to only one cytoarchitectonic area (BA 41), whereas others have came up with two, three or even eleven different types of cortex within Heschl's gyrus, as can be read in the following excerpt from a paper of Morosan and colleagues (2001):

> Although Heschl's gyrus is usually defined as the site of primary auditory cortex, existing cytoarchitectonic parcellation schemes of primary auditory cortex vary among different available maps. While Brodmann (1909) described only one koniocortical area (area 41) as primary auditory cortex, others have identified two areas within the primary auditory cortex (Economo and Koskinas, 1925; Sarkissov et al., 1955; Galaburda and Sanides, 1980) [59,

[i] Quote taken from www.talkingbrains.com

> 61, 62]. Moreover, the size and exact location of areal borders and anterior-to-posterior or medial-to-lateral distribution of these areas differ between the studies, and the intersubject variability of cytoarchitecture, size, and topography was not analyzed (Lashley and Clark, 1946) [63]. It is not known to what degree divergent anatomical patterns may reflect inter-individual variability or only interobserver differences caused by the highly observer-dependent methods of classical architectonic studies. The classical maps do not include data on variations in topography or size of primary auditory cortex between the hemispheres and/or the sexes. Finally, the 2-D presentation in highly schematized maps of the previous cytoarchitectonic parcellations are of limited use for comparisons with functional imaging studies based on 3-D data sets from living human brains. [64]

There remains a lot of work to be done at the cytoarchitectonical level. To facilitate comparison between different subjects and different studies, results should be made available in a common—standardized—reference frame [65]. To circumvent the problems that are associated with visual (i.e. observer-dependent) inspection of histological sections, some authors have turned to a more quantitative cytoarchitectonical approach. Morosan (2001), for instance, analysed ten post-mortem brains with user-independent algorithms and found three different areas within the primary auditory cortex (defined as BA 41), which they labelled Te1.0, Te1.1 and Te1.2 [64]. They also confirmed previous observations that cytoarchitectonic borders do not match gyral or sulcal borders. The picture will likely change again when results of receptor-architectonic and myeloarchitectonic studies are included [65].

The anatomical asymmetries of the end of the Sylvian fissure have initially been interpreted as evidence that Wernicke's area must reside within the left posterior temporal region. However, the frequency of leftward asymmetry (65–80%) is low in comparison with estimates of left hemisphere language dominance in the general population (>90%) [39, 66]. In addition to that, studies have clearly demonstrated that hemispheric asymmetry within this region is not unique to humans and is in fact evident in all great apes. Gannon described in 1998 that the left planum temporale was significantly larger in 17 of 18 chimpanzees [67]. From this we can at least conclude that the planum temporale is not uniquely related to our current language functions, as chimpanzees have communicating abilities that are inferior to the language abilities of humans. Gannon furthermore suggests that:

> an initial distinguishable suite of human-like receptive language area anatomic traits appeared in our common ancestor with early hominoids around 20 million years ago (although functional asymmetries may have been present prior to this). (…) Subsequently, these traits were gradually elaborated until they became anatomically expressed more prominently, likely within our common ancestor with orangutans around 14 million years ago, and arriving at a human condition in our common ancestor with gorillas around 10 million years ago. [68]

As in humans, great apes (including chimpanzees, bonobos and gorillas) also have an enlarged BA 44 in the left hemisphere [66]. These comparative reports do not necessarily change our understanding of purported frontal and temporal human language areas; they merely ground evolutionary interpretations [40].

5.4 Some Concluding Remarks

Anatomical studies, from microscopic cytoarchitecture to macroscopic sulcal topography, have made it clear that there are significant differences between subjects. This makes it a priori impossible to create any reliable, precise anatomical 'standard brain' that can serve as a road map for scientific or clinical investigations of individual subjects. Many researchers in the nineteenth century already appreciated that brain functions are not simply characterized by certain architectural features or the location of a single brain region. Take for instance the more connectionist view of Brodmann (1909):

> Thus to wish to draw conclusions about the level of the organisation of a brain from the high or low cell density of its cortex must be considered in principle as doomed to failure. It is not the quantity of cells per unit volume, but their quality, their detailed intrinsic specialization, their surface area as manifested by the number of dendrites, and the richness of their connectivity, that all form a yardstick for the functional sophistication of the cortex or of a particular cortical region. [13]

Still, connectionism only became the more predominant scientific view at the end of the twentieth century, and it has yet to gain significant clinical implications. Nowadays, scientists consider brain functions to be represented in large-scale brain networks, described in terms of nodes and hubs. However, at the end of the nineteenth century, and for most of the twentieth century, localism and its dogma of a one-to-one correspondence of brain structure and function had the upper hand. This view was supported by the many clinical and experimental lesion and stimulation studies that were done at that time. These studies seemed, at first glance, to confirm localist theories. However, a closer look at the original works certainly gives a much more nuanced view on functional localization, as seen in the next chapter.

References

1. Ecker A. On the convolutions of the human brain (translated by JC Galton). London: Smith, Elder; 1873.
2. Ribas GC, Ribas EC, Rodrigues CJ. The anterior sylvian point and the suprasylvian operculum. Neurosurg Focus. 2005;18:E2.
3. Schiller F. Paul Broca: founder of French anthropology, explorer of the brain. New York: Oxford University Press; 1992.
4. Leuret F, Gratiolet P. Anatomie comparée du système nerveux considéré dans ses rapports avec l'intelligence. Paris; 1839.
5. Rolando L. Della struttura degli emisferi cerebrali. Mem R Acad Sci. 1830;XXV:103–45.
6. Gratiolet LP. Mémoire sur les plis cérébraux de l'homme et des primatès. Paris: A. Bertrand; 1854.
7. Pearce JM. Louis Pierre Gratiolet (1815-1865): the cerebral lobes and fissures. Eur Neurol. 2006;56:262–4.
8. Gould SJ. Wide hats and narrow minds. The panda's thumb: more reflections in natural history. New York: WW Norton & Company; 1992.
9. Zimmerman A. Anthropology and antihumanism in imperial Germany. Chicago: University of Chicago Press; 2001.

10. Finger S. Origins of neuroscience: a history of explorations into brain function. New York: Oxford University Press; 2001.
11. Rapport RL. Nerve endings: the discovery of the synapse. New York: WW Norton & Company; 2005.
12. Zilles K, Amunts K. Centenary of Brodmann's map—conception and fate. Nat Rev Neurosci. 2010;11:139–45.
13. Garey LJ. Brodmann's localisation in the cerebral cortex. London: Smith-Gordon; 1999.
14. Grodzinsky Y, Amunts K. Broca's region. New York: Oxford University Press; 2006.
15. Amunts K, Schleicher A, Zilles K. Cytoarchitecture of the cerebral cortex—more than localization. NeuroImage. 2007;37:1061–5. Discussion 1066
16. Talairach J, Tournoux P. Co-planar stereotaxic atlas of the human brain. New York: Thieme; 1988.
17. Keller SS, Highley JR, Garcia-Finana M, et al. Sulcal variability, stereological measurement and asymmetry of Broca's area on MR images. J Anat. 2007;211:534–55.
18. Jones EG. Cortical maps and modern phrenology. Brain. 2008;131:2227–33.
19. Campbell AW. Histological studies on the localization of cerebral function. Cambridge: Cambridge University Press; 1905.
20. Grünbaum ASF, Sherrington CS. Observations on the physiology of the cerebral cortex of some of the higher apes. Proc R Soc Lond. 1901;69:206–9.
21. ffytche DH, Catani M. Beyond localization: from hodology to function. Philos Trans R Soc Lond B Biol Sci. 2005;360:767–79.
22. Amunts K, Schleicher A, Burgel U, et al. Broca's region revisited: cytoarchitecture and intersubject variability. J Comp Neurol. 1999;412:319–41.
23. Petrides M, Pandya DN. Comparative cytoarchitectonic analysis of the human and the macaque frontal cortex. In: Boller F, Grafman J, editors. Handbook of neuropsychology. Amsterdam: Elsevier; 1994. p. 17–58.
24. Bernal B, Ardila A. The role of the arcuate fasciculus in conduction aphasia. Brain. 2009;132:2309–16.
25. Friederici AD. Broca's area and the ventral premotor cortex in language: functional differentiation and specificity. Cortex. 2006;42:472–5.
26. Eickhoff SB, Heim S, Zilles K, Amunts K. A systems perspective on the effective connectivity of overt speech production. Philos Trans A Math Phys Eng Sci. 2009;367:2399–421.
27. Ono M, Kubik S, Abernathey CD. Atlas of the cerebral sulci. Stuttgart: Georg-Thieme; 1990.
28. Bartley AJ, Jones DW, Weinberger DR. Genetic variability of human brain size and cortical gyral patterns. Brain. 1997;120:257–69.
29. Hervé G. La circonvolution de Broca: étude de morphologie cérébrale. Paris: Davy; 1888.
30. Keller SS, Crow T, Foundas A, et al. Broca's area: nomenclature, anatomy, typology and asymmetry. Brain Lang. 2009;109:29–48.
31. Duvernoy HM, Bourgouin P. The human brain: surface, three-dimensional sectional anatomy with MRI, and blood supply. Wien: Springer; 1999.
32. Zilles K, Armstrong E, Schleicher A, Kretschmann HJ. The human pattern of gyrification in the cerebral cortex. Anat Embryol (Berl). 1988;179:173–9.
33. Tomaiuolo F, MacDonald JD, Caramanos Z, et al. Morphology, morphometry and probability mapping of the pars opercularis of the inferior frontal gyrus: an in vivo MRI analysis. Eur J Neurosci. 1999;11:3033–46.
34. Wada JA, Clarke R, Hamm A. Cerebral hemispheric asymmetry in humans. Cortical speech zones in 100 adults and 100 infant brains. Arch Neurol. 1975;32:239–46.
35. Bogen JE, Bogen GM. Wernicke's region—where is it? Ann N Y Acad Sci. 1976;280:834–43.
36. Geschwind N, Levitsky W. Human brain: left-right asymmetries in temporal speech region. Science. 1968;161:186–7.
37. Eberstaller O. Das Stirnhirn. Urban & Schwarzenberg; 1890.
38. Cunningham DJ. Contribution to the surface anatomy of the cerebral hemispheres. Dublin: Royal Irish Academy; 1892.

39. Shapleske J, Rossell SL, Woodruff PW, David AS. The planum temporale: a systematic, quantitative review of its structural, functional and clinical significance. Brain Res Brain Res Rev. 1999;29:26–49.
40. Gannon PJ, Kheck NM, Braun AR, Holloway RL. Planum parietale of chimpanzees and orangutans: a comparative resonance of human-like planum temporale asymmetry. Anat Rec A Discov Mol Cell Evol Biol. 2005;287:1128–41.
41. Rubens AB, Mahowald MW, Hutton JT. Asymmetry of the lateral (sylvian) fissures in man. Neurology. 1976;26:620–4.
42. Pfeifer RA. Myelogenetisch-anatomische Untersuchungen über das kortikale Ende der Hörleitung. Leipzig: BG Teubner; 1920.
43. Economo C, Horn L. Über Windungsrelief, Maße und Rindenarchitektonik der Supratemporalfläche, ihre individuellen und ihre Seitenunterschiede. Zeitschrift für die gesamte Neurologie und Psychiatrie. 1930;130:678–757.
44. Ide A, Rodriguez E, Zaidel E, Aboitiz F. Bifurcation patterns in the human sylvian fissure: hemispheric and sex differences. Cereb Cortex. 1996;6:717–25.
45. Witelson SF, Glezer II, Kigar DL. Women have greater density of neurons in posterior temporal cortex. J Neurosci. 1995;15:3418–28.
46. Witelson SF, Kigar DL. Sylvian fissure morphology and asymmetry in men and women: bilateral differences in relation to handedness in men. J Comp Neurol. 1992;323:326–40.
47. Geschwind N, Galaburda AM. Cerebral lateralization. Biological mechanisms, associations, and pathology: I. A hypothesis and a program for research. Arch Neurol. 1985;42:428–59.
48. Steinmetz H. Structure, functional and cerebral asymmetry: in vivo morphometry of the planum temporale. Neurosci Biobehav Rev. 1996;20:587–91.
49. Steinmetz H, Rademacher J, Jancke L, et al. Total surface of temporoparietal intrasylvian cortex: diverging left-right asymmetries. Brain Lang. 1990;39:357–72.
50. Jancke L, Schlaug G, Huang Y, Steinmetz H. Asymmetry of the planum parietale. Neuroreport. 1994;5:1161–3.
51. Lattner S, Meyer ME, Friederici AD. Voice perception: sex, pitch, and the right hemisphere. Hum Brain Mapp. 2005;24:11–20.
52. Witelson SF, Kigar DL, Harvey T. The exceptional brain of Albert Einstein. Lancet. 1999;353:2149–53.
53. Geschwind N. Disconnexion syndromes in animals and man. Part I. Brain. 1965;88:237–94.
54. Critchley M. The parietal lobes. New York: Hafner; 1971.
55. Crammond DJ. Motor imagery: never in your wildest dream. Trends Neurosci. 1997;20:54–7.
56. Hadamard J. The mathematician's mind: the psychology of invention in the mathematical field. Princeton: Princeton University Press; 1949.
57. Hickok G. The functional neuroanatomy of language. Phys Life Rev. 2009;6:121–43.
58. Galaburda AM, Sanides F, Geschwind N. Human brain. Cytoarchitectonic left-right asymmetries in the temporal speech region. Arch Neurol. 1978;35:812–7.
59. Galaburda A, Sanides F. Cytoarchitectonic organization of the human auditory cortex. J Comp Neurol. 1980;190:597–610.
60. Mesulam MM. Principles of behavioral and cognitive neurology. New York: Oxford University Press; 2000.
61. de Economo CF, Koskinas GR. Die Cytoarchitektonik der Hirnrinde des Erwachsenen Menschen. Berlin: Springer; 1925.
62. Sarkissov SA, Filimonoff IN, Kononowa IP, et al. Atlas of the cytoarchitectonics of the human cerebral cortex [in Russian]. Moscow: Medgiz; 1955.
63. Lashley KS, Clark G. The cytoarchitecture of the cerebral cortex of Ateles: a critical examination of architectonic studies. J Comp Neurol. 1946;85:223–305.
64. Morosan P, Rademacher J, Schleicher A, et al. Human primary auditory cortex: cytoarchitectonic subdivisions and mapping into a spatial reference system. NeuroImage. 2001;13:684–701.
65. Roland PE, Zilles K. Brain atlases—a new research tool. Trends Neurosci. 1994;17:458–67.

66. Toga AW, Thompson PM. Mapping brain asymmetry. Nat Rev Neurosci. 2003;4:37–48.
67. Gannon PJ, Holloway RL, Broadfield DC, Braun AR. Asymmetry of chimpanzee planum temporale: humanlike pattern of Wernicke's brain language area homolog. Science. 1998;279:220–2.
68. Falk D, Gibson KR. Evolutionary anatomy of the primate cerebral cortex. Cambridge: Cambridge University Press; 2008.
69. Meynert T. Psychiatry: a clinical treatise on diseases of the fore-brain based upon a study of its structure, functions, and nutrition. 1885. Trans B Sachs. New York: Hafner; 1968.
70. y Cajal SR. A new concept of the histology of the central nervous system (1892). In: Rottenberg DA, Hochberg FN, editors. Neurological classics in modern translation. New York: Hafner; 1977. p. 7–29.
71. Nowinski WL, Belov D. The Cerefy Neuroradiology Atlas: a Talairach-Tournoux atlas-based tool for analysis of neuroimages available over the internet. NeuroImage. 2003;20:50–7.
72. Witelson SF, Pallie W. Left hemisphere specialization for language in the newborn. Neuroanatomical evidence of asymmetry. Brain. 1973;96:641–6.
73. Steinmetz H, Volkmann J, Jancke L, Freund HJ. Anatomical left-right asymmetry of language-related temporal cortex is different in left- and right-handers. Ann Neurol. 1991;29:315–9.
74. Aboitiz F, Scheibel AB, Zaidel E. Morphometry of the sylvian fissure and the corpus callosum, with emphasis on sex differences. Brain. 1992;115:1521–41.

Mapping and Lesioning the Living Brain 6

It is now well known that the brain is electrically excitable. Physicians frequently make use of electromagnetic properties to monitor or localize brain functions in patients (e.g. EEG, MRI or electrical stimulation of the brain). However, in the nineteenth century, it was accepted as a fact that the cerebral hemispheres were non-excitable 'by all common psychologic stimuli' [1]. This dogma prevailed to such an extent that studies that challenged this concept initially had to be performed outside of the universities [2].

6.1 Fritsch and Hitzig

In 1870, Gustav Fritsch (1838–1927) and Eduard Hitzig (1839–1907) electrically stimulated the cortex of a dog. Their experiments were done at home, on the dressing table of Frau Hitzig [3]. They were able to demonstrate that stimulation of certain parts of the cortex induced (contralateral) motor responses, and thereby they presented for the first time convincing experimental evidence for localization of function (Fig. 6.1) [1]. The motivation for their experiments followed from several developments in the eighteenth and nineteenth centuries. Galvani, and many predecessors, had already induced movements in animals by stimulation of the muscles, nerves and brain [4]. Animals seemed to contain 'animal electricity', something that was distributed by the nerves and secreted by the brain [4]. Volta constructed the first battery, which proved useful for electrical stimulation (Fig. 6.2). By the early nineteenth century, medical applications of electricity were commonly advocated, and a unique field of study, called galvanism, had developed [4]. Boling describes some of the experiments of Aldini, a nephew of Galvani and professor of physics at the University of Bologna [4]. Aldini used a voltaic pile for his experiments on animals and humans (Fig. 6.2). He could elicit strong muscular contractions with stimulation of the dura and the cortex of a trephined ox. He had also noted that electrical stimulation of the scalp significantly improved the mood in his human patients. In 1802, Aldini stimulated the cortical surface of the left hemisphere of a decapitated

G.-J. Rutten, *The Broca-Wernicke Doctrine*, DOI 10.1007/978-3-319-54633-9_6

Fig. 6.1 Drawing after Fritsch and Hitzig's (1870) figure of stimulation sites on the dog's cortex. Δ twitching of neck muscles, + abduction of foreleg, † flexion of foreleg, # movement of foreleg, ◇ facial twitching (Figure taken from Gross, 2007 [3])

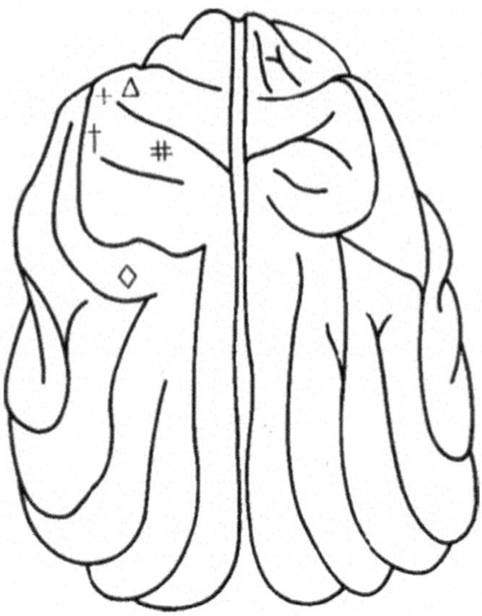

criminal and observed contractions of the right face [5]. Fritsch and Hitzig's experiments thus stand in a particular 'scientific' tradition. Their work also followed from personal experience:

> Hitzig had tried electrical stimulation of the human head for therapeutic purposes and had noticed it caused eye movements. He then tried rabbits and also elicited movements. Fritsch, while working as a battle field surgeon, had apparently noticed that the contralateral limbs twitched while dressing an open head wound. [3]

The dog's cortex was stimulated with platinum wires with brief pulses of monophasic direct current from a battery (i.e. galvanic stimulation). A current was used that was just sufficient to elicit sensations on the experimenter's own tongue (these measurements were at that time seen as the best means for regulation of the stimulation current) [6]. In the animal, usually a muscle twitch or spasm [Zuckung] was observed. According to Gross (2007), who reviewed Fritsch and Hitzig's experiments, the central findings were that:

> (a) the stimulation evoked contralateral movements, the crossed laterality confirming observations dating back to Hippocrates in the 5th century BC [7], (b) only stimulation of the anterior cortex elicited movements, (c) stimulation of certain parts of the cortex consistently produced the activation of specific muscles, and (d) the excitable sites formed a repeatable, if rather sparse, map of movements of the body laid out on the cortical surface. [3]

Fritsch and Hitzig confirmed their stimulation findings by lesioning the areas whereby electrical stimulation had led to muscular responses. They often also observed some recovery of function, which led them to suggest that there was more than one centre involved in motor control [8]. Their experimental results are

Fig. 6.2 Volta's electrolytic battery, as used in a modified form for the stimulation experiments of Aldini and Rolando (circa 1800). (*Upper figure*) Volta's chain of cups. (*Lower figures*) Pile that Volta named the artificial electric organ, columnar apparatus and electromotor apparatus (Text and Figure taken from Boling, 2002 [4])

meanwhile seen as a major step in the development of modern neuroscience [9]. At the time, however, findings were met with great scepticism. Contemporary opinion was that motor functions were controlled from the basal ganglia, in particular the corpus striatum (remember that, for instance, Broca explained Leborgne's hemiparesis by a lesion of the striate body; see Chap. 1). The cortex was thought to be rather insignificant for this purpose (or, in fact, for any purpose). One of the chapters in Young's book gives an excellent overview of the concepts and dogmas on sensorimotor functions that prevailed at the turn of the nineteenth century:

> The corpora striata seem to have held the loyalties of all as the major motor organs. When, in 1865, Luys assigned discrete motor functions to cortical cells on histological grounds, he still held that the corpora striata were the effective motor organs. Carpenter's standard text on Physiology held in 1869 that the corpus striatum was the motor ganglion and the thalamus the sensory. William Carpenter's writings provide a clear picture of the orthodox view, and an opportunity to contrast this with the new approach of Spencer and Jackson. [9]

Young tries to reconstruct why the nineteenth-century investigators failed to see that there were anatomical connections that ran from the cortex to the spinal cord:

> It is perfectly understandable that the investigators of the brain in the nineteenth century related the sensory tracts to the optic thalamus and the motor tracts to the corpus striatum. Todd and Bowman were quite right in tracing the posterior columns of the spinal cord to the thalami and the anterior columns to the corpora striata. But why did they stop there? It appears that their preconception allowed them to see this far and no farther. Neither the thalami nor the corpora striata are the termini of tracts which are seen to pass into them. [9]

The corpus striatum consists of both the caudate and lentiform nucleus. Ontogenetically, these nuclei are single structure, divided by the internal capsule. Modern neuroanatomy teaches that there are direct connections from the motor cortex to the spinal cord, passing through the internal capsule. Sensory information is indirectly passed on to the cortex, via a relay station in the thalamus. This 'passing on' of sensorimotor tracts was not seen or better perhaps not acknowledged by most researchers before 1870. Even after Fritsch and Hitzig's experiments, it took years to convince the scientific community that the cerebral cortex was directly implicated in muscular movements:

> As late as 1886, Jackson indicated that most physicians thought epilepsy to be a dysfunction of sub-cortical and medullary centres [10]. It is not until 1890 that one finds, in Foster's standard *Text Book of Physiology*, the modern view which sees the fibres of the cortico-spinal tract merely passing through the corpora striata, structures whose functions are unknown. [9]

Hughling Jackson's work played an important role in cerebral localization theories, although—as stated by Young—'no claim is feasible that Jackson predicted Fritsch and Hitzig's findings' [9]. Hughling Jackson adopted the view that the nervous system is an aggregate of distinct functional organs and was convinced that all neurological functions were exclusively of sensorimotor origin: 'The psychical, like the physical, can only be functions of complex combinations of motor and sensory nerves' [11]. In 1873 Hughling Jackson wrote that epileptic discharges had their origin in the cortex. Later, Charcot would honour him with the term 'Jacksonian epilepsy', referring to his description of the 'march' of seizures over different body parts, that implicated a somatotopic representation of motor functions [12]. To capture some of the Zeitgeist, it is illustrative to quote a famous footnote that Hughling Jackson wrote in 1870:

> It is asserted by some that the cerebrum is the organ of mind, and that it is not a motor organ. Some think the cerebrum is to be likened to an instrumentalist, and the motor centres to an instrument; one part is for ideas, and the other for movements. It may then be asked, How can discharge of part of a mental organ produce motor symptoms only? I say motor symptoms only, because, to give sharpness to the argument, I will suppose a case in which there is unilateral spasm without loss of consciousness. But of what 'substance' can the organ of mind be composed, unless of processes representing movements and impressions; and how can the convolutions differ from the inferior centres, except as parts representing more intricate co-ordinations of impressions and movements in time and space than they do? Are we to believe that the hemisphere is built on a plan fundamentally different from that of the motor tract? What can an 'idea', say of a ball, be, except a process representing certain impressions of surface and particular muscular movements? What is recollection, but a revivification of such processes which, in the past, have become part of the organism itself? What is delirium, except the disorderly revival of sensori-motor processes received in the past: What is a mistake in a word, but a wrong movement, a chorea? Giddiness can be but temporary loss or disorder of certain relations in space, chiefly made up of muscular feelings. Surely the conclusion is irresistible, that 'mental' symptoms from disease of the hemisphere are fundamentally like hemiplegia, chorea and convulsions, however specially different. They must all be due to a lack, or to disorderly development, of sensori-motor processes. [9]

6.2 Ferrier

In the years following the publication of Fritsch and Hitzig's paper, the literature on cerebral localization became exhaustive [9]. Cerebral centres for various functions were described in various animals, and techniques for stimulation and ablation were further refined. Among the many investigators, Sir David Ferrier (1843–1928) stood out. He was the first to confirm the experiments of Fritsch and Hitzig, and, according to Sherrington, was the main figure to provide the basis for a 'scientific phrenology' [13]. He also paved the way for intracranial surgery (Fig. 6.3). The principal inspiration for Ferrier's research was Hughlings Jackson's ideas on functional localization in the cortex. Ferrier's book *The Functions of the Brain* (1876) was dedicated to Hughlings Jackson, 'who from a clinical and pathological standpoint anticipated many of the more important results of recent experimental investigation into the function of the cerebral hemispheres' [6]. Another motivation was to follow-up on the discovery of the electrical excitability of the hemispheres by Fritsch and Hitzig. Ferrier was initially not very explicit in acknowledging the work of his predecessors, probably because they had not referenced Hughlings Jackson in their paper.[a] Ferrier's method differed considerably from that of Fritsch and Hitzig. He was careful to use the minimal current necessary to obtain responses and to guard against conduction to neighbouring structures 'by insulation of the electrodes, and careful removal of fluid which is apt to collect on the surface' [6]. Still, Ferrier

[a] See Young (1970) for details of this dispute [9].

observed considerable differences between different animals with respect to the excitability of the hemispheres. He also observed a significant variation between different regions in the brain:

> A current sufficient to cause decided contraction of the orbicularis oculi will frequently fail to produce any movements of the limbs. By arbitrarily fixing a standard of stimulation which they thought sufficient, Fritsch and Hitzig failed to elicit the most important positive

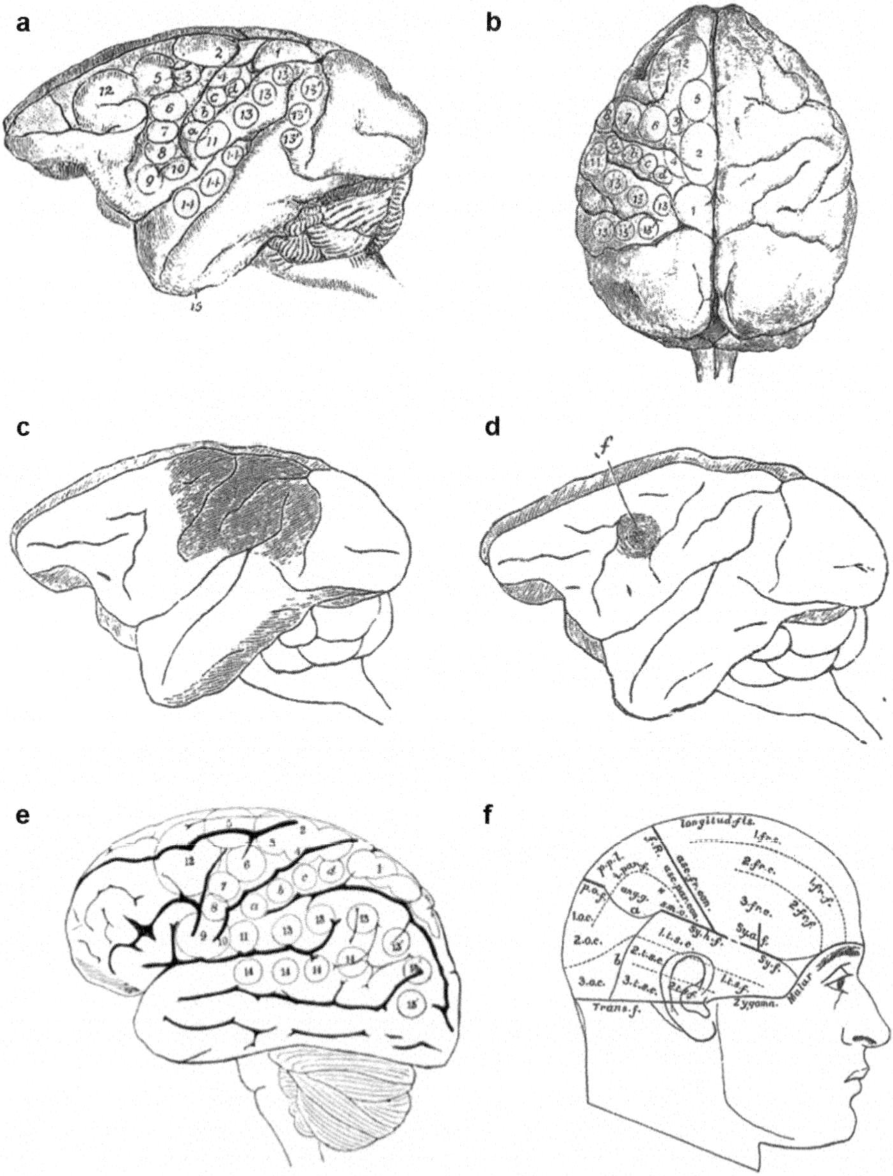

result of deep significance in regions of the brain which they choose to call inexcitable. There is no reason to suppose that one part of the brain is excitable and another not. The question is, how the stimulation manifests itself. [6]

Ferrier used a faradic current (via an induction coil) which resulted in complex muscular movements of longer duration. Remember that Fritsch and Hitzig used direct and monophasic (galvanic) stimulation whereby they induced muscle twitches. Ferrier believed that these more complex movements resembled intentional movements better than the short muscular reactions that were produced by the galvanic method:

The closing and opening shock of the galvanic current, applied to the region of the brain, from which movements of the limbs are capable of being excited, causes only a sudden contraction in certain groups of muscles, but fails to call forth the definite purposive combination of muscular contractions, which is the very essence, and key to its interpretation. Fritsch and Hitzig, in their description of the results of their experiments with the galvanic stimulus, did not, in my opinion, sufficiently define the true character of the movements. If the galvanic current is applied for a longer period than necessary to cause the momentary closing or opening shock, electrolytic decomposition of the brain substance ensues at the points of contact with the electrodes; an objection from which the faradic stimulus is entirely free. I have in my possession the brains of monkeys and other animals, on which experimentation by the induced current was maintained for many hours, which, with the exception of some degree of hyperaemia consequent on exposure as much as stimulation, are entirely free from structural lesion. [6]

Fig. 6.3 (**a**, **b**) Ferrier's composite results of his monkey experiments, showing areas whereby stimulation results in motor responses. Figures and quotes are taken from Ferrier's book *The functions of the brain (1886)* [6]. Note that motor responses are found well behind the central sulcus and also in the temporal region. Ferrier states that there are no exact boundaries for these areas and that stimulation results are dependent on the duration and intensity of the current. For example (compare figure **d**), stimulation of area 6 results in 'Flexion and supination of the forearm—the completed action bringing the hand up to the mouth. The movement is essentially the same as that which occurs on stimulation of the sixth cervical root of the brachial plexus.' (**c**) 'Lesion of the left hemisphere, causing motor paralysis of the right leg and right hand and wrist, and of some of the movements of the right arm, and loss of sight of the right eye.' Ferrier believed that the paralysis that followed a large ablation was permanent. Others observed that there was a considerable recovery when the animals were kept alive for a longer period [8, 162]. (**d**) 'Lesion (f) of the left hemisphere, causing paralysis of the action of the biceps on the right side.' (**e**) Ferrier, following the nomenclature and anatomical scheme of Ecker, translated his results in monkeys to the human brain. He acknowledged that 'An exact correspondence can scarcely be supposed to exist, inasmuch as the movements of the arm and hand are more complex and differentiated than those of monkey; while, on the other hand, there is nothing in man to correspond with the prehensile movements of the lower limbs and tail in the monkey.' (**f**) Diagram showing the relationship between the convolutions and the skull. This sort of information guided surgeons in their trephination and cortical exploration. Macewen was the first surgeon to use Ferrier's data for localizational purposes and operated on a patient guided solely by the motor phenomena of the patient [4]

In his book, Ferrier meticulously presents arguments against the criticism that the elicited movements 'are in reality due to conduction of currents to the real motor centres situated at the base of the brain' [6]. That dispute was certainly still not settled at the time:

> Areas in close proximity to each other, separated by only a few millimetres or less, react to the electrical current in a totally different manner. If there were no functional differentiation of the areas under stimulation the diverse effects would be absolutely incomprehensible on any theory of mere physical conduction, which would, under the circumstances, be practically to the same point in all cases. Movements of the limbs can only be excited from certain points, all others being ineffective. No current applied to the prefrontal or occipital regions will cause movements of the limbs, yet physical conduction to supposed motor centres and tracts at the base is just as easy from these points as from the parietal regions, which react invariably and uniformly. The supposition that it is mere conduction to the corpus striatum and motor tracts which accounts for the movements is further absolutely contradicted by the simple experiment of placing the electrodes on the island of Reil, which immediately overlies the lenticular nucleus. Here we get in nearest proximity to the corpus striatum and internal capsule, and yet no reaction whatever can be induced by currents which are highly effective when applied to the more distant parietal regions. [6]

Ferrier could elicit movements from a rather wide cortical territory in monkeys, including the parietal and temporal lobe (Fig. 6.3). However, he commented that:

> The mere fact that movements result from stimulation of a given part of the hemisphere does not necessarily imply that the same is a motor centre in the proper sense of the term. [6]

For Ferrier, motor reactions could also be secondary related to stimulation of a sensory centre, 'being of the character of associated or reflex indications of sensation' [6]. When, for instance, stimulation of the temporal regions induced movements of the contralateral ear (area 14 in Fig. 6.3), these findings could also point to this region's involvement in hearing [8]. The central sulcus did not form a boundary for motor function in Ferrier's studies. In fact, he (wrongly) assumed that tactile sensation had no representation within the peri-Rolandic region:

> In my earlier experiments, which I have since abundantly confirmed, I could discover no sign of impairment or loss of tactile sensibility after the most extensive lesions involving the convex aspect of the cerebral hemisphere. And yet, considering the definite localization of the centres of sight, hearing, smell, and probably taste, as well as the respective motor centres, no conclusion seems a priori better warranted than that there must be a definite region for the various forms of sensibility included generally under the sense of touch (contact, pressure, temperature, &c.). [6]

This region for tactile sensory functions was, according to Ferrier, located in the falciform lobe (i.e. hippocampal, parahippocampal and cingulate gyrus), a view adopted also by others (see Fig. 6.4). Ferrier rejected, based on experimental and pathological findings, the hypothesis that motor and sensory tracts could become 'jumbled together indiscriminately in the cortical areas' [6]. Instead he conducted several experiments whereby he selectively destroyed areas in the hippocampal region of monkeys to prove that 'beyond all doubt (…) the falciform lobe is the centre of common and tactile sensations.' In order not to damage the brain areas that surround the deeper-lying mesiotemporal regions, which would have undoubtedly

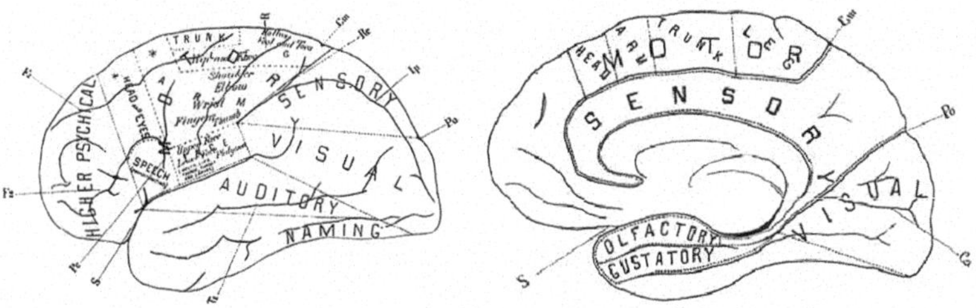

Fig. 6.4 Diagram and quotes taken from Mills' *Cerebral localization in its practical relations* (1888). Mills's findings were 'based upon the investigations of Ferrier, Horsley and Schäfer, and others, and upon a study of cases, personal and collected from the literature of the subject' [163]. 'By the sensorial area is meant that for the senses of touch, pain and temperature, and modification of these senses, and it has been made to include the gyrus, fornicatus, hippocampal convolution, precuneus, and also portions of the superior and inferior parietal convolutions. This sensorial area has therefore been extended to the external surface of the cerebrum so as to include the general postero-parietal region'. Note the naming centre in the posterior part of the third temporal convolution

occurred during surgical exploration, Ferrier inserted a wire underneath or through the occipital lobe to cauterize the deeper parts of the brain. In his chapter 'The sensory centres', Ferrier concluded that the falciform lobe is the cortical centre for the sensory fibres in the internal capsule. He was unable to differentiate the lobe further into various different sensory centres, as he had observed for cortical motor function. Still, he assumed that the 'various motor centres are each anatomically related by associating fibres with corresponding regions of the falciform lobe. This association would form the basis of a musculo-sensory localization' [6].

Although the evidence against Ferrier's view gradually mounted, it took considerable time before the modern view was accepted that sensory functions were located in the postcentral region. See, for a historical review on this topic, the paper of Boling (2002) [4].

6.3 Sherrington and Grunbaum: The Primate Motor Cortex

Ferrier's observations that motor areas were located both anteriorly and posteriorly to the central sulcus were initially widely supported by other clinical and experimental studies. But this view changed, albeit only gradually. Horsley and Beever, for instance, observed that electrical stimulation of the postcentral region was usually not very successful in evoking motor responses [14]. Then again, they had also observed recovery of function from small lesions in the precentral gyrus and concluded that 'the pre-central gyrus is not in man the only out-going motor centre for voluntary movements of the upper limb' [8]. Convincing evidence for the modern view that motor functions are located exclusively in the precentral gyrus was given in the animal studies of Charles Scott Sherrington (1857–1952) and Albert Grunbaum (1869–1921) [15, 16].[b] Their paper, published in 1917, had a lasting

[b] Grunbaum changed his name into Leyton in 1915.

influence for many years. The various observations and conclusions are still well worth reading and remain among the most informative investigations to date. It is worth mentioning that Cushing and Campbell cooperated in some of the experiments and were acknowledged in the paper. Lemon (2008), in a historical perspective on the paper, summarizes the findings as follows:

> Leyton & Sherrington (1917) provided the first detailed proof that there was indeed localization of function within the cerebral cortex. The durability of their report probably owes most to the fact that Leyton & Sherrington (1917) were the first to establish precisely the true extent of the motor area, and to provide the first detailed 'motor map' of the primate motor cortex. In addition, they showed that surgical extirpation of the cortical tissue that, when stimulated, gave rise to movement of a particular body part, resulted in a widespread weakness and loss of use of that same body part. There was, however, substantial recovery in the weeks that followed, recovery that was not lost on lesioning either the adjacent tissue in the same hemisphere or the equivalent cortical area of the opposite hemisphere. Finally, they were able to trace the course of the degenerating corticofugal and corticospinal fibres. They observed widespread degeneration in the cervical cord after a lesion of the hand and arm cortical area and noted that after such a lesion in the chimpanzee (p. 185), 'the whole of the cross-area of ventral horn has scattered through it many degenerating fibres…', which I think is the first report of the direct cortico-motorneuronal projection, a projection whose existence was confirmed physiologically by Bernard & Bohm (1954) and one that appears to be unique to primates (Porter & Lemon, 1993). [16]

The fact that Leyton and Sherrington did not evoke motor responses from post-central areas contrasted with the observations of previous investigators such as Ferrier. Ferrier seems to have been very careful in his stimulation procedures, and yet Leyton and Sherrington obtained different results and conclusions. This discrepancy must be explained by their further refinement of experimental methodology [16]. It is likely that Ferrier and previous experimenters used currents that were 'too strong' so that motor responses were produced from areas outside the true motor cortex [4, 17, 18]. Leyton and Sherrington had realized that several factors could affect stimulation results, notably 'the depth of narcosis, freedom of blood supply, local temperature, and such effects of experimental exposure of the cortex as "drying".' They went to great lengths to control experimental conditions as much as possible:

> For stimulation of the cortex we have used faradisation, applied for the most part by the unipolar method [15, 19]. For this a broad copper plate was strapped over a pad wetted with strong sodium chloride solution lying against the sole of the foot contralateral to the hemisphere under examination. The pattern of electrode used was that figured in the Journal of Physiology, vol. xxviii. p. 16 [19]. It has the advantage of being easily applied with a light and fairly constant pressure against the cortex surface without risk of pricking the cortex or its pia; also of being easily sterilised by the flame, and of being readily bent to any appropriate curve when surfaces not otherwise easily reached have to be explored. The inductorium was of the usual physiological pattern, worked by a single Daniell cell. In many instances we have used also the bipolar method, the electrode tips being 2 mm. apart. The unipolar method is preferable, and gives minuter localisation. Especially where, as in certain experiments, a cut surface is to be explored for fibres running at right angles to that surface.
>
> The animals were in all cases deeply anesthetised with chloroform and ether mixture for the whole of the operation by which the cortex is exposed. During the actual exploration with faradism the anaesthesia was lightened, since in profound anesthesia the cortex becomes inexcitable.

> After the dura mater was opened it was always necessary to prick or tear some small holes in the arachnoid to let out the subarachnoid fluid. If that is not done, localization in the neighborhood of the sulci is almost or quite impracticable.
>
> A precaution found necessary for success in a prolonged examination of the cortex is prevention of a fall in temperature of the exposed cortical surface. The temperature of the room was therefore always kept high, usually fully 30 °C; and the cortex was kept as far as possible covered with cotton-wool swabs wrung out after being soaked with Locke's fluid at 38 °C. [20]

The peri-Rolandic cortices of 22 chimpanzees, three orang-utans and three gorillas were probed and studied with electrical stimulation. Stimulation usually lasted 1–2 s [21]. The results are extensively described in a paper that would nowadays probably not be accepted for publication in such a lengthy format (88 pages). However, the amount of detail and data that is provided is well justified given the many insightful observations and conclusions. More than 400 (!) different movements are listed. These movements often consist not only of a first but also a second, third or even fourth movement. Each unique response is numbered and indicated on a scale drawing of the cortex. For instance, observation no. 187 consisted of flexion of the fingers without the thumb (first movement), wrist flexion (second movement), wrist supination (third movement) and elbow flexion (fourth movement). Movement no. 192 consists of flexion of all fingers and the thumb (first movement), thumb adduction (second movement) and elbow flexion (third movement). In seven animals, precentral areas where specific movements had been elicited with electrical stimulation were lesioned to study the resulting neurological deficits. This resulted in severe motor deficits, as expected from the stimulation results. However, these animals made a remarkable and fast recovery, up to the point where there was almost no deficit detectable. One animal (see Fig. 6.5 for an extensive description) was studied with stimulation mapping on six different occasions. In consecutive sessions, surgical lesions were made in both hemispheres specifically to test whether these areas were involved in recovery of motor functions:

> Improvement in the willed actions of the limb set in very early, and progressed until the limb was finally used with much success for many purposes even of the finer kind. Thus after destruction of the greater part of the arm areas of both hemispheres the two hands were freely and successfully used for breaking open a banana and bringing the exposed pulp of the fruit to the mouth. And again, after considerable destruction of one leg area the foot was successfully used for holding on the bars when climbing about the cage. [20]

Leyton and Sherrington were impressed by the fact that it took the animals many hours to 'realize' that their arm or leg had lost its particular function. They kept on using the limb as if it was not affected. This made them wonder whether the function of the ablated motor cortex could be 'infra-mental':

> The impression given us was that the fore-running idea of the action intended was present and as definitely and promptly developed as usual. All the other parts of the motor behaviour in the trains of action coming under observation seemed accurate and unimpeded except for the role, as executant, of the particular limb whose motor cortex was injured. [20]

The paper has 'a great many key points of lasting value' [16]. Of course, there is the first convincing somatotopic representation of motor functions along the

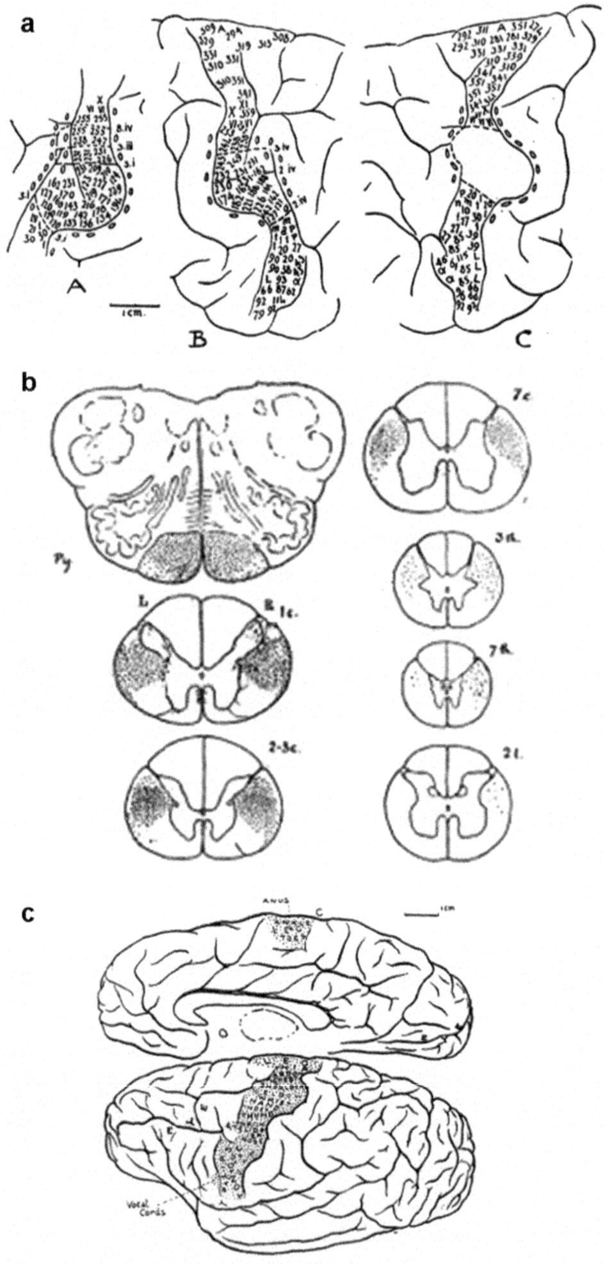

Fig. 6.5 (a *A*) **First operation, January 3—partial resection of left precentral gyrus**. Shown is part of the exposed left hemisphere of a chimpanzee during surgery. Numbers denote the various motor responses of the wrist and hand that were obtained along this part of the precentral gyrus. 'The part indicated by the enclosure within the dotted line was then extirpated: care was taken to include the whole anterior wall of the sulcus centralis, i.e. down to the bottom of the sulcus.' What followed was a partial paresis of hand musculature. 'Fifteen days after the operation great improvement had occurred in the use of the limb; a cursory examination would hardly detect any defect of movement in it; the wound had completely healed.' **Second operation, March 3—extension of**

previous resection in left hemisphere. 'Faradisation of the cortex along the lower edge of the old lesion evoked no movement in hand, but retraction and raising of right angle of mouth, and at one point quite regularly a brisk turning of neck and head toward the opposite side. Faradisation by plunging the unipolar electrode into the soft scar, even when the penetration amounted to 1 cm, failed to evoke movement. Precentral gyrus above the lesion was explored up to the upper genu; it gave the same results as at the previous examination 2 months before.' 'The old scar was then entirely cut away, and the old lesion deepened everywhere by further ablation; and the old lesion was increased upward by removing part of the gyrus previously uninjured as far as the line marked 3. iii. in the map.' 'Next day, the animal doing very well and being very active, the movements of right arm were thoroughly examined. No difference was detected between its existing motility and that obtaining before the last operation.' (**a** *B*) **Third operation, April 2—resection of homologue motor area in right precentral gyrus.** 'Careful search was made for evidence of movement in the right arm on stimulation of this cortical area for the left arm, in order to test the supposition that the recovery of the right arm movements might be explicable by supplementary functions for right arm taken over by the cortical field for left arm. Even with very strong and diffuse (widespread bipolar electrodes) stimulation, let alone moderate and weak with the unipolar electrode, never was any trace of movement of right arm evoked by excitation of the motor arm area of the right hemisphere. The movements elicited in left arm were, however, very various and vigorous. Finally, the whole of the area which under faradisation had provoked "leading" (primary) movement in fingers, thumb, wrist, and elbow was then extirpated by the knife to a depth of about 8 mm, and the floor of the ablated area cauterised superficially with the electro-cautery.' Lines of lesion shown in figure *AB*. This resulted in a clear impairment of left hand and wrist functions. 'Not the slightest recrudescence of symptoms of paresis and clumsiness in the right arm was detected.' **Fourth operation, April 3—extension of previous resection in right hemisphere.** Given the good condition of the animal and the fact that it 'climbed actively about the cage', it was decided to ablate more of the arm area of the right hemisphere. 'The area of cortex indicated on the figure as bounded by fissura centralis and the dotted line above (…, limit line marked 3, iv.,) was then excised to the same depths as yesterday.' Directly after surgery, the paresis of the left arm had increased. However, improvement was already seen the day after surgery. Four days after the last operation, there is 'Further improvement in motility of left arm. Since the right cortex operation, which impaired the motility of left arm, the motility of right arm has notably increased, and right arm has been much more frequently employed than before. No paresis remains detectable in it; and hand, and the whole arm, are now repeatedly employed for all their usual purposes. This morning, after its breakfast, the animal sat and picked its teeth with the isolatedly extended index finger of right hand. It was seen also to pick and scoop out the furrows of the pinna of the left ear with right index finger. When making an effort to take with the left hand a small object, e.g. maize-grain, there occurs frequently an accompanying strong contraction (flexion) of the fingers of right hand. The converse has not been noticed to occur.' (**a** *C*) **Fifth operation, April 8—third resection in left hemisphere.** 'Gyrus post-centralis was then tested by faradisation to see if, especially at the levels opposite the excised portion of arm area, it had acquired motor responsiveness to the electric stimuli, but no motor responses could be elicited from it. From precentralis above the lesion, from the edge of the lesion right up to the trunk area, between arm area and leg area, repeated excitation elicited, and elicited easily, movements of shoulder, but of no other part of arm. Movements thus evoked in shoulder were never on any occasion accompanied by or followed by movements of elbow, wrist, fingers, or thumb. (…) The strip of cortex above the former lesion was then excised to the limit shown (…) by the broken line marked 8, iv. From the old lesion the electrodes never obtained responses, although plunged deeply into the tissue, and although both the single and double electrodes with strong stimuli were used. On recovery from the operation narcosis the animal showed no impairment in the motility of right arm. (…) May 4.—The animal now uses both hands and arms well. Employs either hand in feeding himself with banana or grapes. Peels banana, holding it in one hand and stripping off the peel with the other.' Left and right hemisphere were then again examined with similar stimulation results as in previous sessions. The animal was then killed. (**b**) Microscopic examination of bulb and spinal cord revealed degeneration of pyramids both on the right and left side. (**c**) Stimulation responses from one gorilla 'grouped diagrammatically' (Figures and quotations taken from Leyton and Sherrington (1917) [20]) precentral gyrus (Fig. 6.5c). The drawing of the gorilla's brain is an important precursor of the 'homunculus' images that—20 years later—would become one of the most pervasive pictures in the neurological and neuroscientific literature. Leyton and Sherrington emphasized that their image was only a simplified and diagrammatical representation of their observations; it was not an accurate depiction of an individual's functional topography. Despite these cautions, these visually appealing images are frequently cited and (mis)interpreted outside the context of the original publication. Consequently, important details and nuances are 'lost.' For instance, these diagrams do not show the anatomical and functional variability between animals, a variability that was much larger than what Leyton and Sherrington expected:

> The dissimilarity of the convolutional pattern of the hemispheres even in individuals of the same species (…), and the seemingly variable relation of analogous functional points to sulci of corresponding name, makes it practically impossible to decide with sufficient exactitude what point on the hemisphere of one individual is identical with a given point upon another hemisphere. [20]

The animal studies showed that there was no strict anatomico-functional correlation and certainly no invariant somatotopic order of functions along the precentral gyrus. Of equal importance was Leyton and Sherrington's observation that the areas from which responses of a particular body part were evoked overlapped with those that controlled other body parts, a fact that would later also be confirmed in humans [22]. The fact that there is no pointlike representation for muscles or movements and that functional borders are diffuse and not strict is also something that simply cannot be adequately visualized by a drawing.

Then there was another important issue that was broadly questioned at the time: what exactly was *represented* by the cortical motor areas? If one would again simply look at Fig. 6.5c, one is apt to think that each part of the motor cortex controls the muscles of the body parts that are schematically written on the different parts of the cortex. However, Leyton and Sherrington stressed that movements, and not muscles, were represented in the primary motor cortex (an opinion that was also favoured by Hughlings Jackson and Ferrier). Sherrington's observations are in line with current neurophysiological findings that each part of the motor cortex is involved in the control of multiple muscles and that, conversely, individual muscles are controlled from a wide cortical territory. What the exact role of the primary motor cortex is in movement control, and how the information is coded in the cortex, remains to be determined. The fact is that motor control involves a great many more areas outside the primary motor cortex, and there are multiple functional representations, probably also within the primary motor cortex.

Leyton and Sherrington noted in particular that the movements that resulted from electrical stimulation were *fractional* in their nature, consisting of more or less elementary movements from which other and more complex movements were 'constructed', or as Leyton and Sherrington eloquently put it themselves:

> that the individual movements, elicited by somewhat minutely localized stimulations, are, broadly speaking, fractional, in the sense that each, though co-ordinately executed, forms, so to say, but a unitary part of some more complex act, that would, to attain its purpose, involve combination of that unitary movement with others to make up a useful whole. In evidence of this 'fractional' character it is only necessary to note the predominantly unilateral character, as elicited from the cortex, of movements that under natural circumstances are symmetrically bilateral. [20]

Electrical stimulation thus cannot evoke 'natural' movements. Then there is another very important aspect of electrical stimulation mapping, one that is surprisingly seldom discussed in the modern (clinical) literature, that:

> the cortical motor point, or many of them, are within limits functionally unstable. The chart obtained from a motor region examined at one time and by one series of stimulations may not agree in detail with that obtained from the same motor region at another time and under another series of stimulations. [20]

So the responses that result from stimulation of a certain cortical area are not 'fixed' and can be altered by preceding stimulations of the adjacent cortex. This phenomenon of 'facilitation' was also studied by others, notably by Brown [23, 24]. Leyton and Sherrington assumed that facilitation of responses was a physiological phenomenon that reflected the 'rich mutual associations of the cortical motor points' [20]. Facilitation was needed to compose more purposeful movements out of the partial and fractioned ones:

> Phenomena, such as (…) the functional instability of cortical motor points, are indicative of the enormous wealth of mutual associations existing between the separable motor cortical points, and those associations must be a characteristic part of the machinery by which the synthetic powers of that cortex is made possible. The motor cortex seems to possess, or to be in touch with, the small localized movements as separable units, and to supply great numbers of connecting processes between these, so as to associate them together in extremely varied combinations. The acquirement of skilled movements, though certainly a process involving far wider areas (cf. v. Monakow) of the cortex than the excitable zone itself, may be presumed to find in the motor cortex an organ whose synthetic properties are part of the physiological basis which renders that acquirement possible. [20]

Leyton and Sherrington's observational work is truly a landmark in the study of functional brain topography. This was not only because of the ordered motor maps that were the precursor of later human findings, but in particular because of the careful practical approach and more theoretical considerations of the underlying physiological processes. The latter is often forgotten, or, even worse, not known.

6.4 Krause, Foerster and Penfield: The Human Motor Cortex

We know that at the end of the nineteenth century, several surgeons began to use electrical stimulation to map motor areas in human patients. Feodor Krause (1857–1937) reported stimulation of the central area to map function and identify seizure foci as early as 1893 [25]. He may have been the first to do so, although Horsley, Sherrington and Keen are also named [26, 27]. What is beyond dispute is that Krause and Schum's map of 1911, based on monopolar (faradic) stimulation in 142 patients, is the first detailed representation of the human motor cortex [28]. In an impressive series (even measured by current standards), they systematically found all motor areas to lie on the precentral gyrus and in a somatotopic order that resembles that of the well-known homunculus (see Fig. 6.6). A similar map of the precentral gyrus, albeit more schematic, was published in 1930 by Foerster and Penfield [29]. Both Krause and Foerster made monumental contributions to neurosurgery

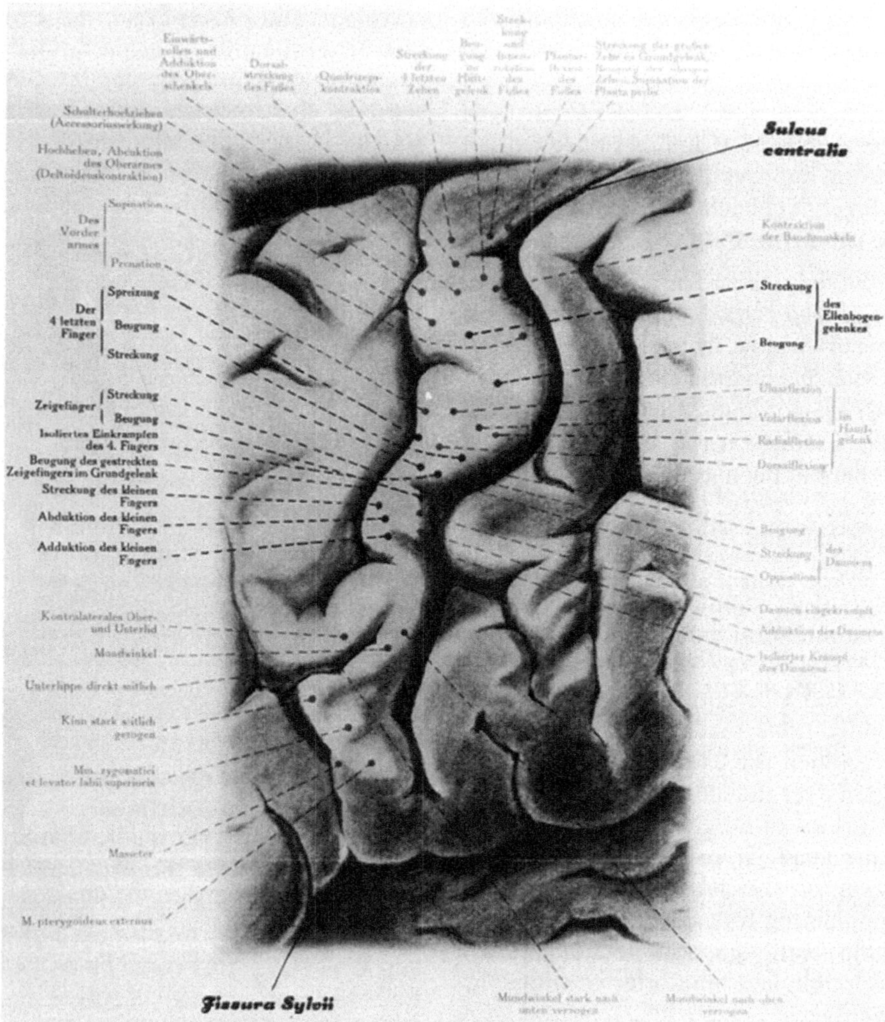

Vergrößerung 2 : 1.
Linke Zentralregion mit F. Krause's durch einpolige faradische Reizung an 142 Menschen gewonnenen Ergebnissen.
Alle Foci liegen in der vorderen Zentralwindung.

Fig. 6.6 Krause's map of motor functions as a result of (faradic) stimulation in 142 patients, published in 1911. Note the amount of detail and the resemblance to the latter maps of Penfield. (Figure taken from Devinsky, 1992 [21]). 'All the foci which are found belong to the precentral convolution. They lie on the cortex, so arranged that the centers for the lower extremities are situated above, near the longitudinal sinus, and, as has been determined on experimental animals, they extend down to the median side of the hemisphere also. About the upper fourth of the precentral convolution is taken up with the lower extremity of the opposite side of the body. About half of the middle portion contains the foci for contractions in the upper extremity, from shoulder to fingers. In the lower fourth are situated the foci for the muscles of the face and the muscles of mastication; here the centers for muscles of the larynx, the platysma myoides and the tongue should also be found' (Krause 1934) [164] including the introduction of surgery for epilepsy. They greatly contributed to the development of electrocorticography [30] as well as intraoperative electrocortical stimulation for localization of functions and localization of the epileptogenic focus (see for details the historical chapter in Lüders and Comair, 2001) [31]. Foerster was strongly influenced by Wernicke, with whom he cooperated and published an anatomical atlas. Following Wernicke's suggestion, he went to Paris to study with Dejerine, Marie and Babinksi, before returning to Breslau [32].

The somatotopic order of the primary motor cortex was only to become famous, however, with the introduction of the homunculus by Wilder Penfield (1891–1976), first published in *Brain* with Boldrey in 1937 [33] and later, in a more final graphical form, in the seminal monograph *The cerebral cortex of man* in 1957 [34]. More than 30 years after Leyton and Sherrington's landmark paper, Penfield and Rasmussen devoted six pages of their monograph to their studies of the anthropoid cortex [34]. At that time, Penfield was already a world-renowned neurosurgeon and famous for his awake surgical procedures. The monograph describes Penfield's extensive experience with electrocortical mapping of the human cortex and begins with an overview of historical studies on brain mapping. Penfield was a great admirer of Sherrington, whom he knew as a teacher from his (under)graduate studies (1915–1919) in Oxford (see Fig. 6.7). Sherrington had a lasting influence on his clinical and scientific work

Fig. 6.7 In 1915, Penfield (*middle row, third from the left*) enrolled in the course on mammalian physiology, directed by Sherrington (*top row, left*) at Oxford University. The photo shows the graduating class of 1916. For Penfield, Sherrington always remained his 'scientific hero.' In a tribute (1952) he stated: 'It was not the example of Horsley or Cushing that led me into surgery of the nervous system. It was the inspiration of Sherrington. He was, so it seemed to me from the first, a surgical physiologist, and I hoped then to become a physiological surgeon' [35]

[35]. In the monograph, the various types of 'instability of a motor point' that were mentioned by Leyton and Sherrington are summarized by Penfield. Penfield had previously shown himself that the rules of facilitation and deviation of response also applied to motor and sensory responses in humans [33, 36]:

> **Facilitation**
> Suppose stimulation is carried out at any given point on the precentral gyrus, for example at point A, which produces finger flexion. If now the stimulation is regularly repeated, advancing the electrode step by step across the cortex anteriorly to A, the same response continues to follow each stimulus until the electrode is a considerable distance anterior to what was otherwise the anterior limit of motor response.
>
> **Reversal of Response**
> If the electrode stimulates point A and then, after time is allowed for the movement to subside, the stimulation is carried step by step downward along the precentral gyrus, flexion of the digit continues to result from each stimulus. Thus when a point B is reached from which at a previous time extension of the digit had been produced, flexion instead of extension results. Consequently, the response from B has been reversed by antecedent stimulation.
>
> Graham Brown and Sherrington (1912) [24] found reversal of response to occur so frequently that they concluded that reversal is "one of the specific offices of the cortex cerebri."
>
> **Deviation of Response**
> If the electrode begins stimulating again at point A and progresses step by step along the motor cortex, producing finger flexion each time, it may happen that a point C is reached which had previously moved the wrist. Stimulation of C now produces finger flexion instead of wrist movement. This is deviation of response at point C. After little time has elapsed, the points B and C will go back to their former state and will yield their original response and not that which they were caused to yield by facilitation. [34]

Modern clinical opinion agrees with Leyton and Sherrington's work that in humans the central sulcus is the border between primary motor and sensory representations. However, the matter has never been completely settled, and questions remain to what extent both gyri are conjointly involved in sensorimotor functions. Sherrington was unable to evoke motor responses from the postcentral gyrus in his animals. However, there was one exception—when postcentral stimulation was 'facilitated' by an immediate previous and positive stimulation of the precentral gyrus. In this way Sherrington—indirectly—demonstrated a functional connection between the peri-Rolandic areas:

> When the centralis posterior near to the central fissure is faradised immediately after elicitation of a motor response from centralis anterior at a point in the latter lying about opposite the point faradised in centralis posterior, the motor response obtained from the centralis anterior may reappear, and this even a few times in succession, though not for many unless centralis anterior be restimulated. This 'echo-response' is a phenomenon of considerable constancy. Our observations on it were made chiefly in the region of the inferior genu and below that, and with motor responses in lips, thumb, or index finger. Graham Brown [23, 24] has, independently of us, observed the phenomenon in regard to flexion of the arm, and in small monkeys macacus and cercopithecus as well as in chimpanzee. [20]

From an anatomical perspective, the precentral and postcentral gyri have direct connections via short, U-shaped fibres. Intergyral connections were first described by Meynert in the second half of the nineteenth century [37]. In 1906, Jakob specifically described connections between 'homologous' parts of the precentral and postcentral gyrus (a 'brachial centre' and a 'facio-lingual centre') [37]. Jakob's work was published in Spanish and therefore at the time had scant diffusion in the English literature [37]. Foerster described motor disturbances with lesions in the postcentral region (afferent paresis), whereby in severe cases these resembled pareses [38, 39]. Although power was preserved, the required movements could not be performed or were insufficient because of diffuse contraction of agonists and antagonists. Peri-Rolandic anatomico-functional connections are nowadays well established in animals and human [40–42]. With non-invasive MRI techniques, it has very recently become possible to visualize the various connections of the precentral and postcentral gyrus, as well as several other connections to neighbouring cortical and subcortical areas (see the beautiful images of Catani, 2012, in Fig. 6.8). Penfield, who systematically confirmed Sherrington's findings in humans, already considered the

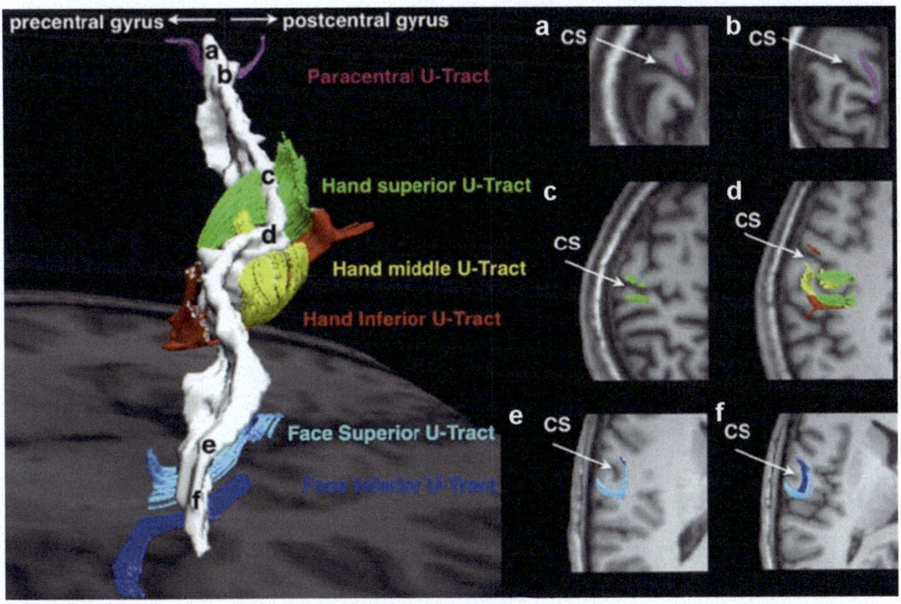

Fig. 6.8 MRI-based fibre tractography enables non-invasive visualization of white matter connections in the brain (i.e. virtual dissection). The method requires (manual) placement of at least two regions to select individual tracts. The white region corresponds to the central sulcus (cs). Letters *a–f* indicate the level of the axial MRI slices as shown on the right side of the figure. (Figure taken from the paper of Catani and Stuss, 2012 [37]). These authors state in the discussion of their paper that 'The exact functional role of the short U-shaped connections remains to be explained. Overall our study suggests that the distribution of the U-shaped fibres follows a functional division rather than a purely anatomical pattern. The three tracts of the central sulcus, for example, whose distribution and relative volume have a precise correspondence with the homunculus regions (Penfield, 1937) [33], are probably in relation to the importance of sensory information for motor control of skilful movements of the hand, mouth/tongue and foot' [165]. Given the dense peri-Rolandic anatomical connectivity, it is 'surprising that direct connections between primary sensory and motor cortices are not considered to play a significant role in current models of sensory-motor integration, for example, in relation to grasping' [37]

primary sensorimotor cortex a 'functional unit' (Fig. 6.9). Although he demonstrated that the 'primary representation of movement' was to be found in the precentral gyrus, he stated that:

> the study of the cerebral cortex of man indicates (…) that there is a subordinate motor representation in the postcentral gyrus. Conversely, the primary somatic sensory representation is postcentral but there is a corresponding representation of sensation in the precentral gyrus. [34]

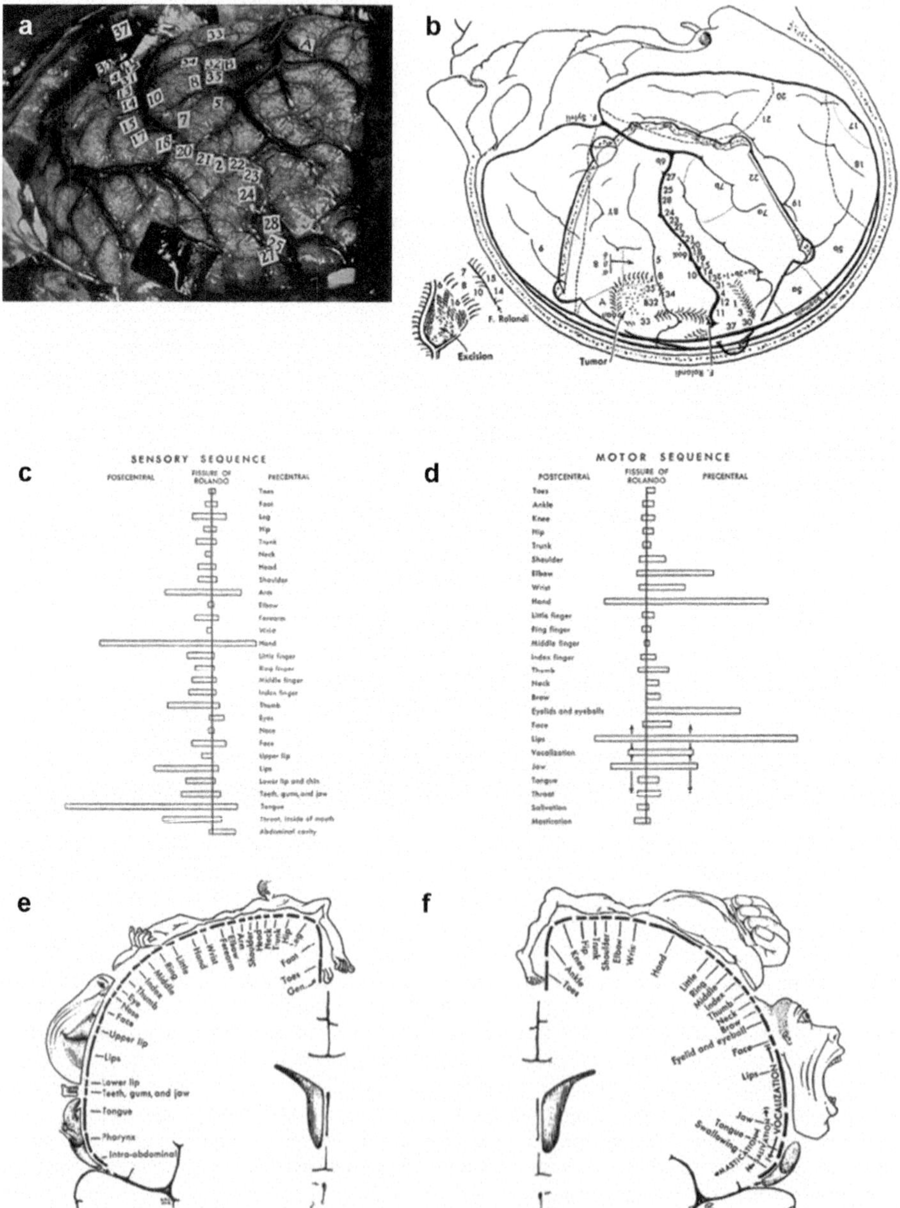

Penfield observed in his patients that as much as 20% of motor response was evoked from stimulation of the postcentral gyrus. Vice versa, sensory responses could be evoked from the motor cortex in 25% of stimulations (see Fig. 6.9). Further evidence of a more intimate functional connection between the peri-Rolandic areas came from resections within this area. Penfield observed that, when part of the (diseased) postcentral gyrus was surgically removed:

> Stimulation of the exposed precentral gyrus may still cause the patient to feel a sensation in the part that has lost its postcentral representation. The reverse is true for precentral excision. Paralysis follows the removal of the precentral gyrus alone. But this is followed in turn by partial recovery, and after recovery stimulation of the postcentral gyrus produces limited movement. [34]

Penfield's sensory and motor homunculi have become iconic images that are consistently cited in medical and scientific textbooks (see Fig. 6.9). In some respects, however, these images have done more harm than good, in a manner similar to the Broca–Wernicke models [33, 43]. As was the case with Sherrington's drawing of the gorilla's motor functions, the complexity of Penfield's experimental findings was greatly simplified to obtain the graphical representations. This was clearly acknowledged by Penfield and his co-authors. They warned their readers that 'the exact position of the parts must not be considered topographically accurate. They are aids to memory, no more' [43]. Penfield was well aware of the inconsistencies of previous mapping studies and certainly realized the complexity of sensorimotor representations in the brain. He denied a simple one-to-one mapping of structure and function:

> The cortical motor sequence of man shows little preservation of the segmental representation of muscles found in the spinal cord and brain stem. There was no evidence of separation of the movement of primitive flexors and extensors. Movements produced by cortical stimulation are gross, awkward. They involve multiple joints and numerous muscles. [44]

Fig. 6.9 Figures and quotations from Penfield and Ramussen's monograph, *The cerebral cortex of man* (1957) [34]. (**a, b**) Photograph and corresponding drawing (upside down) of the exposed cortical right hemisphere of an 18-year-old boy with a history of focal epilepsy. The epilepsy started with sensation in the left side of his body, followed by clonic movements of the left arm and leg. During the awake procedure, electrocortical recordings were made. Tickets A and B mark abnormal spontaneous cortical activity. A small tumour was seen anterior to ticket B. Motor and sensory areas were mapped out by stimulation. Observed motor responses or patient's reported subjective experiences are recorded, and the site of stimulation is indicated with a numbered ticket. 'For this purpose a bipolar electrode was used with the points separated about 3 mm. Occasionally we find it useful to employ a monopolar electrode. The current used was from a stimulator built by Rahm [166] and modified by Jasper. It is our custom to begin stimulation with a frequency of 60 cycles per second and a voltage of 1/2 a volt. The voltage is gradually increased until the first response is obtained.' In the book, clinical and electrocortical mapping results of numerous cases are described and illustrated with photographs and drawings, similar to the example given above. Chapters are devoted to specific functions (e.g. Head and Eye Movement, Vocalization, Arrest of Speech) and end with general inferences on brain function and related cortical areas. (**c**) Penfield observed that as much as 25% of sensory responses was elicited from stimulation of the precentral gyrus. Conversely, motor response (**d**) was obtained in 20% of stimulations of the postcentral gyrus. (**e, f**) The sensory (*left*) and motor homunculus show the order and comparative extent of functional cortical regions. The homunculus is laid upon a (coronal) cross section of the hemisphere. The bars denote more accurately the relative proportion of the area from which responses in the corresponding body part were evoked. These maps show the evidence that Penfield collected from a large number of neurosurgical patients

> In the sensorimotor strip there is an orderly succession of responses to electrical stimulation, but physiologically speaking there is no representation in points or centres. Instead, there is a succession of nerve circuits in which precentral and postcentral gyri are closely related to each other. [34]

Several authors elaborated on the dangers of cartoons such as the homunculus. Some considered it a misleading model of cortical functions that persisted for many decades [45–47]. Others, for instance, Schott (1993), even worried that the homunculus had impeded scientific advance:

> Penfield's homunculus was a deceptively simple and yet naive concept. This type of illustration, a form of map, was a highly original attempt to portray graphically the observations of brilliant and painstaking research and one which has had a lasting influence as a mode of representation. It is memorable and useful. It has, however, been of limited and even doubtful scientific value, since fact and fancy have been confused. Illustration of brain function by projected drawings may best be reserved for those rare instances where true images can be derived and recorded. (…) Representation of everything else may best be served by an unambiguous diagram or words. [43]

Somehow this critique never made it to mainstream science or clinical practice. As of today, the view prevails that the order of primary cortical motor functions is fixed and invariant for every individual, either healthy or diseased.

6.5 Bartholow and Cushing: First Experiences from Conscious Patients

The first record of the use of electrical stimulation in an awake patient dates from 1874 [48]. The case was published by Robert Bartholow (1831–1904) as 'Experimental investigations into the functions of the human brain' in the *American Journal of the Medical Sciences* [49]. The patient was a young woman with a carcinoma that had eroded the skin and skull beneath it. The lesion encompassed both hemispheres and unfortunately had led to brain abscesses that required surgical drainage. Given these circumstances, Bartholow applied faradic stimulation to the dura and the brain of the left hemisphere:

> (…) when the circuit was closed, distinct muscular contractions occurred in the right arm and leg. The arm was thrown out, the fingers extended, and the leg was projected forward. The muscles of the neck were thrown into action, and the head was strongly deflected to the right. [21]

No pain was noted upon stimulation. Bartholow then repeated his experiments on the right side with similar results [21]. He used higher currents to produce 'more decided reactions.' This resulted in a generalized seizure (with focal onset in the left hand) that lasted 5 min. Following the experiments, there were several recurrent seizures; the patient died 4 days later. Autopsy revealed 'needle tracts from the electrodes, extensive thrombus in the longitudinal sinus, and a thick layer of pus covering the left hemisphere' [21]. Bartholow's experiments, which obviously did not serve any medical purpose, met with fierce criticism. He responded in a letter to the editor with an explanation of the case and his considerations. The (dying) patient had given consent, he wrote, and he had expected that the small electrodes would have caused no injury and that the procedure would be safe [4]. He regretted that his experimental results, which he hoped would progress knowledge, 'were obtained at the expense of some injury to the patient' [21]. Bartholow acknowledged that injury was done, but that this was not the cause of the fatal outcome.

It can be difficult to judge whether or not ethical borders are encountered when treating patients. This is obviously always a concern when doing clinical research. Progress inevitably means the use of new methods that have yet to prove their clinical effectiveness and safety. Such methods may be labelled 'experimental' and are especially sought for when patients are suffering, and conservative treatments do not provide significant relief. Good examples can be found in the practice of Harvey William Cushing (1869–1939), who is considered by many the 'father of neurosurgery' and honoured both as a neurosurgeon and a physiologist. When he founded a school of neurosurgery in Johns Hopkins Hospital, he was keen to integrate laboratories and post-mortem examinations in clinical practice. He was convinced that he needed to be both a clinician and a scientist and that medical progress required both laboratory experiments and surgical experiments [50]. His innovations demonstrate that there is really no clear-cut border between conservative and new 'experimental' treatment of patients. One of many examples was his approach to patients with trigeminal neuralgia. The facial pain that is caused by this disease can be excruciating, as was demonstrated by the first patient to be operated on by Cushing and Walker. This former sea captain felt 'a devil twisting a red-hot corkscrew into the corner of the mouth.' He was:

> very near the end of his rope after years of seeking relief from the malady. (…) The slightest movement of his face or beard could set off an attack. Drugs were useless. His teeth had long been extracted [still today, a dental or mandibular origin for these complaints is frequently suspected first, GR]. He could barely eat or talk, and in the summer of 1899 he appeared at Johns Hopkins threatening suicide if the surgeons couldn't help him. Walker was emaciated and shrunken, unwashed and red-eyed from sleeplessness, drooling and writhing and crying out in pain. Two previous operations for his nerve trouble had given him only short-term relief. Now he did not much care if he died on the table. [51]

In Cushing's time several surgeons had attempted to relieve the pain by cutting out parts of the peripheral or central nervous system. Cushing modified a surgical technique that was developed by Hartley and Krause and significantly contributed to the safety and effectiveness of the procedure (i.e. extirpation of the trigeminal ganglion via a craniotomy and subtemporal approach). As always, Cushing was well prepared. He first built on his observations by studying the literature and by practicing on cadavers. Exemplary for his determination and level of preparation is the fact that he found practice on ordinary anatomic material unsatisfactory and therefore performed a great many operations on fresh cadavers; the toughening of the bodily structures gave him markedly different sensations than those of fresh tissue [51]. Cushing carefully documented clinical findings in order to better understand the pathophysiological basis of the disease and to further improve his surgical procedures. He published extensively on his experiences in the literature [52]. Clinical documentation meant that the patients were subjected to time-consuming, daily investigations for a period of weeks. These rigorous and sometimes uncomfortable examinations were of no direct benefit to the patient. Nevertheless, patients participated in the experiments because they thought they were part of their treatment or perhaps felt obliged to cooperate to their surgeon who had relieved them of their pain. Contemporary surgeons such as Halsted and Keen stressed that surgeons had a moral obligation to perform experiments and to pioneer new techniques. Hospitals like the Johns Hopkins Hospital had an important function in education and research. In Cushing's time antiseptics and anaesthesia greatly improved, and this substantially lowered the surgical risks [53, 54]. As a consequence, surgery became more elective, and surgeons routinely gained access to the human body which revealed information that could not be extracted from laboratory experiments or clinical bedside teaching.

In 1900–1901 Cushing made a tour visiting several of Europe's leading clinics and surgeons, an exercise that was at that time frequently done by (young) American doctors. He visited, among others, Kocher, Kroneck, Horsley and Sherrington. On the instigation of Kocher, Cushing studied the relationship between intracranial pressure, vascular dynamics and respiration. Cushing's name was eventually given to the phenomenon ('reflex') of increased systolic pressure, bradycardia and irregular respiration in patients with elevated intracranial pressure. Although he carefully studied the brain's reaction to compression and made significant contributions, the pathophysiological mechanisms were in fact already known for decades [55]. At the University of Pavia, Cushing saw Riva-Rocci's machine for measurement of blood pressure. Cushing was given an inflatable armlet to take home, and 4 months later he would introduce blood pressure measurements for his anaesthetized patients in the Johns Hopkins Hospital [56]. These are some of the more beneficial examples of his tour around Europe. However, in his opinion, there were also disappointments. Cushing was generally not very impressed by the quality of the surgical and anaesthesiological procedures. Cushing's meticulously precise and perhaps even neurotic style of operating was in strong contrast with most contemporary surgical procedures. He witnessed Horsley's operation on a trigeminal ganglion that was done within an hour. Cushing claimed he saw nothing more than 'blood and swabs' [57]. This all made him even more determined to develop new methods for his own patients.

Despite the criticism that had followed the Bartholow case (and other cases), electrocortical mapping gradually gained acceptance as a clinical tool for localization of both the epileptic focus and sensorimotor functions. In 1909 Cushing was the first to demonstrate that the human postcentral gyrus contained sensory representations and that upon electrical stimulation patients reported some kind of sensation [58]. Although it had been repeatedly confirmed in animal experiments that the precentral gyrus was involved in motor functions, the role of the postcentral gyrus was still a matter of debate at that time. Animals obviously could not report their subjective feelings and were also anaesthetized during the stimulation procedure.

Cushing had already used electrical stimulation for motor mapping in several patients, a technique he had learned during his stay with Sherrington in 1901 [59]. In his 1909 publication, which was widely lauded for his surgical and electrophysiological accomplishments, Cushing reported on two patients with epilepsy [60]. After the craniotomy had been performed under morphine and chloroform, these patients were awoken. In the first patient, a 15-year-old boy, mapping along the postcentral gyrus resulted in sensation in various body parts. In his second patient, Cushing was also able to map the adjacent gyri, to confirm that these gyri did not induce sensations (see Fig. 6.10). With Horsley and Krause, Cushing became one of the pioneers of electrical stimulation in humans.

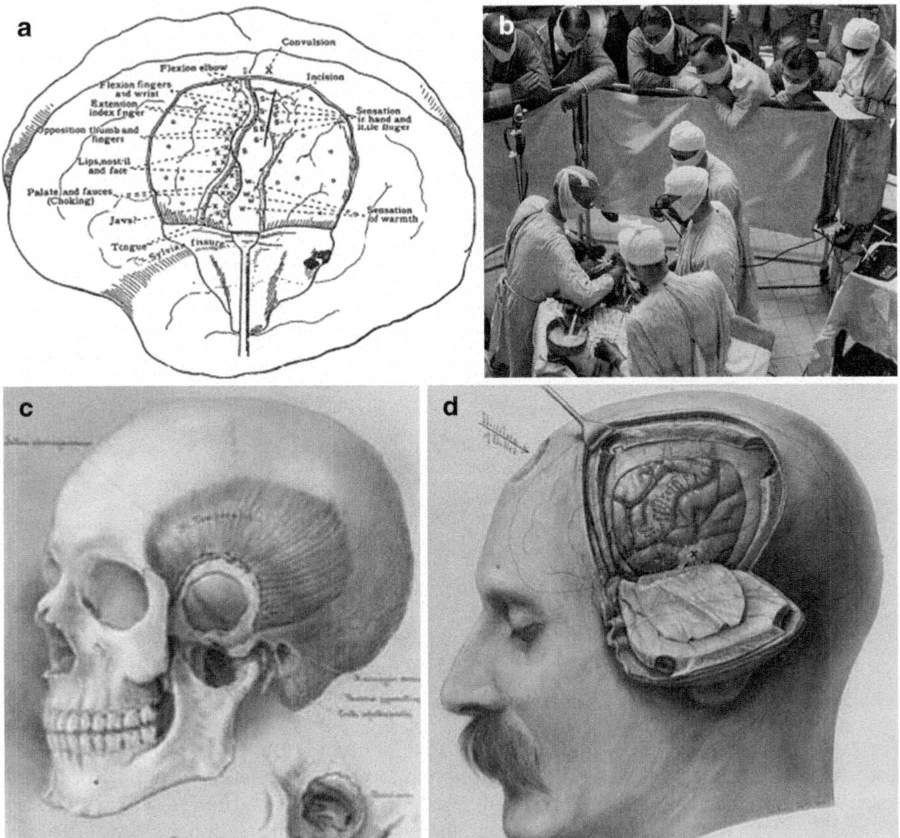

Fig. 6.10 (*Top left*) Cushing's drawing of his results of electrocortical mapping in one of the first patients that were studied under awake conditions. Results from both motor and sensory mapping are shown, as well as 'negative' mapping results of posterior parietal areas. (*Top right*) Cushing operating in the Brigham and Women's Hospital, Boston. (*Bottom*) Cushing was also a gifted artist and often made drawings of his anatomical studies or intraoperative findings

6.6 Penfield's *Speech and Brain Mechanisms*

Penfield was the first to study systematically cortical language organization from the perspective of both electrical brain stimulation and resection of cortical areas (Figs. 6.11 and 6.12). He did so in patients with traumatic brain lesions or tumours who suffered from epileptic seizures. Penfield was much indebted to Foerster, whom he had visited and worked with for 6 months in Breslau in 1928. Foerster was an extremely innovative neurophysiologist, neurologist and neurosurgeon. According to Tan and Black (2001), 'he published more than 300 scientific monographs encompassing every aspect of the nervous system, including tabes, movement disorders, spasticity, extrapyramidal diseases, dermatomes, epilepsy, cortical localization, brain tumors, peripheral nerve injuries, and pain' [61]. Penfield learned Foerster's method of cortical stimulation under local anaesthesia that was aimed at localization of both functional (motor) cortex and epileptogenic tissue. Electrocorticographic recordings and galvanic stimulation of the cortex helped Foerster to delineate the epileptogenic region during surgery [32]. Whenever possible from a functional point of view, Foerster performed a radical excision of traumatic 'scar tissue' to cure the patient from invalidating epileptic seizures. The many veterans from World War I with cerebral injuries and resulting epilepsy gave him enormous experience. Foerster and Penfield conjointly performed a number of studies on this subject [29, 62]. Penfield

Fig. 6.11 Wilder Penfield (1891–1976) as a student at Princeton University in 1913. Two years later, Penfield received a Rhodes scholarship and went to Oxford University to study medicine. There his clinical and surgical thinking was inspired by men like Osler, Holmes and Sherrington. In 1924 Penfield went to Spain to investigate the histological aspects of brain cells with Rio Hortega and Cajal, culminating in publications on oligodendroglioma [167]. Later, he would write and edit a textbook on neuropathology, *Cytology and Cellular Pathology of the Nervous System* (1932) [168]. In 1928 Penfield visited Foerster in Breslau, who educated him on epilepsy surgery and surgical procedures in awake patients. With Foerster he published a topographical functional map that was based on 100 patients; they also studied damaged brain tissue under the microscope in order to understand better the process of scar formation [62]. Penfield was the founder and first director of the Montreal Neurological Institute, which he modelled on Foerster's institute in Breslau. Photograph taken from Wikipedia (https://en.wikipedia.org/wiki/Wilder_Penfield)

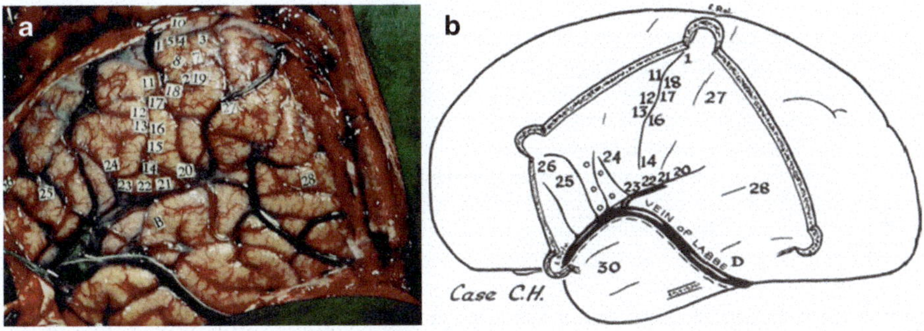

Fig. 6.12 Penfield's case C.H. (*left*) Photograph showing a large craniotomy that exposes parts of the frontal, parietal and temporal lobe. The order of (positive) stimulations is indicated by the numbers on the tickets. Patient was a 37-year-old male with focal and secondary generalized seizures that had started 3 months after a head trauma. Clinical semiology, as well as preoperative electroencephalography and pneumoencephalography, suggested pathological changes within the anterior part of the temporal lobe, and operative cortical excision was recommended. During surgery, a traumatic scar and dense gliosis were found under the tip of the temporal lobe (i.e. a lesion had resulted from the head trauma). Electrocorticography found a focus of high-voltage sharp waves under the surface of the anterior end of the temporal lobe. (*Right*) Schematic drawing of intraoperative results. *Dotted line* indicates the part of the anterior temporal that was eventually removed. Language disturbances were produced at points 26, 27 and 28. Anarthria (motor speech arrest) was produced at points 23 and 24. In the book, all the patients' responses are documented. Here is an excerpt: 24—patient tried to talk and mouth moved to the right, but he made no sound. 25—the patient hesitated and then named 'butterfly' correctly. Stimulation was carried out then below this point and at a number of points on the two narrow gyri that separate 25 from 24, but the result was negative—no interference with the naming process. The points of negative stimulation are shown by the small circles in the figure. 26—the patient said, 'Oh, I know what it is. That is what you put in your shoes.' After withdrawal of the electrode, he said, 'foot.' 27—unable to name tree which was being showed to him. Instead he said, 'I know what it is.' Electrode was withdrawn and then he said, 'tree.' Stimulation at point 28 and 30 led to a naming problem and speech arrest, respectively, but the electrograph also recorded (widespread) afterdischarges; thus, the stimulation results were not considered of localizing value. There was no evidence of aphasia until 20 h after operation. Following that, there was progressive development of profound aphasia. This began to improve at the end of 2 weeks and cleared up finally several weeks later (Figures and text taken from Penfield and Roberts's *Speech and Brain Mechanisms* (1959) [36])

eventually established the Montreal Neurological Institute, which was modelled on Foerster's institute in Breslau [4]. Penfield's perioperative approach became known as the Montreal or Penfield procedure and is still the basis for modern awake surgical procedures [63]. He refined contemporary methods that were needed to study his patients under local anaesthesia and carefully documented operative and clinical findings. Penfield's many experiences and ideas on the neurophysiological underpinnings and localization of language functions culminated in Penfield and Roberts' book *Speech and Brain Mechanisms* (1959). Here is the beginning of the first paragraph of a chapter entitled 'Forbidden Territory':

> Twenty-five years ago we were embarking on the treatment of focal epilepsy by radical surgical excision of abnormal areas of brain (Foerster and Penfield, 1930 [29, 62]; Penfield 1930 [64]). In the beginning it was our practice to refuse radical operation upon the

> dominant hemisphere unless a lesion lay anteriorly in the frontal lobe or posteriorly in the occipital lobe. Like other neurosurgeons, we feared that removal of cortex in other parts of this hemisphere would produce aphasia. The left temporal lobe and the fronto-centro-parietal areas were considered to be devoted to mechanisms of speech, and aphasia literature gave no clear guide as to what might and what might not be removed with impunity.
>
> But patients continued to present themselves in increasing numbers with focal epilepsy that had followed scars and atrophic lesions and small tumors, placed by chance within this general area. Many of these patients were not aphasic. Some had not been aphasic at the time of a well localized previous injury. And so we were emboldened gradually to make more and more excisions within this forbidden territory. [36]

The book includes chapters on the general brain anatomy, functional topography and history of language and is illustrated with many figures from historical publications. It is interesting that Penfield and Roberts chose to reproduce all four of Wernicke's figures from his 1874 monograph (see Fig. 2.3) and not only the one that is nowadays generally associated with the Broca–Wernicke model. They discussed the concepts of agnosia and apraxia, and stressed the intimate and inextricable relationships with aphasia. They were fairly critical when they reviewed the available evidence that had led certain authors to conclude that there was a strong relationship between focal lesions and specific cognitive disorders:

> No discrete localization of lesions producing various types of agnosia and apraxia has been found. It seems, as Jackson (1931) stated that any acute lesion to any gross part of the left hemisphere will produce some disturbance in speech. [36]

In their overview, Penfield and Roberts narrowed down the location of language areas to certain parts of the left hemisphere without becoming too specific:

> In summary, much has been learned about aphasia since Broca's time. It would seem that most authors agree that lesions in specific localities produce definite types of aphasia. The closer the lesion is to Broca's area (the posterior part of the third frontal convolution) and the adjacent precentral face area, the more the motor components of speech are involved. The nearer the lesion is to the vicinity of the junction of the parietal, temporal and occipital lobes, the more reading and writing are affected; and the more the posterior superior temporal region is involved, the greater the difficulty in the comprehension of spoken words. Head's (1926) warning that there are no fundamental individual faculties of speech, reading, and writing should be heeded; but until a better classification of the dysphasias is forthcoming, the disorders of speech, reading, and writing should be recorded.
>
> Lesions restricted to small areas such as Broca's area are extremely rare. The best example of the so-called 'pure' disturbances such as 'pure motor aphasia' are clinical descriptions without pathological correlations (eg, Nielsen, 1936) [65]. In most if not all of the reported cases, discrete lesions have resulted in only transient aphasia.
>
> As far as the recorded literature is concerned, the difference in the clinical syndrome produced by lesions of the precentral motor face area as compared with lesions of Broca's area is extremely difficult to ascertain because of the scarcity of cases with a lesion limited to one or the other area, or insufficient clinical data. [36]

Penfield made detailed recordings of his surgical cases and in many patients also acquired information on long-term outcome. This information is really one of the pillars upon which his work rests. It allowed him to draw general conclusions from his large population. For the book, 190 case records were reviewed in which electrical stimulation had been applied for the purpose of localization of language functions (121 of these cases involved the left hemisphere); 72 of these patients were studied with an extensive (neuropsychological) investigation before and after surgery for evidence of dysphasia or other neurological problems. The book lists many cases, documented with findings from stimulation and excision.

With the patient under local anaesthesia, the brain was typically exposed by means of a large craniotomy (see Fig. 6.12). The brain was then electrically stimulated to identify functional areas. Over the years, changes were made to the stimulation methodology. Penfield did so in close collaboration with Herbert Jasper (1909–1999), the neurophysiologist who was in charge of electrocorticographic recordings and always present during surgery (see also their book *Epilepsy and the Functional Anatomy of the Human Brain*, 1954) [66]. Eventually, a square wave generator was built that produced rectangular unidirectional pulses. Pulses were between 0.2 and 0.5 ms in duration and the frequency was usually set at 60 Hz. Stimulation was done with bipolar electrodes. This setup is still widely used in modern neurosurgery and is often referred to as the Penfield method. From 1951 onwards a monopolar electrode was used that was coated except for the tip, 'so that it may be passed into the brain and stimulation carried out deep to its convexity when desired.'

Stimulation usually started at the postcentral gyrus to determine a minimum threshold. This was done by gradually increasing the voltage until a sensory response was obtained:

> It was often necessary to increase that slightly to obtain movement from the precentral gyrus. Response from the supplementary motor area might be obtained with the same voltage or after a minor increase. Responses from the auditory and visual areas of the temporal and occipital lobes usually called for double the threshold voltage required on the postcentral gyrus; and for the localization of speech areas of the cortex, the voltage was likewise set at double the threshold intensity.
>
> The invariable effect of simple electrical stimulation in any area of cortex is to produce interference with the normal employment of that area. In some areas the stimulus produces activation. In other areas this seems to be impossible and only interference is produced. [36]

Stimulation of either the left or right Rolandic or supplementary motor areas sometimes resulted in vocalization (i.e. a sustained or interrupted vowel cry) but never elicited any intelligible word; it only interrupted ongoing language processes. This was why patients had to perform language tasks during stimulation, usually counting or naming of a series of pictures of objects. Incidentally, but too infrequently to draw general conclusions, tasks for reading or writing were employed. Stimulation led to a number of different types of language errors that were categorized as follows: (a) arrest of speech, (b) hesitation or slurring of speech, (c) distortion of words or syllables, (d) repetition of words or syllables, (e) confusion of numbers while counting, (f) inability to name with the retained ability to speak, (g) misnaming with evidence of perseveration and (h) misnaming without perseveration. When plotted on composite maps, these different language disturbances (a–h) failed to show clear regional preferences, indicating that the type of language disturbance is not indicative for the region that is electrically stimulated. The most interesting map is the one that displays all the speech areas that were found in all patients (Fig. 6.13a). This figure is a representation of the raw

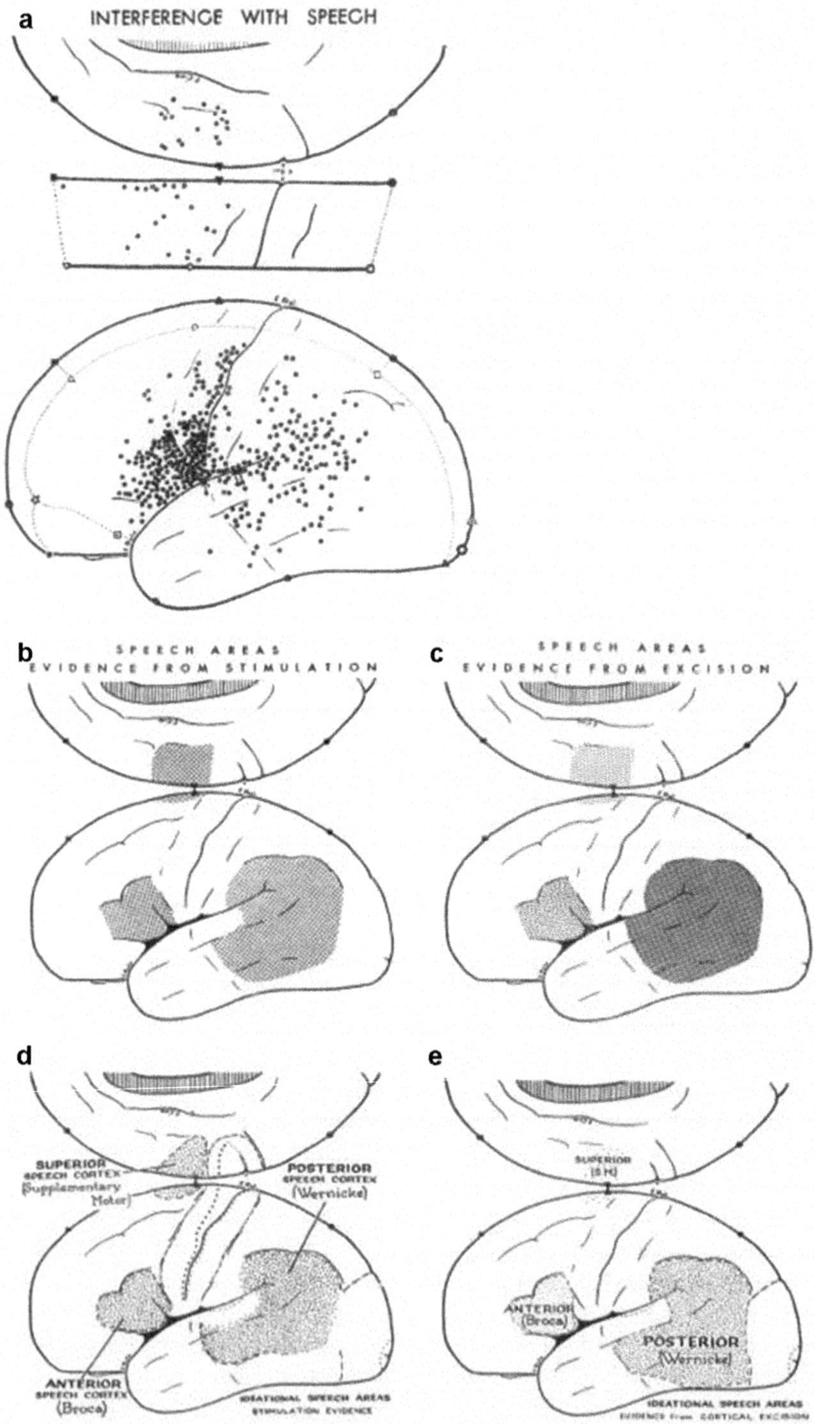
a
INTERFERENCE WITH SPEECH
b
SPEECH AREAS
EVIDENCE FROM STIMULATION
c
SPEECH AREAS
EVIDENCE FROM EXCISION
d
SUPERIOR
SPEECH CORTEX
(Supplementary
Motor)
POSTERIOR
SPEECH CORTEX
(Wernicke)
ANTERIOR
SPEECH CORTEX
(Broca)
IDEATIONAL SPEECH AREAS
STIMULATION EVIDENCE
e
ANTERIOR
(Broca)
POSTERIOR
(Wernicke)
IDEATIONAL SPEECH AREAS

stimulation data, so to speak. New was the involvement of the supplementary motor area (i.e. the posterior part of the superior frontal gyrus) in language functions (although a first suggestion of the importance of the parasagittal region for speech can be found in the work of Schwab and Foerster, who reported that 14 of 21 patients with parasagittal excisions developed a transient aphasia that usually started around the third postoperative day) [67–69]. One of the other striking findings was that the location and the number of language sites strongly differed between patients. Penfield speaks of an 'area localization and not a point localization', meaning that language sites can be found in one of three fairly confined regions but that within these regions there is a large inter-patient variability. Within the individual patient, there were usually only a number of relatively small language areas, a finding that strongly contrasted with the large language areas that were depicted in the literature at that time:

> For example, electrical arrest in the temporal regions five centimeters, seven centimeters, and nine centimeters from the tip may interfere with speech; whereas, at other points between them no interference is produced. Also, stimulation of the first, second, and third gyri anterior to the left precentral face area may produce effects upon speech, though stimulation of the same gyri at other points does not.

Bogen and Bogen (1976) in their brilliant review paper noted that Penfield and Roberts's book provides us with a number of figures in which the language territories are all slightly different; they remarked that in this respect 'the book is a little confusing' [70]. So why was that? We do not know exactly. When discussing

---

Fig. 6.13 (**a**) Penfield's composite map of stimulation sites where electric stimulation led to an interference with speech tasks. (**b, d**) Penfield deduced three speech areas from the results of electrical stimulation: inferior frontal, supplementary motor and temporoparietal. All three areas seemed to him of equal value from the perspective of electrical stimulation. 'So far as can be determined there is no difference between the effects of the electric current when applied to the dominant Broca's area, supplementary motor area, or parieto-temporal region as regards the various alteration in speech. The reason for this lack of difference could be that these three areas are connected by transcortical and subcortical pathways in a single system. An electrical disturbance set up in any part of the system might disrupt the function of the whole system.' If we assume that the horizontal and anterior rami of the Sylvian fissure are indicated on the scheme (this is not explicated in the text), then Broca's area consists of the pars triangularis and pars opercularis of the left inferior frontal gyrus. Penfield explicitly defines Broca's area as 'the three gyri in front of the precentral gyrus.' (**c, e**) These figures were composed after studying possible language impairments after cortical excisions. Note the dissimilarity between figures (**c**) and (**e**); figure (**e**) shows less involvement of the posterior part of the superior temporal gyrus, but more of the middle and inferior temporal gyrus. 'The three speech areas, we believe, are of different values. The posterior, or parieto-temporal, area is the most important. The anterior, or Broca's, area is the next most important but is dispensible in some patients, at least. The superior, or supplementary motor, area is dispensible but probably very important after damage to one of the other speech areas.' (**e**) 'Summarizing map of the areas of the cortex in the dominant hemisphere which are normally devoted to the ideational elaboration of speech. These conclusions are derived exclusively from the evidence from cortical excisions made *around* [italics GR] the speech areas. (…) The size and number of dots suggest the order of dispensability. Removal of the superior speech area produces aphasia of a few weeks' duration; removal of the anterior area, an aphasia of longer duration' (Figures and text taken from Penfield and Roberts's *Speech and Brain Mechanisms* (1959) [36]) the stimulation results, Penfield and Roberts provide us with a more abstracted map (Fig. 6.13b) that significantly differs from that of Fig. 6.13a in two aspects. First, the inferior part of the sensorimotor cortex is completely left out. Penfield considered this part of the brain a pure motor area. He believed that the speech arrest that was obtained with stimulation was caused by an interference with the muscles that are necessary for speech and not with speech mechanisms per se. Hence such a response was classified as an anarthria and not an aphasia. Note that this classification relies on a priori assumptions of the functional character of the cortex and is not only solely based on the behavioural response (a complete speech arrest). Note also that the supplementary motor area was *not* left out in Fig. 6.13b–e. One could think of arguments to do so, because the supplementary motor area resembles the face motor area in stimulation results: its most prominent behavioural response was a speech arrest, whereas this was the only area—together with the face motor cortex—where positive motor phenomena were elicited. In other words, the supplementary motor area may indeed only be a motor area. A second striking difference between the stimulation data and its more schematic counterparts (Fig. 6.13b, d) is that only a small part of the posterior temporal gyrus is included in Wernicke's area, whereas this gyrus clearly shows several positive stimulation sites [70]. In Fig. 6.13d, which is presented in the *concluding discussion* of the book, it is for unknown reasons, even almost completely left out. Thus, the figure that was to become one of the classic pictures in neurology and neuroscience was already an interpretation of results and not a pure reflection of observations.

In addition to the stimulation data, Penfield and Roberts also provided their estimation of language areas from the perspective of cortical excisions (Fig. 6.13c, e). But what exactly do these figures tell us? At first glance, and without any background information, one is apt to think that these figures point to the areas where surgery has a high chance of inducing severe and permanent language deficits, that the figures thus represent a surgical risk map. But that interpretation is not correct. On the contrary, when discussing the *evidence from cortical excision*, Penfield and Roberts write:

> We have shown that any limited, previously damaged area of the left cerebral hemisphere may be excised with transient aphasia, but without immediate or permanent aphasia, so long as the remaining brain functions normally. (…) if the outlines of excisions [from all patients] are combined, all parts of the left hemisphere are included in the map. [36]

Later in their book, they add that removals within language areas, and particularly the posterior language area, should be 'small' in order for aphasia to recover. Still, if excision of *any* part of the brain, including parts of Broca's and Wernicke's territories, is possible without lasting deficits, what do Fig. 6.13c, e tell us? First, it is important to note that, obviously, Penfield did not remove 'normal' brain areas. These figures thus do not represent language organization in a healthy brain, something Penfield and Roberts were of course well aware of: 'We do not know what immediate disturbance in speech would have occurred if there had been normal brain.' Penfield also did not remove the areas that he detected with cortical stimulation, as he considered these areas critical for normal language function. This is the probable reason why the excision data largely resemble the stimulation data.

Generally speaking, the closest evidence that a functional area has been resected (or damaged) comes from patients that have immediate language disturbances after surgery. However, this did not occur very often in Penfield's series; only in 22 of 273 operations (and never in the right hemisphere) did patients have language disturbances immediately after surgery. Most of these patients had only transient deficits. Only three patients had a lasting aphasia and all of these had a tumour. Penfield considered the non-tumour patients his 'best' patients to judge whether functional cortex had been resected, because the subcortical involvement and mass effect of a tumour were likely to overestimate the contribution of cortical function:

> After reviewing the location of the excisions in all these cases without tumor, it seems surprising that dysphasia did not occur more often immediately after operation. (…) There were only five patients whose immediate dysphasia could not be explained on the basis of fatigue, post-ictal phenomena, or probable vascular occlusion. The difficulty in language rapidly cleared in these five individuals. The recovery occurred in a period of time too short to consider that some other area had taken over the function of the area excised. [36]

Thus, permanent language deficits after surgery were rare. For Penfield this meant that his electrocortical stimulation procedure was adequate for identification of functional cortex and preservation of language functions. But the number of patients with lasting deficits was too small to justify conclusions. There was, however, another and larger group of patients who experienced language disturbances. Strangely, in these patients the deficits did not develop immediately after surgery but only after some hours or days. Penfield lists examples of many patients who had intact language functions during and immediately after surgery, but developed dysphasia after a delay of hours or days (Fig. 6.14). He suggested that these transient deficits were locoregional phenomena due to 'neuroparalytic oedema':

> Most patients classified as having aphasia after operation began to have difficulty in speech one or more days after operation. Frequently, disturbance in speech would be noted first on the fourth day after operation, would increase almost to global aphasia within the next day or so, would begin to lessen a week later, and would disappear after several weeks. It was assumed, then, that following prolonged exposure to air and ultraviolet rays, as well as numerous electrical stimulations of the brain, physiological or pathophysiological changes (see Prados et al., 1945 [71]) had occurred which were different from those seen after ordinary brain trauma. Therefore, we have used the word 'neuroparalytic' to describe the edema of those patients. (…) This time course was considered most unusual for brain edema.

Penfield and Roberts assumed a local cortical effect of the neuroparalytic oedema and hence suspected that the dysphasia was caused by a temporary malfunction of language areas that were close to the area of resection. They did not consider the resected area to play a role in language functions (any more): 'It is permissible to assume that the adjacent areas of cortex were temporarily not functioning normally at that time.' It is via this indirect line of evidence that they must have deduced the location of the language areas that were eventually graphically depicted as Fig. 6.13c, e.

Let us rethink this further and study some of Penfield's individual cases, which are shown in Fig. 6.14. The fact that these patients did not show any immediate (new) deficits after resection of left perisylvian areas strongly suggests that language functions had already relocated to other brain areas prior to surgery, probably because of

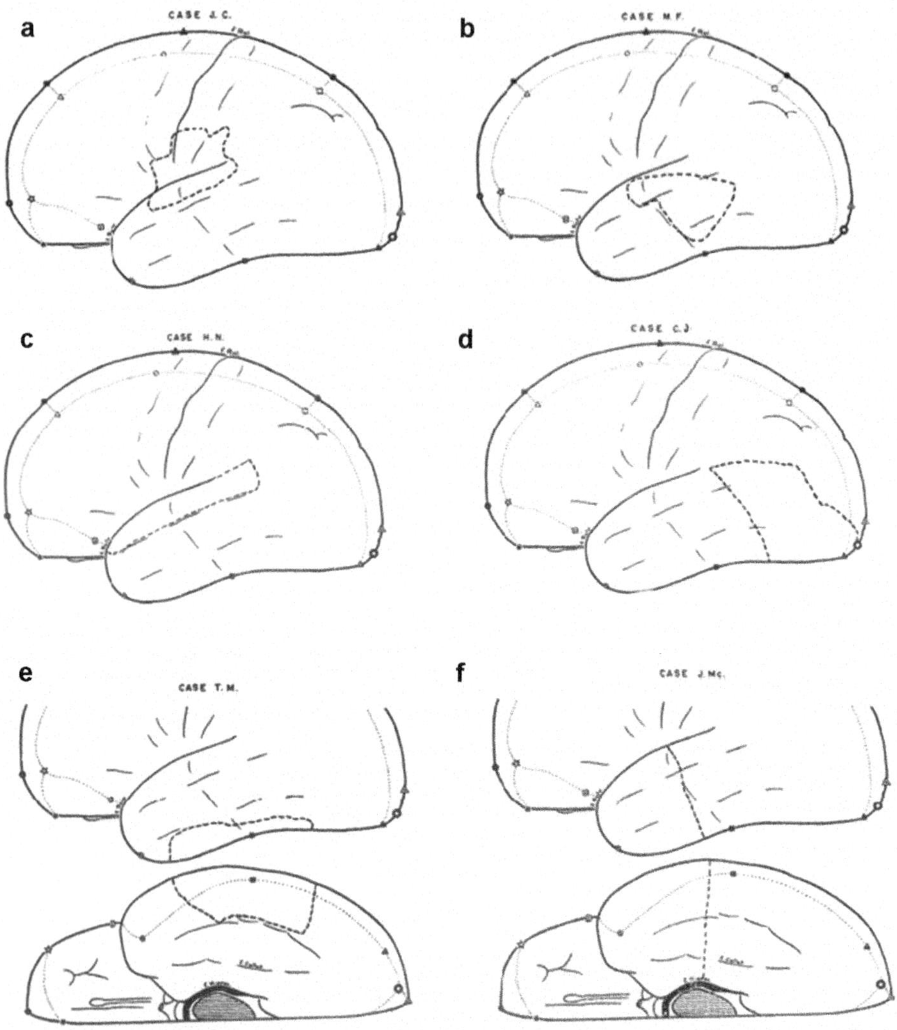

Fig. 6.14 Six epilepsy surgical cases whereby the extent of the cortical resection in the left hemisphere is indicated with dotted lines. Three patients had no language deficits prior to the operation (cases **a**, **b** and **d**). The other patients had slight (**c**, **e**) or moderate (**f**) problems with language during testing. Except for case E, there were no language deficits immediately after surgery. A severe but transient secondary aphasia developed later. (**a**) Twenty-one-year-old right-handed man. Seizures starting 2 years after head trauma with skull fracture. Scar lesion was excised. A practically complete aphasia developed 12 h after surgery. After 3 weeks only language deficits were apparent with testing of more complicated speech functions. (**b**) Thirty-two-year-old right-handed man with left parietal and skull base fractures after a car accident. Scar lesion was excised. Seizures started 9 months after. Speech difficulties started 1 day after operation and lasted 30 days. (**c**) Nineteen-year-old boy with seizures since the age of 11. No neurological abnormalities except for slowness of speech. During surgery, a 'yellow and tough' first temporal convolution was excised. One day after operation, he had a slight difficulty in spontaneous speech and seizures. Four days postoperatively, there was a marked dysphasia. 'When tested seven months after operation, he was slow and hesitant in speaking—perhaps, a little more than before operation. He had difficulty reciting the alphabet, getting to "k" and then becoming confused, but finally doing it correctly. He had

seizures that caused functional impairments of this area (there is another explanation, namely, that there is some form of immediate reorganization during surgery, but that is very unlikely given the time course of the language disturbances). The fact that there were transient deficits argues that perilesional areas are involved in this reorganizational process. Modern studies have confirmed Penfield's clinical observations of delayed and transient neurological deficits following surgery, but have also demonstrated that functional recovery usually takes months, and not weeks. These studies speak of brain plasticity, but this term is perhaps just as uninformative as Penfield's neuroparalytic oedema. Penfield and Roberts did accept the fact that brain functions could relocate to homologous areas in the contralateral hemisphere after traumatic or vascular lesions, in particular, in the case of large lesions and young age. They also speak of 'replacement' of speech by areas within the ipsilesional hemisphere:

> If one of the speech areas is destroyed, then adjacent areas of cortex and the other speech areas function during speech.
>
> It is clear that in cases of cortical destruction some degree of replacement of speech function is possible within the same hemisphere. But this substitution of one area for another does not seem to take place when there are continuing local epileptic discharges in the cortex. [36]

Penfield had his own theory to explain 'plasticity of the brain' (his own term) [36]. In his view, all the different cortical areas were somehow functionally

no, or only questionable, difficulty in all other tests. It is believed that his residual difficulty is related to the frequent minor seizures.' (**d**) Eighteen-year-old, left-handed boy with a right hemiparesis as a result of birth injury. Seizures started at the age of 11. No speech difficulties prior to operation. On the third day after surgery, a transient 'and rather severe aphasia' developed. (**e**) Twenty-four-year-old left-handed man that had a mastoiditis and left epidural abscess as a child. Seizures (with postictal dysphasia) started at age 22. Presurgical testing revealed slight difficulties with reading and spelling. At operation 'abnormal brain' under the surface of the temporal lobe was removed. 'After the excision and before closure of the wound he had no difficulty in spontaneous speech or in reading words and short sentences, but he did have moderate difficulty in naming. (…) The next day he was practically speechless, though emotional speech, particularly swearing, was present. He was unable to name, read, or obey oral or written commands.' Patient started improving 2 weeks after the operation. Two months later testing showed no difficulties as compared with presurgical findings. (**f**) Fourteen-year-old right-handed girl who had had measles and a generalized seizure at the age of 9 years. Three months later habitual seizures started. No neurological abnormalities. Testing showed that she had slight difficulty in obeying oral command and moderate difficulties in reading and spelling. After surgery there were initially no deficits. 'Thirty-nine hours after operation she was speechless and perseverated on the single sound "owl". There was also a slight weakness of the right side of the face and the hand.' On the fourth day after operation, she said nothing that was intelligible but was able to obey simple oral commands or point to objects after she read the word on a card. 'Three weeks after operation she showed definite improvement. Twenty-five days after operation she had no abnormal neurological signs.' Testing still revealed deficits 'probably due to abnormally functioning brain associated with numerous seizures after operation' (Figures and text taken from Penfield and Roberts's *Speech and Brain Mechanisms* (1959) [36]) integrated via subcortical connections. As he had observed in many cases that cortical areas, independent of their location, could be resected with no, or no permanent, loss of function, he posed on clinical grounds that there is 'evidence of a level of integration within the central nervous system that is higher than that to be found in the cerebral cortex.' Already in 1938 he had published about this centrencephalic system, whereby each functional area of the cortex 'forms a unit with some portion of the diencephalon of which it is a developmental projection' [36]. Penfield was convinced that the higher brain stem, and in particular the thalamus, was the 'central coordinating and integrating mechanism' that served the purpose of inter- and intra-hemispherical integration (Fig. 6.15) [36]:

> For example, the anterior frontal cortex might be thought of as an elaboration from the dorso-medial nucleus of the thalamus, and much of the temporal cortex as an outward projection of the pulvinar and posterior part of the lateral nucleus of the thalamus (...). This is

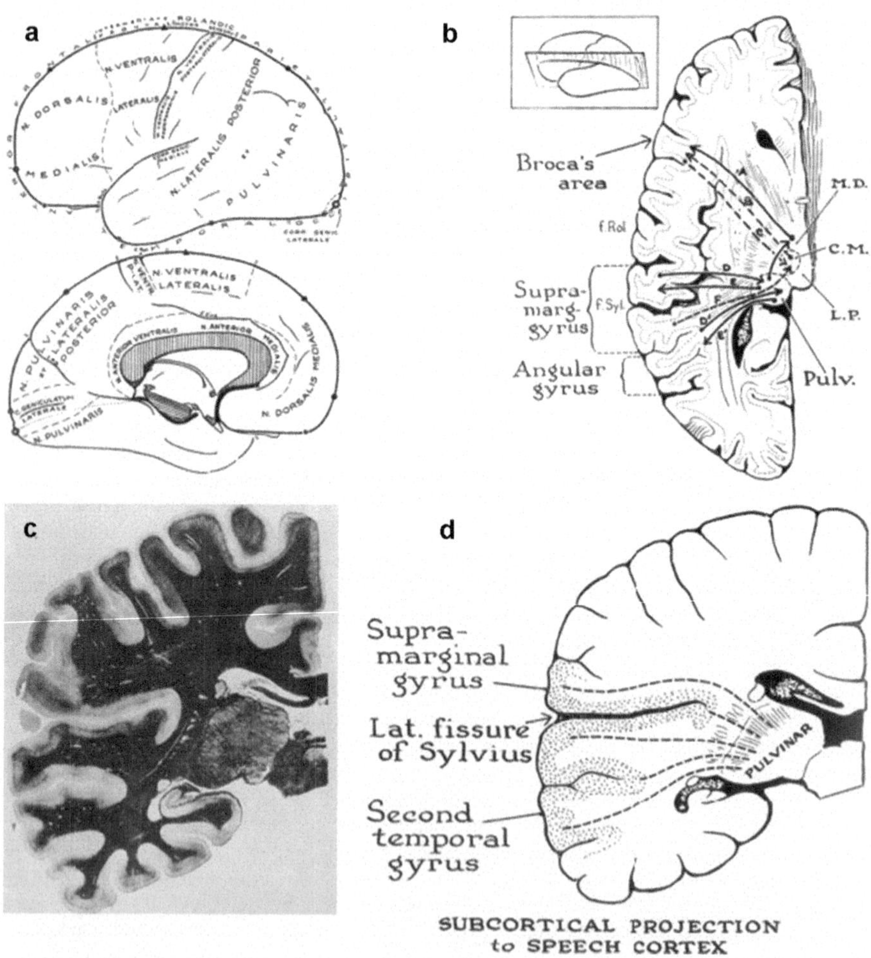

> in many ways a surer guide by which to predict functional subdivisions of the cortex than the cyto-architectonic parcellation of cortex (eg, Brodmann's area). [36]

Penfield and Roberts further specified their proposed 'speech hypothesis' whereby:

> the functions of all three speech areas in man are coordinated by projections of each to parts of the thalamus, and that by means of these circuits the elaboration of speech is somehow carried out. Support for such a conception is given by the fact that removal of the gyri all about the two major cortical speech areas does not produce aphasia. Indeed the map in Figure X-10 (our Fig. 6.13d) was drawn from the negative evidence provided by successful excisions of gyri close to the speech area, which resulted in no more than transient postoperative aphasia that began several days after operation. Such removals were carried to the bottom of each fissure but never deeper than the gray matter of the gyrus. The removals would not, therefore, interrupt the connections between other gyri and their own subcortical structures under any circumstances. They would also not ordinarily interrupt the more deeply placed transcortical connections within the white matter. [36]

Figures 6.15 and 6.16 illustrate the proposed cortical–subcortical connections of the speech areas. At Penfield's request, Klingler performed (his now famous) white matter dissection on post-mortem brains to show fibre connections between the language cortex and the thalamus. The role of subcortical structures in the thalamus only gained (some) interest many years later, when technology was able to localize subcortical lesions (i.e. from the 1970s with the introduction of computer tomography) [72]. Interestingly, direct cortico-thalamic circuits that involve Broca's area have only recently been visualized in more detail with MRI tractography [73]. Penfield acknowledged the importance of transcortical association tracts, but considered these 'certainly of less importance than subcortical integration' [36]. Figure 6.15c, d depicts a cross section of the brain that Penfield used to illustrate the depth of the sulci and to show the 'close relationship of the posterior cortical speech area, with this basal gray nucleus and the adjacent nucleus lateralis-posterior' [36].

Fig. 6.15 (**a**) Projection of thalamic nuclei and geniculate bodies to the cerebral cortex, as suggested by thalamo-cortical connections. The figure is based on monkey studies by Earl Walker (1938). Penfield was convinced that each functional area in the more recently evolved cortex was connected to portions in the older brain. He wrote that 'the subcortical areas of gray matter, by means of their projection fibers, serve to coordinate and to utilize the functional activities of cortical areas and to integrate that activity with the rest of the brain.' Penfield's ideas are in line with our modern view that the thalamus and the cortex are two closely interconnected structures [169]. (**b**) Cross section through the brain showing hypothetical cortical–subcortical connections between the speech areas, mediated by the basal ganglia. 'The connections that are indicated by solid lines have been established for the monkey by anatomical studies, and by electrographic recording measurements as well. The broken lines of connection to the centrum medianum (C.M.) have been established for the monkey by electrical recording methods only.' (**c**, **d**) Anatomical coronal section of the brain (*left*) and corresponding drawing through the pulvinar and posterior language areas, indicating Penfield's proposed cortical–subcortical connections of the speech system [170] (Figures and text taken from Penfield and Roberts's *Speech and Brain Mechanisms* (1959) [36])

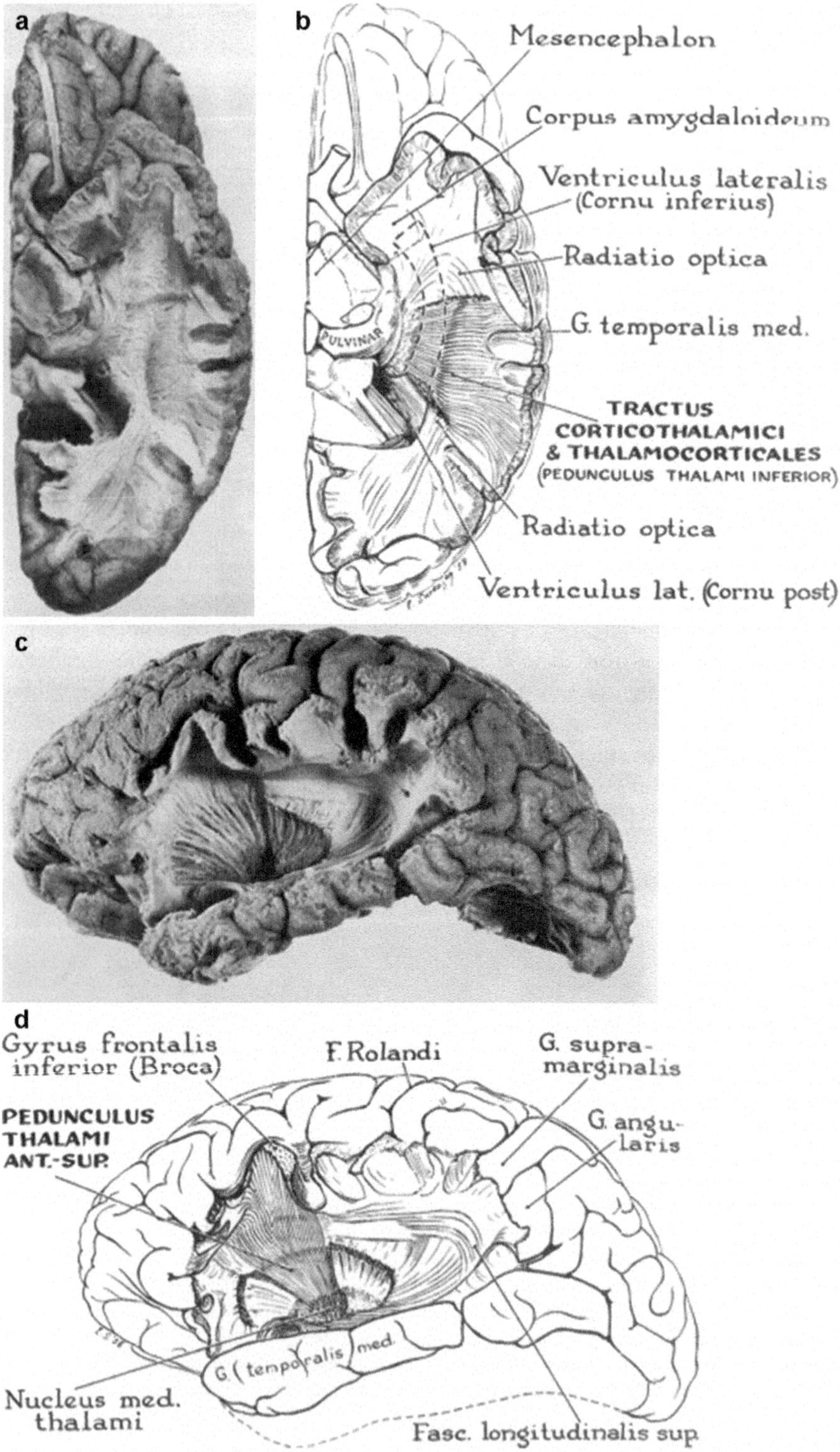
a
b
Mesencephalon
Corpus amygdaloideum
Ventriculus lateralis
(Cornu inferius)
Radiatio optica
PULVINAR
G. temporalis med.
TRACTUS
CORTICOTHALAMICI
& THALAMOCORTICALES
(PEDUNCULUS THALAMI INFERIOR)
Radiatio optica
Ventriculus lat. (Cornu post)
c
d
Gyrus frontalis
inferior (Broca)
F. Rolandi
G. supra-
marginalis
PEDUNCULUS
THALAMI
ANT.-SUP.
G. angu-
laris
G. (temporalis) med.
Nucleus med.
thalami
Fasc. longitudinalis sup.

So what Penfield proposed was in fact a strong and dynamic connectionist view, whereby a network of cortical areas is able to compensate for a cortical lesion within classical language areas. Penfield rejected the view of single areas with specific functions:

> Lesions in particular localities may result in specific clinical syndromes. Lesions in the region of the precentral face area and of Broca's area may cause dysphasic disorders which are predominantly expressive in type. This does not mean that a center for eupraxia, and another center for movements of the lips, etc., have been destroyed. There is no specific site where what Nielsen (1946) calls the motor engrams of speech are stored. A large part of the cortex and sub-cortex appears to be active during the production of a proposition. The transmission of impulses from the precentral gyrus to all of the complex musculature necessary for speech is certainly occurring; and there is activity in Broca's area or another speech area. There is, however, no localized area for articulate language in Broca's convolution. Broca's convolution is only part of the whole. [36]

This view is in strong contrast to the implicit message conveyed by the often reproduced (static) schemes of Fig. 6.13. For Penfield, even these areas were to some extend 'expandable.' The superior language area (i.e. the supplementary speech area) was the area that was:

> The most easily dispensable: The evidence derived from cortical ablations in the dominant hemisphere indicates that removal of the supplementary area produces an aphasia that disappears within a few weeks.

Penfield was not sure about the dispensability of Broca's area: 'our evidence leads us to believe that Marie was probably correct when he asserted that this area of cortex could be sacrificed without eventual loss of speech in the adult.' However, Penfield considered his evidence too meagre to generalize this to all cases (he had himself removed Broca's area only once, in a case of an indolent brain tumour, a hamartoma). His advice to neurosurgeons was:

> Despite the suspicion of indispensability of the anterior speech area of Broca, we still advise that this area, which can be outlined so clearly by stimulation, should be carefully avoided during surgery. No excision should be carried out in the posterior speech area of adults, unless the removal is small. (...) On the contrary, in the posterior speech area any large destruction that involves cortex and the underlying projection areas of the thalamus would certainly produce the gravest aphasia.

Fig. 6.16 (**a, b**) 'Photograph (and corresponding drawing) of the left hemisphere seen from below to show the nerve fiber projection connections between the posterior speech area on the left middle temporal convolution and the pulvinar. The inferior portion of the left temporal lobe has been removed, together with the inferior horn of the ventricle. (...) Dissection by Professor Klingler.' (**c, d**) 'Photograph (and corresponding drawing) of dissection of the left hemisphere seen from the lateral surface to show connections between thalamus and anterior speech cortex (Broca). Note the superior longitudinal fasciculus that connects posterior and anterior language areas. (...) Dissection by Professor Klingler' (Figures and text taken from Penfield and Roberts's *Speech and Brain Mechanisms* (1959) [36])

6.7 Ojemann: Expanding the Language Territory

Since Penfield's comprehensive studies, several other authors have reported their results with electrocortical stimulation in awake patients. In 1974, Fedio and van Buren studied 19 patients and found language areas spread over a larger territory than described by Penfield [74]. These findings were confirmed 15 years later by George Ojemann in his landmark study of 117 patients (Figs. 6.17 and 6.18). Ojemann found that language errors could indeed be evoked from a far wider region than the traditional Broca and Wernicke regions. This was in particular the case in the frontal lobe, where language areas were found in up to 50% of patients within the middle and the superior frontal gyrus. Ojemann analysed his stimulation results in a more systematic way than his predecessors; he documented not only the areas where 'positive' stimulations were found but also the 'negative' sites. That allowed him to draw a probabilistic map that indicated the chance to find a language site in any given region of the left fronto-temporal cortex in his population of patients. Much to his surprise, there was no single area where language functions were present in every individual patient. The area where language disturbances were most frequently found was the inferior frontal gyrus immediately in front of the motor cortex. Still, even in this classic Broca's area, language areas were absent in one of five patients [75]. In the posterior part of the superior and middle temporal gyrus, the chance of finding language disturbances with electrical

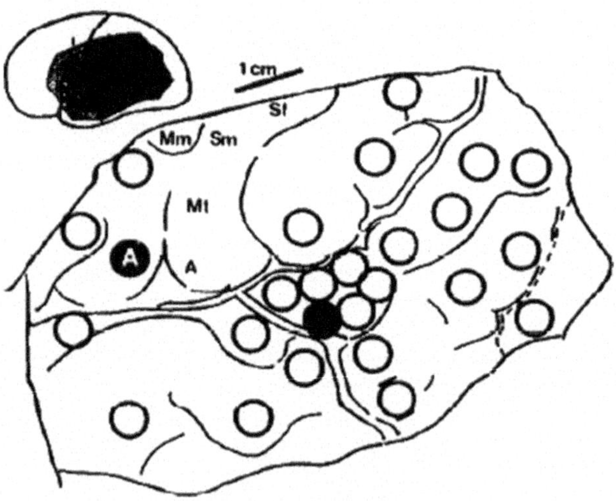

Fig. 6.17 Location of two sites at which electrocortical stimulation mapping disturbed a naming task in the left hemisphere of a 36-year-old patient. *Circles*, sites of bipolar stimulation; *open circles*, no errors; *solid circles*, sites of repeated naming errors; *label A*, predominantly speech arrest. Letters outside of the circles indicate evoked motor (*Mm*, *Mt*, *A*) and sensory (*Sm*, *Sf*) responses identifying Rolandic cortex. The shaded area in the inset indicates the location of the intraoperative cortical exposure. Repeated naming errors were found at one location in the superior temporal gyrus; note the lack of errors at the immediately surrounding sites. Note also that the inferior frontal gyrus has not been fully exposed (only its most posterior part) (Figure taken from Ojemann, 1991 [75])

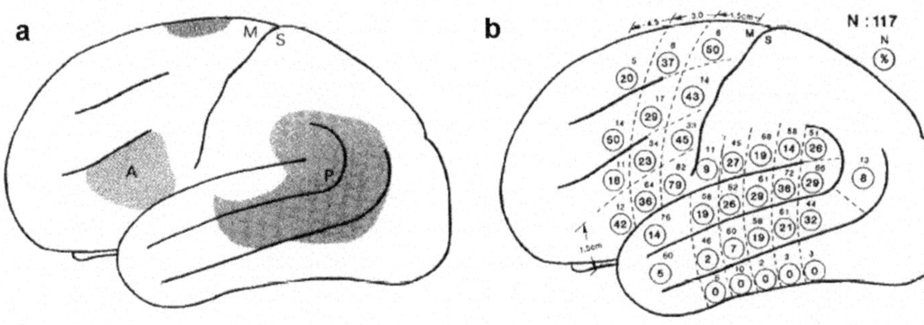

Fig. 6.18 Composite results from intraoperative electrocortical stimulation mapping. (*Left*) Location of essential cortical language areas as typically depicted in traditional textbook models; image taken from the book of Penfield and Roberts's *Speech and Brain Mechanisms* (1959) [36]. *A*, frontal (Broca's) language area; *P*, posterior (Wernicke's) language area; *M* and *S*, motor and sensory cortex. Note that results look entirely different when individual results are examined, compared to language localization in an individual subject (Fig. 6.17) and to the probability map from Ojemann that indicates large interindividual variability (figure on the *right*). (*Right*) Variability in localization of sites essential for naming, based on electrical stimulation mapping in the left, language-dominant hemisphere in 117 patients. Patients were operated on because of medically intractable seizures and tested during surgery with an object naming task. Individual maps, such as Fig. 6.17, were aligned with reference to Rolandic cortex and the end of the Sylvian fissure. The cortex was then divided into zones represented by intersecting solid and broken lines. The upper number in each zone indicates the number of subjects in whom a site was tested in that zone; the lower circled number indicates the percentage of those subjects in whom naming errors were evoked. M and S indicate motor and sensory cortex, respectively. Although it looks as if Broca's area (the zone where in 79% of patients' language functions are found) is located on the precentral gyrus, Ojemann clarifies in his papers that this zone represents the posterior part of the inferior frontal gyrus (Figure and text (modified) taken from Ojemann (1991) [75])

stimulation was even smaller; there was no region where language sites were found in more than one-third of patients.

A probabilistic map is a better way to deal with interindividual variability than simply adding up all the patients' results in a single drawing. It weights each individual's functional contribution to a certain brain area. What is important to realize is that the observed variability is the result of several different contributing factors, some of which are unknown or difficult to control for. One major factor is the nature of the patient's brain disease. Is it a long-standing benign lesion or a rapidly growing malignant tumour? Is there epilepsy within a seemingly normal anatomical brain or are the seizures related to a congenital abnormality (i.e. a dysplasia)? And in the case of a tumour, what is its location, size, growth pattern and so forth [76]? Slowly growing brain lesions, for instance, have been associated with compensatory mechanisms to relocate functions to ipsilateral or even contralateral brain areas [77]. Other factors that may contribute to language variability are age and premorbid cognitive abilities of the patients. A general finding is that younger patients or patients with greater 'intellectual enrichment' are more tolerant of the effects of pathology [78, 79]. They are thought to be more 'cerebrally efficient' and to have more 'cognitive reserves', although a good explanation for these phenomena is lacking [78]. Another source of variability lies in the technical and methodological shortcomings of electrocortical stimulation, which—despite its status as gold standard technique—suffers from false-negative and false-positive findings. A final important factor is the significant anatomical differences between subjects, as already discussed in Chap. 5. Ojemann captured each individual's cortical anatomy with a drawing (see Fig. 6.17). This variability was, however, not really accounted for in the overall results. The anatomical sketches from each individual patient were projected upon a standard schematic brain, with the Rolandic and Sylvian fissures as orientation points. Once aligned, gyri were arbitrarily subdivided into 'zones', and for every zone the percentage was shown whereby patients had naming errors during stimulation. Anatomical variations between patients are inevitably lost with such an 'average' brain representation. In Ojemann's words: 'The functional variability we have noted may also be a consequence of considerable variability in the detailed anatomy of cortex' [80].

It remains unclear why in 25 of 117 patients no language areas were found in Broca's area, but both anatomical and functional variability are likely important reasons [81]. The pars opercularis, for example, can be enfolded and not visible at the cortical surface, making it inaccessible to the effects of electrocortical stimulation [82]. More recent studies argue that neuroplasticity can also account for these results and have demonstrated that cortical mapping can be 'negative' in patients with a lesion in the classic Broca's area. Subsequent resection of the lesion can be achieved without any lasting language deficit (see for a patient case Fig. 4.9) [83].

Like Penfield, Ojemann had good postoperative neurological outcome after surgical procedures near or within classical language areas. He warned his fellow neurosurgeons not to consider solely the anatomical location of a lesion in presurgical decision making, especially when the lesion was present since early life, but advised them instead to use stimulation mapping to assess the risk of aphasia from any proposed resection.

Ojemann confirmed Penfield's earlier findings of substantial differences in the patients' functional anatomy. Another important and remarkable finding was that the language areas themselves were usually very small (1–2 cm^2) and sharply demarcated from 'non-essential' language cortex (something that was also noted by Penfield). Ojemann himself spoke of 'individual mosaics' [84]. In the meantime, these observations have been reproduced by many others [85]. Ojemann criticized figures and models in textbooks that showed language areas that were significantly larger than he had found in his series. He argued that any such representation (e.g. Fig. 6.18) 'is likely to be based on an artifact of pooling data from subjects whose essential areas are in different locations, and thus an overestimate of the extent of essential modules for one language function in an individual brain' [75].

Ojemann, and others, also started to use different language tasks during surgery and quickly found out that this led to new 'eloquent' areas that could not always be detected with the traditional object naming task. When, for instance, both a reading task and an object naming task were used, a large proportion of stimulated frontotemporal sites only showed a positive response during one of the two tasks [86, 87]. Several studies confirmed a similar dissociation with other language tasks (e.g. verb generation or writing [88, 89]) or when language was tested in different modalities (e.g. auditory versus visual naming [90]). Functional specialization seems to be the rule rather than the exception when functions are tested with electrocortical stimulation [75, 91, 92]. Both distinct and shared cortical sites have also been consistently found when different languages were tested in patients who were proficient in more than one language, i.e. in bilingual or multilingual patients [93–95]. Thus, at least from the perspective of electrocortical mapping, language functions seem to be organized in parallel operating subsystems, something that Ojemann named 'compartmentalized.' He found this theory was supported by findings in stroke and other lesion studies, where patients frequently show only a selective loss of language functions. In general, however, stroke patients are not very well suited to research on the functional role of smaller areas, as the ischemic area is usually much larger than those that are found with electrocortical stimulation mapping. It likely includes several 'positive' and 'negative' stimulation sites as well as subcortical structures. One of the advantages of electrocortical mapping is that its 'virtual lesions' are temporary and focal and that multiple and different cortical areas can be studied within the same patient.

Some electrocortical mapping procedures offer the possibility to obtain knowledge about more fundamental aspects of language processing (i.e. knowledge beyond what is considered necessary for clinical decision making). This is particularly the case in patients who have grid electrodes implanted over their cortex in preparation for a surgical procedure for epilepsy. Placement of such a grid is done for diagnostic purpose and aims at localization of the epileptogenic region and/or identification of critical functional cortex. The grid typically stays in for a number of days to optimize the chance of finding the epileptogenic region. The patients are fully awake and are often willing to participate in scientific experiments (they are hooked up to neurophysiological recording machines and are literally tied to their beds). Boatman and colleagues performed several electrocortical stimulation experiments in grid patients and specifically studied the cortical organization of speech perception. They used tasks that ranged from low-level auditory discrimination of tones or syllables to rhyming, picture–word matching or the token test. Their experimental results gave them a framework 'for modeling the organization of functional circuits in the left hemisphere that are critical for speech perception' [96]. Note that grid recording can only be made from regions that are located on the lateral surface of the temporal lobe. These are the so-called belt and parabelt regions that represent secondary and associative auditory areas; primary (core) auditory areas lie buried within Heschl's gyrus and cannot be measured with cortical electrodes. Still, the authors identified several different functional regions within the lateral parts of the temporal lobe. Both the anterior and the posterior part of the left superior temporal gyrus (STG) housed different speech perceptual functions. These regions seemed hierarchically organized (i.e. from lower level to higher level functional processing). Within the posterior part of the STG, 'electrocortical mapping studies identified an anterior region associated with acoustic-phonetic processing, a more posterior region associated with phonological processing, and the temporoparietal junction associated with access to lexical-semantic information for comprehension'

[96]. The anterior part of the STG could be subdivided further into two regions: 'a posterior region associated with phonological processing and an anterior region associated with sentence processing' [96].

Cognitive functions other than language have also been studied with electrocortical mapping. Again, Ojemann and his group were at the forefront of these investigations:

> Ojemann and his colleagues have studied short-term memory (STM) in the neocortex during awake craniotomies (Ojemann 1978 [97], Ojemann 1982 [98]). Their paradigm consisted of stimulus display (photographs of common objects, nonverbal stimuli such as faces or line orientations) via slide presentation, a distractor phase (reading, phoneme identification, orofacial movements, mental arithmetic), and recall of the original stimulus. Ojemann equated stimulus display with acquisition, distraction with consolidation, and recall with retrieval in relating the paradigm to models of STM processing. Stimulation was applied during one of these three phases for three or more trials at multiple cortical sites. (…) For dominant hemisphere studies, the initial stimulus display for acquisition also required naming, so that the linguistic functions of naming (and, in many studies, reading) were tested in conjunction with the acquisition, storage, and retrieval process of memory.[c]

Stimulation studies thus argued that language and verbal memory functions have predominantly distinct representations. Memory sites were more widespread and usually surrounded the language sites. Distinct sites for visuospatial perception and working memory have also been detected [21, 99, 100]. Some authors have advocated that cortical mapping of these higher cognitive functions should have clinical implications (i.e. to tailor the resection), but there remains a lot of discussion whether or not these sites are really 'critical' for normal function [101, 102].

Some authors also investigated whether neurological syndromes, characterized by a combination of functional disturbances, could be better understood with cortical mapping findings. An example is the Gerstmann syndrome, whereby patients typically have problems with calculation, finger recognition, writing and left–right orientation. Roux and colleagues studied the angular gyrus and found specific sites for several of these functions, including also naming and reading. They concluded that symptoms of the Gerstmann syndrome could be found with electrocortical mapping [103]. In another elegant electrocortical mapping study, Boatmann and colleagues found evidence for a cortical substrate of transcortical sensory aphasia. This syndrome is characterized by impaired auditory comprehension, but intact repetition and fluent speech. In all six of the studied patients, multiple sites in the posterior superior and middle temporal gyrus were found where auditory comprehension was impaired, but the other two functions remained intact during stimulation [104]. At a subset of these sites, naming and word reading also remained intact.

Although these studies yield fascinating scientific information, they pose a problem to clinicians: how many language functions or tasks should be tested during surgery to assure that language deficits are prevented after surgery? And, even more challenging, what about all other cognitive functions? There are not yet definite answers to these questions. Mapping for cognition other than language is currently

[c]Quotation taken from Devinsky (1993) [21]

not routinely performed during neurosurgical procedures. For language, some authors advocate using both a naming and a reading task in all patients [87]. Others use only one task (naming) and refer to other language tasks on indication. Then there are neurosurgeons who advocate testing during surgery *all* languages in which the patient is proficient. However, such advice is certainly biased, and there is no strong evidence for it [94]. Penfield, who worked for 30 years in a bilingual city where most people spoke both English and French, said he never had seen a patient in whom one of the two languages was selectively lost; neuropsychological investigation almost always showed an impairment in the other, seemingly non-affected language. To my knowledge he never tested his patients in more than one language during surgery.

6.8 Duffau: Subcortical Pathways and Hodology

Wernicke noted that damage to subcortical structures could cause language dysfunction, and he proposed that a clinical syndrome could be the result of a subcortical lesion alone. But Wernicke's theories went much further, and he attributed a critical role to white matter pathways in general cognitive functioning (see Chap. 2). Both Meynert and Wernicke postulated that connections between different brain regions were part of the distributed system that represented conceptual knowledge. Modality-specific information was thereby stored in specialized brain regions, but the 'bigger picture' only emerged when all this information was somehow connected at the right moment. This view still largely reflects our modern theories on semantic memory [105, 106]. Despite Meynert and Wernicke's foresight, the role of the subcortex in speech and language has always been downplayed. This attitude had already biased Broca's conclusions; remember his patient Leborgne, whose extensive subcortical damage was a priori not taken into account when he discussed the cause of his patient's dysphasia. In the century that followed, the subcortex was simply not considered important for cognitive functioning and remained a sort of 'terra incognita.' The basal ganglia, thalamus and cerebellum were regarded primarily as motor structures. Only late in the twentieth century were these views challenged and shifted towards a more prominent role in cognition [107].

With the advent of CT in the late 1970s, the number of reports that described subcortical vascular lesions in patients with aphasia increased. Gradually, the concept of subcortical aphasia emerged, although its aetiology, and even existence, remains controversial until today. Subsequent new language models arose from neurological–radiological correlation studies (see, for an example of such a model, Fig. 6.19). Advanced imaging techniques (in particular MRI-based fibre tractography) further contributed to a renewed acknowledgement of fibre pathways. Interestingly, electrical stimulation of white matter and subcortical structures gained clinical interest rather late in the twentieth century. Language mapping techniques were historically developed in the context of epilepsy surgery and focused on the identification and removal of epileptogenic cortical

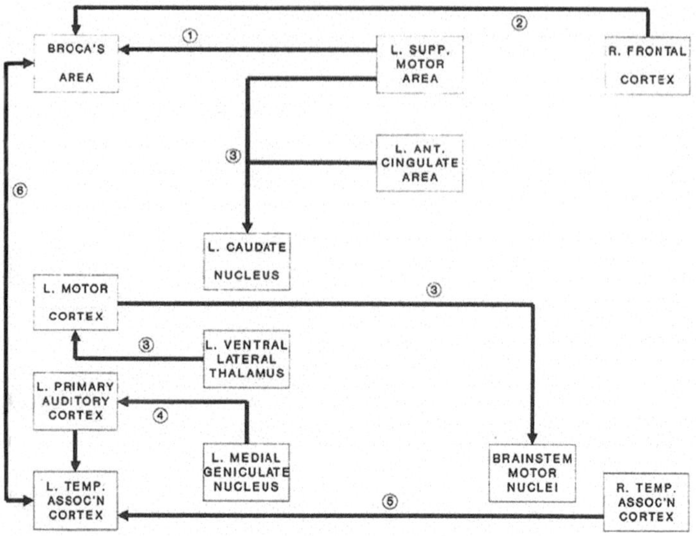

Fig. 6.19 Example of a model for subcortical language functions and aphasia, in this case from Alexander (1987) [171]. These models largely evolved from lesion studies and theories of motor control. They consist of complex circuitry that includes basal ganglia, thalamus and cortex [172]. Key features of these models are feedback loops that connect various regions that are themselves poorly defined in anatomical terms. Note transcallosal connections from right frontal and temporal areas (Figure taken from Crosson, 1992 [72])

areas [108]. Penfield had no real clinical interest in subcortical structures, although he was aware of their critical role and tried to avoid going beyond the depth of a sulcus whenever possible (see Fig. 6.15). When neurosurgeons started to operate more frequently on brain tumours because of the improved abilities for detection and localization of these tumours with CT scanning, it was found (and described in 1996 by the group of Ojemann and Berger) with electrical stimulation that:

> functional cortex and subcortical white matter may be located within the tumor or the adjacent infiltrated brain. Therefore, to safely maximize glioma resection in these functional areas, intraoperative stimulation mapping may be used to identify functional cortical or subcortical tissue within, as well as adjacent to, the tumor, thus avoiding permanent injury. [109]

Many brain tumours infiltrate the white matter and a safe subcortical delineation is thus important, especially because subcortical lesions generally have a higher chance of inducing permanent deficits than cortical lesions [110]. But much is still unknown: sometimes a small subcortical lesion causes severe and lasting functional deficit, whereas in other cases the white matter is very ‘forgiving’ and functions will recover after surgery. The work of Hugues Duffau (1966), a French neurosurgeon, has particularly contributed to the revival of the use of subcortical stimulation mapping and to the development of scientific theories regarding the functional role of white matter tracts. Duffau has systematically studied the function of the cortex and subcortex during surgery of low-grade gliomas. These slowly growing brain tumours (they grow on average a few millimetres per year) invade and gradually destroy normal brain tissue. Remarkably, they can reach a size of several centimetres before becoming symptomatic [111]. Even then, neurological or cognitive deficits are usually not very prominent. The majority of patients with low-grade glioma present with an epileptic seizure that results in detection of the tumour. Some of these patients, when asked in retrospect, will tell of rather 'vague' complaints they experienced, which they themselves attributed to stressful periods in their job or social life (but are presumably related to the presence of the tumour). Still, the fact that large tumours can be virtually asymptomatic for a long time remains a puzzling phenomenon.

Treatment of low-grade gliomas is difficult. To date, there is no cure and median survival roughly lies between 6 and 13 years [112, 113]. Although still a matter of debate, the role of surgery is becoming increasingly important as a first line of treatment. There is strong empirical evidence that the extent of resection increases survival [114, 115]. The surgeon is therefore given the challenging task of maximizing tumour removal while preserving brain functions. The main difficulty with these tumours is that they have no strict anatomical borders, and total resection is thus virtually impossible. MRI scans may give the impression that a low-grade tumour is sharply demarcated, but on a microscopic scale, tumour cells have already crossed the radiological border by several millimetres, or perhaps centimetres, and invaded normal functional brain tissue (see for examples Figs. 6.20 and 6.23) [116, 117]. Duffau is a strong proponent of a 'maximal' surgical resection. In achieving this goal, he does not stop at macroscopic tumour borders, but removes tissue until he reaches functional boundaries as defined with electrical stimulation. The picture that emerges from these surgical procedures is foremost one of an enormous cortical functional flexibility. In many cases, tumours can be resected from presumed eloquent cortical areas with only transient neurological deficits, including the classically defined areas of Broca and Wernicke (Figs. 4.9 and 6.20) [83, 118, 119]. What most frequently limits the extent of resection, though, are not cortical areas but subcortical pathways. Duffau consequently detected and described several locoregional and long-distance fibre pathways that appear to play a critical role in normal speech and language functions (Fig. 6.21). His observations support the existence of not one but two main fibre pathways between frontal and temporoparietal language areas. One is the arcuate fasciculus, the known classic connection between Broca's and Wernicke's areas. Subcortical stimulation of this pathway predominantly induces phonemic paraphasias and repetition errors, suggesting a role for the arcuate fasciculus in phonological processing and articulation. In addition, there is a second pathway that runs within the temporal lobe. Stimulation of this inferior fronto-occipital fasciculus (IFOF) elicits semantic disorders (see Fig. 6.22). This is a significant modification of the clinical language model (see also Chap. 7). The

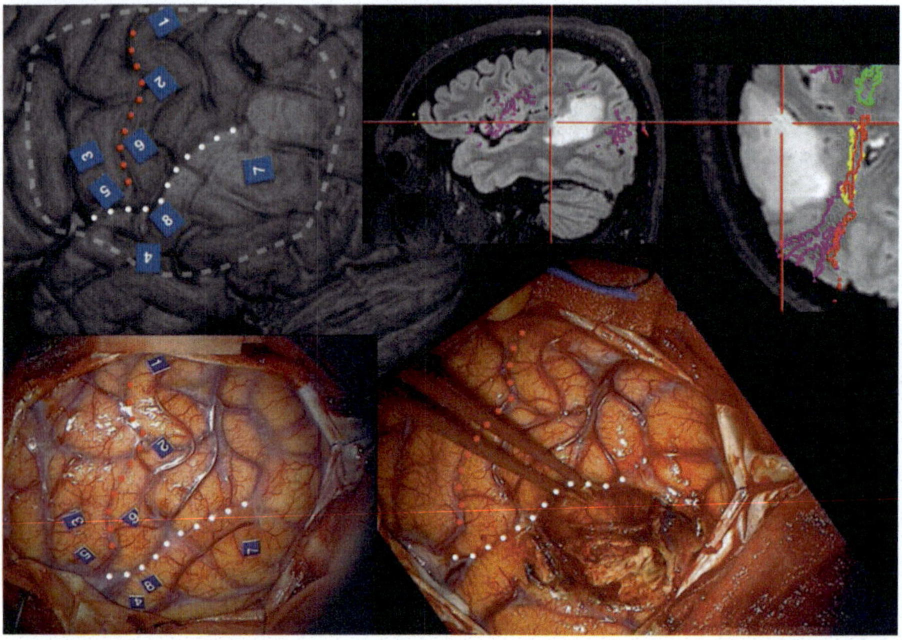

Fig. 6.20 A 30-year-old patient with sudden onset of language disturbances that (completely) resolved after several hours. At time of admission to the hospital, there were no language or other impairments upon neurological examination. In retrospect, the patient reported that he experienced predominantly difficulties in word finding. In later days, the patient experienced auditory hallucinations. These temporary functional disturbances were considered epileptic phenomena. MRI (*top images*) revealed a left temporo-parietal brain tumour (low-grade glioma). MRI-based tractographic information is superimposed on anatomical images (*purple*, arcuate fasciculus; *yellow*, inferior fronto-occipital fasciculus; *red*, optical tract; *green*, corticospinal tract). The deepest part of the tumour is closely related to language and optical tracts. *Dotted lines* show the Sylvian fissure (*white*) and central sulcus (*red*). Photographs during surgery (*bottom images*) show the results of cortical stimulation and the extent of resection. Several tasks were used during the mapping procedure: counting, picture naming and auditory comprehension. (*1*) Sensation in the right hand; (*2*) sensation within the mouth; (*3*) number skipped during counting (31 … 50); (*4*) hesitation and speech arrest; (*5*) speech arrest, "strange feeling" in the mouth; (*6*) sensation in the tongue; (*7*) hesitation. During five subsequent stimulations, no language impairments were found, and the area was considered noncritical for language; (*8*) hesitation, paraphasia and anomia. Subcortical stimulation in the resection cavity yielded visual sensations (in both eyes) and language disturbances (speech arrest, paraphasias). During tumour resection, there were mild language impairments that worsened in the days after surgery. Three months after surgery, at follow-up visit, these had been resolved

ventral and dorsal pathways have obvious parallels to the 'what' and 'where' pathways that have been described for the visual system (and run into the temporal and parietal lobes, respectively) [120]. The results of intraoperative functional mapping also nicely fit with more recently described dual-stream models of auditory language processing (see also Chap. 7).

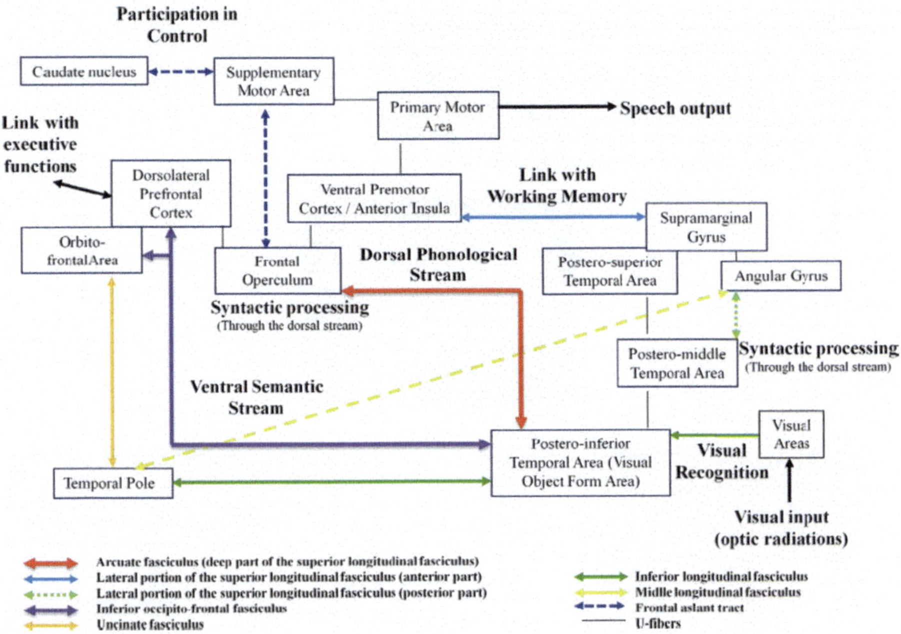

Fig. 6.21 Duffau's 'hodotopical' model of language that is based on structural–functional correlations provided by brain stimulation mapping during a visual picture-naming task (2013) [161]. The model is strongly rooted in anatomy, with a core that consists of two parallel processing pathways: a ventral semantic stream (via the inferior fronto-occipital fasciculus, *purple*) and a dorsal phonological stream (via the arcuate fasciculus, *red*). In addition to the direct ventral route, an indirect pathway is hypothesized with a relay at the level of the temporal pole (i.e. via inferior longitudinal fasciculus and uncinate fasciculus). Such an additional pathway would allow for functional compensation and subcortical plasticity [121]. There is also an indirect dorsal route, via the angular gyrus. These pathways are subserved among others by lateral parts of the superior longitudinal fascicles

Duffau systematically studied various white matter pathways and combined the results of stimulation with those of non-invasive MRI-based tractography [121, 122] and modern post-mortem fibre dissections [123, 124]. His conjoint research with different researchers and institutions has resulted in many papers and a recent book *Brain Mapping: From Neural Basis of Cognition to Surgical Applications* [125]. His work has led him to propose an alternative view on functional language organization, namely, a 'hodotopical (interconnected) and plastic (dynamic) view of brain organization' [110]. The model is strongly rooted in (sub)cortical functional anatomy and describes interactions with other networks for working memory and executive control. It better accounts for functional reshaping before or after a brain lesion or operation than previous models and helps to explain why surgical resections can be performed in presumed eloquent brain areas without neurological deficits. One of

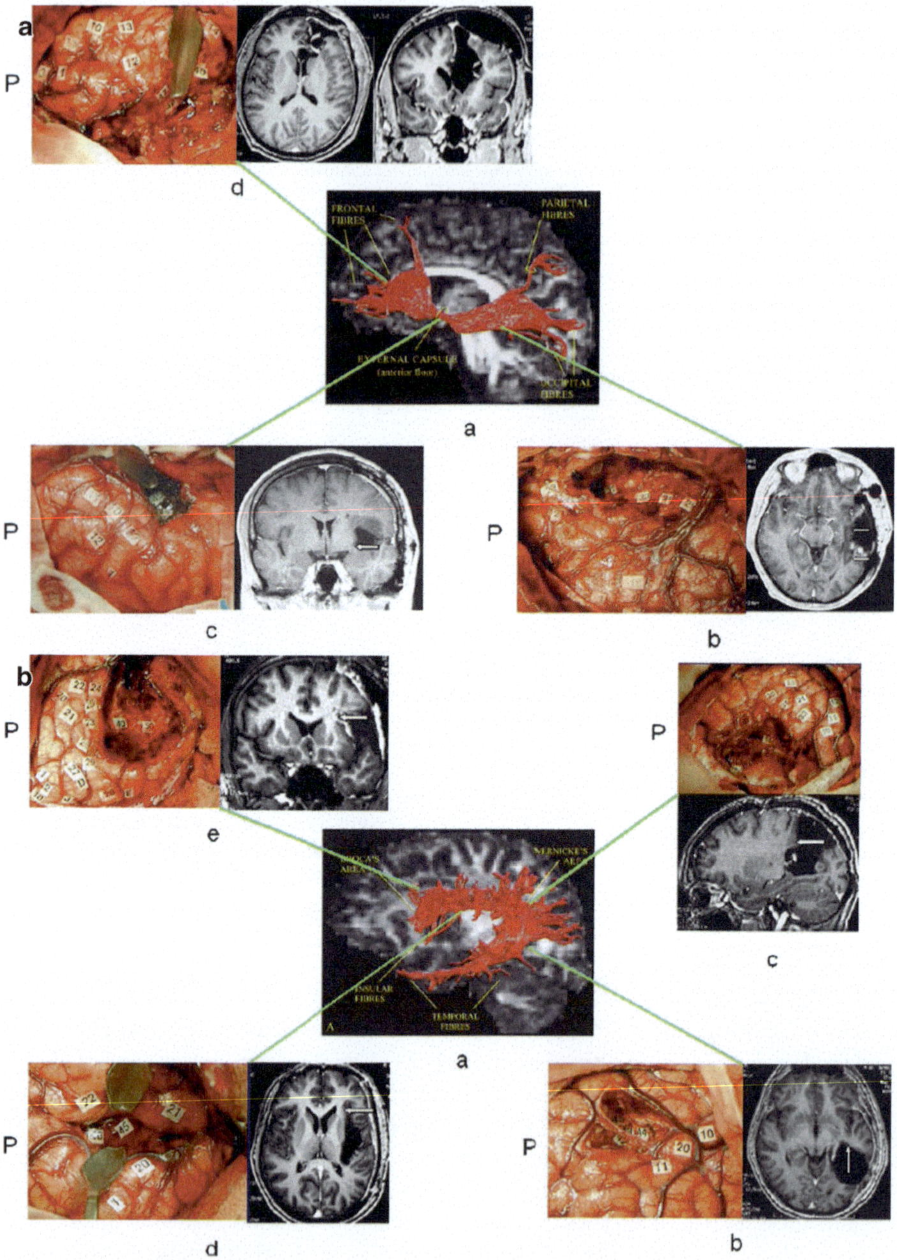
FRONTAL FIBRES
PARIETAL FIBRES
(anterior floor)
INSULAR FIBRES
TEMPORAL FIBRES

its important properties is that functions are *delocalized* and that lesions at various different locations can lead to similar symptomatology. This holds not only for language but also for cognitive functions in general. For instance, when discussing the origin of frontal symptoms, Duffau (2012) states that, although:

> it is still possible to link 'frontal symptoms' to a frontal lesion, it is also possible to link similar "frontal symptoms" to a non-frontal lesion. In addition, extensive frontal lesion (even a total left dominant lobectomy) can occur with neither 'frontal syndrome' nor consequences on the daily familial, social and professional life. Such data show that an improvement of the knowledge of dynamic anatomofunctional connectivity is crucial for both neuroscientists and neurologists/neurosurgeons, based on a more distributed view, preventing to rigidly relate a syndrome to the injury of a specific cerebral area—but rather to analyse each symptom without a priori concerning the location of the damage. [110]

As was mentioned before, the areas that are found with direct electrical stimulation (DES) are usually smaller than the language territories that are claimed from classic neurological studies. In other words, DES frequently yields negative stimulation results in inferior frontal and posterior temporal language regions. In many cases, these areas are anatomically or functionally abnormal due to the presence of a tumour or epileptogenic tissue. The absence of language areas is then hypothesized to be caused by plasticity, which is the obvious explanation given the circumstances [126]. However, neurosurgeons occasionally resect normal brain tissue to

Fig. 6.22 (*Top box*) 'Inferior fronto-occipital fasciculus: ventral semantic stream. (**a**) The anatomical trajectory of the white matter bundle studied by DTI. (**b–d**) The surgical field and post-surgical MRI of different patients operated on for a tumor within various brain locations; (**b**) temporal; (**c**) insular; (**d**) frontal. In all cases, the deep functional boundary of the resection was given by a part of the inferior fronto-occipital fasciculus, identified by subcortical mapping. Electrocortical stimulation of this tract systematically induced semantic paraphasias. The precise location where these language disorders have been induced were marked intraoperatively by number tags in the depth of the cavity. These sites are shown by the arrow on the postoperative anatomical imaging. P: the posterior part of the brain.' (*Bottom box*) 'Arcuate fasciculus: dorsal phonological stream. (**a**) The anatomical trajectory of the white matter bundle studied by DTI. (**b–e**) The surgical field and post-surgical MRI of different patients operated on for a tumor within various brain locations; (**b**) temporal; (**c**) parietal; (**d**) insular; frontal (**e**). In all cases, the deep functional boundary of the resection was given by a part of the arcuate fasciculus, identified by subcortical mapping. Electrocortical stimulation of this tract systematically induced phonological paraphasia. The precise location where these language disorders have been induced, were marked intraoperatively by number tags in the depth of the cavity. These sites are shown by the *arrow* on the postoperative anatomical imaging. P: the posterior part of the brain' (Text and figures taken from Duffau (2008) [122])

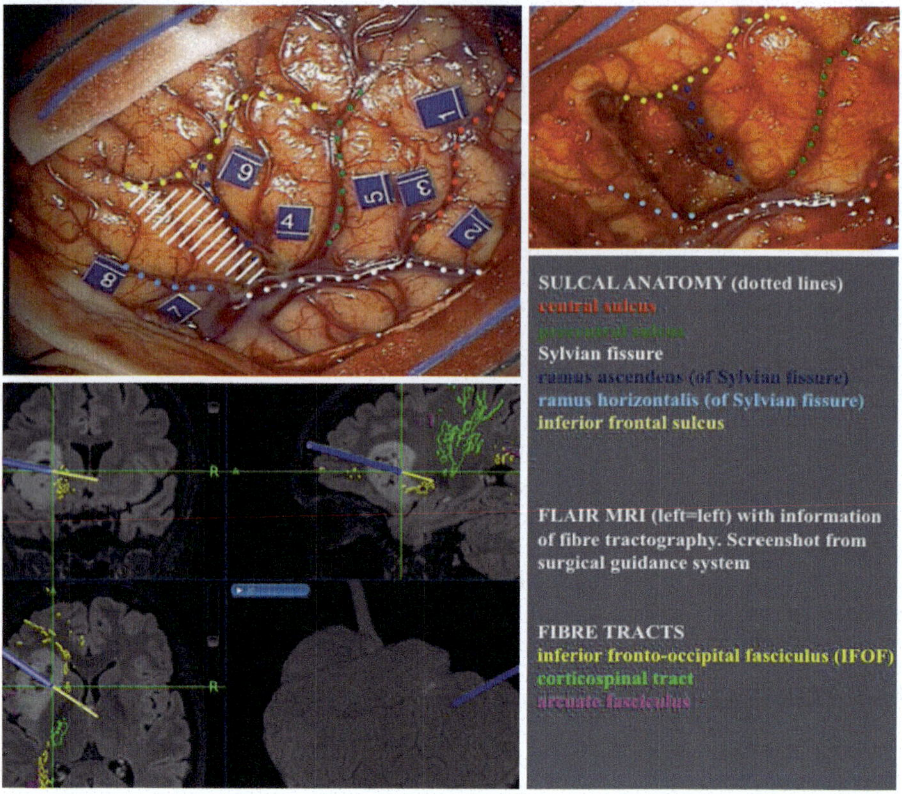

Fig. 6.23 A 25-year-old patient with a clinical debut of focal epileptic seizures (speech disturbances and motor impairment of the right arm) and a left fronto-insular glial tumour. There were no language or other cognitive impairments. An approach through the frontal operculum was planned as the preferred surgical route (we considered an alternative route via splitting of the Sylvian fissure to be less optimal in this case). Awake surgery was performed to map functional areas in the left fronto-central region. Primary sensorimotor cortex and ventral premotor cortex were identified with direct electrical stimulation and low stimulation settings (*1*) extension of fingers, (*2*) "strange feeling" in the right jaw, (*3*) dysarthria, (*5*) speech arrest). Responses over the pars opercularis (*4*, *6*) and pars orbitalis (*7*, *8*) suggested language involvement, but were less convincing ((*4*) 'opening of the hand' but also three times negative stimulation result, (*6*) slight hesitation, (*7*) anomia and four times negative stimulation result, (*8*) hesitation. The pars triangularis was repeatedly stimulated, but no responses were obtained with relative high stimulation parameters (60 Hz, bipolar stimulation, 4 mA, 1 ms pulse width). Consequently, the corticotomy was performed in this part of the inferior frontal gyrus (hatched area). This did not result in any speech or language impairments. Speech remained fluent and formal language tests (e.g. picture naming) remained normal. The depth of the resection border was determined by positive responses obtained from subcortical stimulation of the IFOF (semantic paraphasias, indicated in the lower image by the *blue–yellow line* that represents the pointing device of the surgical guidance system; the *yellow line* represents its virtual extension). When the resection encroached on this functional border, the patient had subtle semantic problems. After surgery there was a mild dysphasia (word-finding difficulties, frequent hesitations and pauses) and patient complained of memory loss. Six weeks after surgery nearly all of these problems were resolved, and after 3 months they had disappeared. Postoperative MRI showed a small remnant in the insula

expose a deep-lying tumour. These areas are used as a corridor to reach the tumour [127]. For insular tumours, these transcortical windows are located in the perisylvian gyri: the inferior frontal, lower central, supramarginal or superior temporal gyri [127, 128]. It is remarkable that this kind of surgery in the left hemisphere hardly ever results in permanent language deficits [128]. Our surgical team had a number of cases where the pars triangularis was used as an entry to the anterior part of the insula. There were no speech or language disturbances during the corticotomy or removal of this part of the language-dominant inferior frontal gyrus; see Fig. 6.23 for an example. Why is this possible? Is this area truly non-functional, or are we not picking up the more subtle results of our iatrogenic damage? In our cases, the pars triangularis appeared normal on MRI scans, as well as on macroscopic inspection during surgery (white and grey matter were distinguishable through the surgical microscope and had normal tissue consistency). Possibly, the area was already disconnected from the normal language network due to infiltrative growth of the underlying insular tumour. Then again, DES-positive sites were found in adjacent gyri that potentially suffered from a similar problem. Perhaps future studies can shed light on this problem, by determining the connectivity of these seemingly non-functional areas with more recent MRI techniques (fibre tractography and functional MRI). Alternatively, DES may have yielded false-negative information. Although it is the gold standard, it has its methodological and practical drawbacks (see next paragraph). In any case, I find it astonishing that a seemingly normal part of Broca's area can be resected without noticeable behavioural deficits. This again demonstrates that the individual functional anatomy can be significantly different than what is normally assumed in the medical and neuroscientific literature, and adds to the discussion of the functionality and indispensability of the left inferior gyrus.

6.9 The Wada Test and Electrical Stimulation Mapping: Gold Standards by Default

6.9.1 Language Dominance

Ever since the pioneering observations of Dax and Broca, discussion has continued about the contribution of each hemisphere to essential language functions. Is the left hemisphere indeed 'dominant' in most subjects and the right hemisphere the 'minor' or 'subordinate' hemisphere? Or are the various language functions to a certain extent represented in both hemispheres? Clinical observations have played a major role in this discussion and led to the dichotomized approach that is still commonly practised by clinicians. In this view, a hemisphere is either necessary or not necessary for performance of language functions. Most of the evidence for this view stems from stroke studies where language deficits are predominantly found after lesions of the left hemisphere.

Most clinicians, and probably also many scientists, will argue that our left hemisphere is biased to harbour essential language functions in the absence of early brain damage. Still, even in healthy subjects, the predisposition of only one dominant (left) hemisphere for language seems unjustified, as some patients become aphasic after a right hemispheric stroke (Fig. 6.24) [129, 130]. Lesion-deficit studies strongly suggest that a small percentage of the normal adult population has atypical language representation, whereby language is represented in the right hemisphere (estimates range from 0.4 to 4%) [131–134]. The incidence is somewhat higher in left-handed patients and in those with early-life brain damage. As these patients generally make a better recovery than aphasic patients with left hemisphere

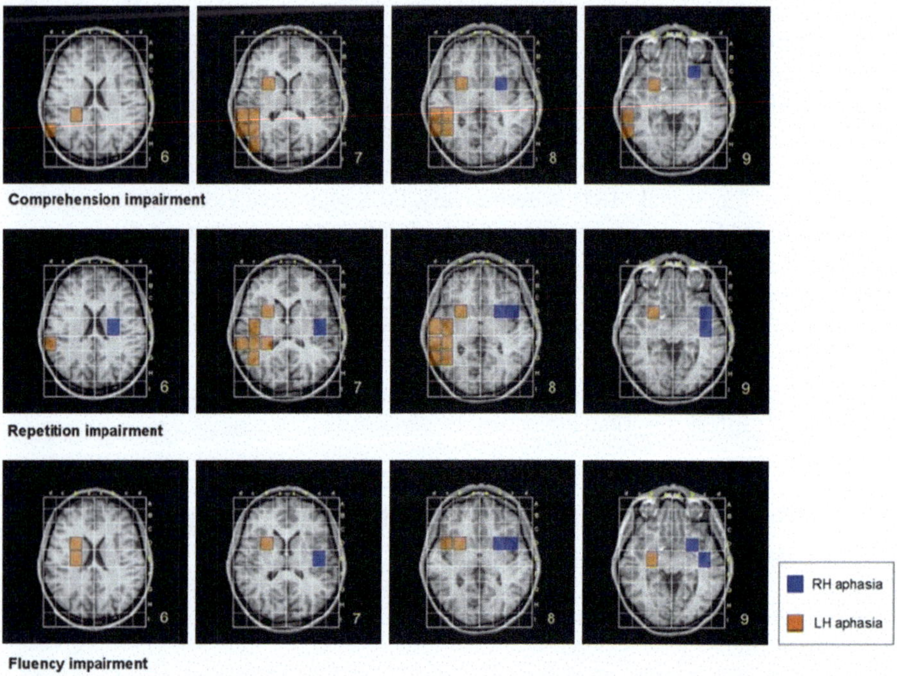

Fig. 6.24 Of all patients that suffer from a stroke, there is a small number with right hemisphere damage who fails on clinical tests for aphasia. This is generally seen as strong evidence that these patients have a language-dominant right hemisphere. This study of Dewarrat (2009) compared lesion sites in patients with aphasia after a stroke in the right hemisphere (RH) or left hemisphere (LH). Results are shown for three types of language disturbances in 16 patients with RH damage and a control group of 25 patients with a stroke in the LH. Lesions that in more than 50% of patients resulted in deficits are indicated with *orange* or *blue cubes*. The results suggest that there may be differences regarding the severity of language impairments (less severe in the right hemisphere group) and precise location of the lesion between both groups. Repetition and comprehension impairments in patients with RH stroke are more often associated with an anterior lesion (Figure taken from Dewarrat, 2009 [129])

damage, it has been suggested that many of them may in fact have bilateral language representation.

It should be noted that there are obvious confounders that influence the discussion on hemispheric language dominance. An important confounder is the operational definition of language, or better of *language deficits*, that any investigator uses, as it determines the choice of language tests. In particular in the clinical situation, there is a real danger that the investigation becomes a sort of self-fulfilling prophecy, as clinical tests are rather simple and biased towards detection of typical left hemisphere functions (e.g. confrontational naming, repetition, the occurrence of paraphasias in spontaneous speech, etc.). If we broaden our view, and step away from this traditional language view, there is convincing evidence that the right hemisphere is not the non-language hemisphere, but that it plays an important role in normal communication [135]. Studies of so-called split-brain patients (where the corpus callosum has been surgically sectioned) indicate that the right hemisphere is capable of elementary linguistic functions, in particular related to the meaning of words. Experiments with these patients have shown that the right hemisphere can process verbs or nouns. As this hemisphere seems to lack speech and syntax, it communicates in a nonverbal manner [136]. There is also a fair amount of evidence from stroke studies that some aspects of normal language function are mediated more by the right than the left hemisphere. Many of these deficits will normally not be noticed in regular conversation or standard neurological examination, which is why they are typically underreported in the clinical literature. But language is much more than just speech and a vocabulary; it is also importantly based on prosody and kinetics:

> Prosody and kinetics constitute the paralinguistic elements of language and play an equally prominent role in the organization of human communication and discourse. The right hemisphere appears to exert a major influence on the organization of the paralinguistic features. Prosody is the suprasegmental feature of language that conveys information beyond that transmitted by word choice and word order alone. The acoustic features associated with prosody include pitch, intonation, melody, cadence, loudness, timbre, tempo, stress, accent, and timing of pauses. (…) Kinetics refers to limb, body, and facial movements that normally accompany discourse and serve to modulate the verbal message being communicated. [135]

Patients with damage to the right hemisphere often have difficulty with integration of information or cannot adequately use contextual information. Although they can interpret the literal meaning of language, they fail to get jokes or metaphors or the irony, that is, the actual message of a sentence. But even on typical 'left hemisphere' comprehension tasks, patients who suffer a lesion in the right hemisphere do not necessarily do better than those with left hemisphere damage when they are carefully examined, as is shown in Fig. 6.25.

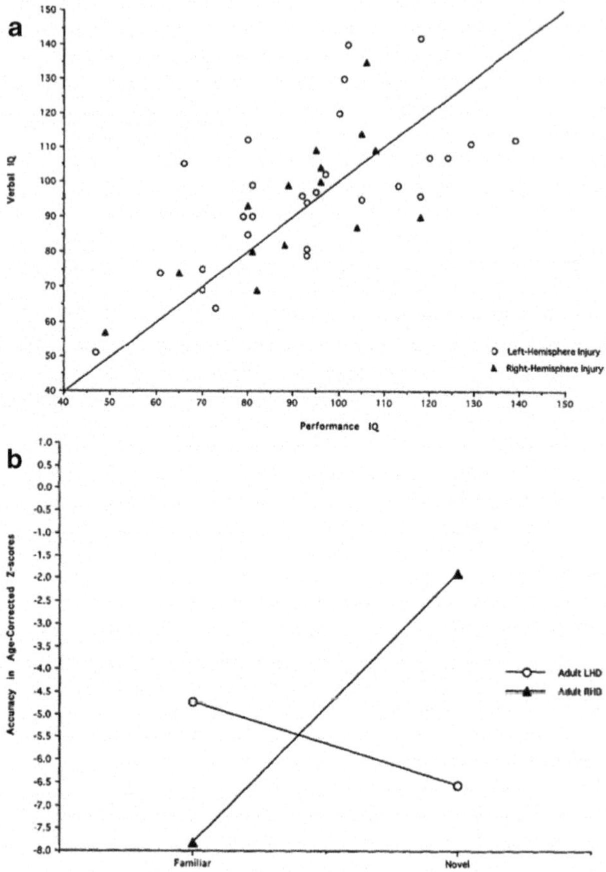

Fig. 6.25 (*Top*) Comparison of verbal and performal IQ in children with congenital damage in either the left or right hemisphere. Results show no relationship between the affected hemisphere and either verbal or performal IQ score in later years. Children were tested at various ages between 3 and 10 years. Full-scale IQ was 93 (i.e. within the normal range, but below the population mean of 100). Note the wide range in scores, with some children having IQs above 120. (*Middle and bottom*) Scores on a task where patients had to match either novel or familiar sentences to one of four pictures. Tested groups were patients with adult-onset lesions or children (6–12 years) with left or right hemisphere damage (respectively, LHD and RHD). A double dissociation is seen in the adult group: adults with LDH do better on the familiar phrases, whereas adults with RDH do better on the novel phrases. This implies that the right hemisphere is important for language functions, but in a qualitatively different manner than the left hemisphere. The bottom figure shows that children do far better than adults with comparable brain damage. Note also that the double dissociation is lacking here. RHD children actually performed worse on comprehension of novel sentences than the LDH group (Figures and text (modified) were taken from the chapter by Bates in Broman and Fletcher (1999) [173])

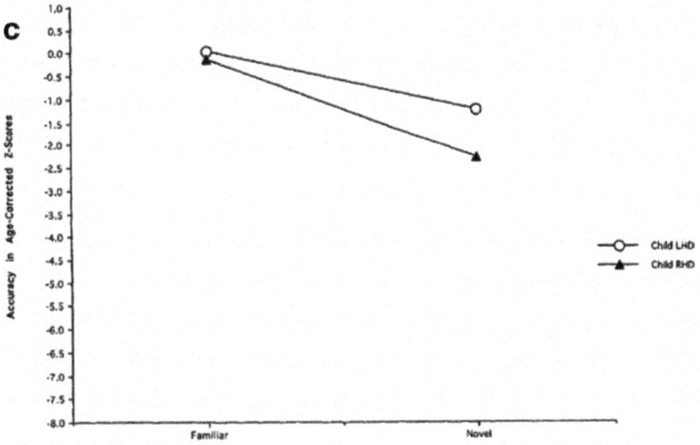

Fig. 6.25 (continued)

6.9.2 Wada Test

There is only one clinical test to determine hemispheric language dominance: the sodium amytal or Wada test (named after its inventor, Juhn Wada) [137]. The Wada test is an invasive and also somewhat controversial procedure. It is predominantly performed in epilepsy patients who are considered candidates for surgery and aims to test if the contralesional hemisphere has sufficient capacity for normal language and memory functions. It does so by temporarily disabling a large part of one hemisphere (effectively simulating surgery within this hemisphere). It is of note that the Wada test is seldom requested for brain tumour patients. Apparently, neurosurgeons are not very concerned about abnormal language organization in this group of patients.

The Wada test has been called the 'least ambiguous method' to test for hemispheric language dominance [138]. Nonetheless, it is the gold standard technique for this purpose. The procedure itself is fairly simple. A short-acting barbiturate is injected via a catheter that is positioned in one of the carotid arteries. As the barbiturate spreads in the vascular territory of the carotid artery—which usually supplies a large part of one hemisphere—there is an immediate hemiplegia and unilateral slowing of the EEG.[d] Because several cognitive or behavioural functions are disrupted, there is a danger that neglect or somnolence will influence or even invalidate patient testing [140]. Once one of the hemispheres has been put 'to sleep', the investigator has approximately 5–10 min

[d] The Wada test cannot specify laterality of all language functions, in particular functions that are classically attributed to the postero-temporal and infero-parietal region (such as verbal comprehension) [139]. Possibly, these regions are spared after injection of sodium amytal.

to investigate the patient's language (or other) functions. The patient will be asked to identify and name objects and to read words or answer questions. On the basis of a series of fairly simple tests, the examiner judges whether or not language functions reside in the tested hemisphere. The outcome of the Wada test is therefore usually dichotomic [141].

Results of the Wada test in patients that suffered from a brain lesion in later life (and thus had normal language development) probably reflect the incidence of language dominance in the normal population. At least, the incidence resembles that found in stroke studies, whereby 96% of right-handers and 70% of left-handers have a left language-dominant hemisphere for simple language functions [142]. The incidence of atypical language representation is higher in patients with early brain damage (see also Chap. 9). Estimates in the literature range from 63 to 96% for right-handers and 48 to 75% for left-handers or ambidextrous patients. In rare bilateral cases, representation of different language functions in different hemispheres has been reported [143–145]. It is of note that Wada test results are not always concordant with another gold clinical standard: electrocortical stimulation [146, 147].

6.9.3 Electrocortical Stimulation Mapping

Electrocortical stimulation mapping is the current gold standard to localize so-called eloquent brain structures. It has a good track record in neurosurgery, and most surgeons consider it a valuable technique to maximize safe tumour resection [148, 149]. The method of direct electrical stimulation (DES) is very convincing for those that have been able to observe it in the operating room. It is hard not to acknowledge the functional importance of an area when a patient suddenly gives a wrong answer or stops speaking when this area is electrically stimulated. However, the principles and behavioural effects of stimulation on the human brain are poorly understood [150]. There is no convincing proof that areas that are found with DES are truly functional in a sense that resection of these areas will leave the patient with permanent deficits.

DES relies on the principle that a particular brain area or fibre pathway can be functionally disabled for several seconds during electrical stimulation, thus temporarily eliciting a 'virtual lesion.' At first glance, the technique seems very intuitive and valid [151]. When a particular region is stimulated and the patient has difficulty performing a task, there must be a close and essential relationship between that brain area and the disturbed function. Consequently, areas in which ESM is positive are considered to be indispensable for normal function and are not included in the resection. Duffau (2012), for instance, formulated that:

> if a structure of the brain is still detected as functional by the mapping (i.e., eliciting the same reproducible and transient deficit during each stimulation performed at its level), it means that this structure is not only involved in a large-scale network, but is an 'epicentre' crucial for function of the whole network—and thus cannot be compensated. [110]

However, such a straightforward inference is not always justified. Premotor areas, in particular the supplementary motor area (SMA) proper, can be resected with little, if any, permanent deficit despite occasional positive responses after stimulation. After (or sometimes even during the course of) surgery, there is an immediate and often severe akinesia or mutism, as expected from the stimulation results [152]. However, these deficits typically resolve in several weeks or months. The fact that an area tests positively with electrical stimulation therefore does not necessarily imply that it is indispensable (i.e. eloquent) for that particular function *in the long run*. Note that in this case an eloquent area is a posteriori defined as an area that, when damaged, leads to permanent deficits. This finding calls into question the clinical usefulness and even the validity of electrical stimulation for its purpose, as it is unable to account for functional reorganization or compensation after surgery. Stated otherwise, the technique is not predictive of permanent loss of function. In case of resection within the SMA region, it has been shown that secondary motor areas in the healthy hemisphere are recruited in response to functional deficits. Unmasking of new motor areas is correlated with (partial) motor function recovery and has been demonstrated in humans by comparison of fMRI brain maps before and after surgery [153].

It is very likely that such a redundancy of positive stimulation sites is not only present in the motor domain but also holds for other (cognitive) functions. There is indirect evidence for this in the language domain. Surgery within the left fusiform gyrus never results in a lasting dysphasia, although language errors may be elicited with electrical stimulation [154]. Perhaps in these cases stimulation of anterior or basal temporal areas interferes with more distant critical areas via subcortical connections [155]. Long-term functional compensation could also account for this redundancy. Although no randomized studies have been performed to resolve these issues (for obvious reasons), it seems that with electrical stimulation 'some areas are more equal than others' [156]. Clearly, there are methodological concerns besides practical limitations. Not much is known about the local and global effects of electrical stimulation [157]. With bipolar stimulation, the area that is locally depolarized is approximately 5 mm in diameter, and spatial resolution therefore seems high [158]. In addition, there is distant current spread via physiological or biological propagation [155]. As formulated by Mandonnet (2009):

> So, in essence, DESs [direct electrical stimulations] are highly non-local: they enter the whole network that sustains a function. The stimulated point (axonal or cortical) is only an input gate to the whole network. (…) It is worth noting that the perturbation induced by DES in a functional network is small enough that it propagates only in a 'sub-circuit', thus inhibiting solely a specific component of the tested function. This is why for cortical mapping of language function, depending on which 'sub-circuit' is disturbed, one may observe phonological errors, speech apraxia, semantic paraphasias, anomia or syntactic mistakes. Of course, when increasing the stimulation intensity, one ultimately generates a speech arrest [154]. Similarly, when stimulating axonally, one generates a dysfunction of a specific network. While in a connectionist point of view this dysfunction is mediated by a mere transient disconnection between two still effective areas (i.e., only the link is disturbed, corresponding to a connectionist point of view), the non-local theory assumes that one or both of the areas linked by the stimulated pathway will be disturbed. Anyway, the resulting effect mimics disconnection syndrome, as observed in lesions of white matter. [158]

Another drawback of intraoperative electrical stimulation mapping is the relatively short duration of the applied stimulus (usually not more than 4 s) and the obvious constraints due to the surgical setting and stressful circumstances (Fig. 6.26). As a consequence, tasks are relatively simple, and extensive testing of a wide array of different functions is not possible. These methodological and practical issues limit thorough investigation of higher-order cognitive functions such as emotion or discourse. False-negative results can be expected when a function is not, or not correctly, tested for an appropriate area. For instance, it has been shown that patients who undergo awake surgery for a left frontal lobe tumour may exhibit postoperative working memory deficits, despite extensive intraoperative language mapping [158, 159]. Possibly, the use of a working memory task can prevent similar postoperative impairments in other patients. However, such a task

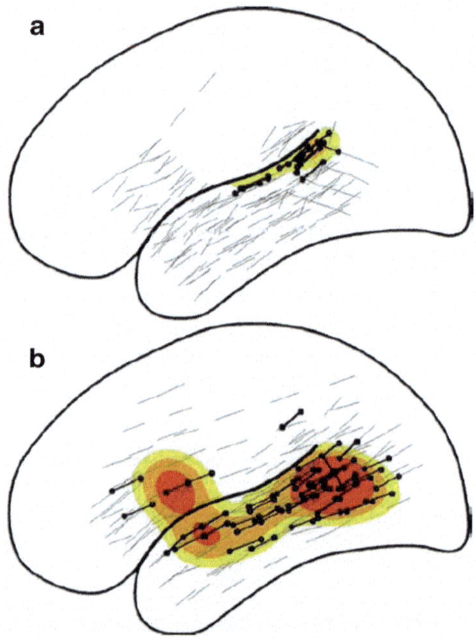

Fig. 6.26 Results of an electrocortical stimulation study in patients with implanted cortical grid electrodes (Boatman 2005) [174]. Patients had chronic medically refractory epilepsy and were candidates for surgical removal of a possible epileptogenic region. The grid was implanted for diagnostic reasons for a period of several days in which extra-operative testing took place. This situation allowed for much more extensive testing than during an operation. In a scientific study, patients were tested with three auditory discrimination tasks (syllables and tones). The figures show the probability densities for the distribution of auditory speech discriminations in normal listeners (**a**) and impaired listeners (**b**). *Black dots* show the locations where task performance was impaired. It was found that patients who were already impaired at a particular listening task used a much larger part of the temporal cortex than patients who were unimpaired listeners for that particular task [174]. One possible explanation for this phenomenon is that impaired listeners used more cortical resources to achieve similar performance on the task

has not yet been developed or validated, and it is questionable whether higher cognitive functions can be adequately tested in an awake surgical environment. Then there is the paradoxical observation that many patients do experience language (or other) impairments during or after a surgical procedure where DES was used and that therefore DES itself is not always able to prevent functional deficits. This is known by an experienced surgical team and seen as part of the surgical strategy. Patients are told before surgery that there is a chance that they will have motor or language impairments after the operation, but that these will usually disappear after a period of weeks or months (a process that is, by the way, largely unexplained).

In conclusion, electrical stimulation mapping seems a reliable technique to assess the immediate consequences of removal of a part of the brain in terms of sensorimotor and language functions. It is currently the best technique available for this purpose. Of course, the cortical or subcortical area that is temporarily disabled by stimulation is part of a much wider brain network. It is therefore in itself never solely responsible for any given function that is identified with DES. As a consequence, DES cannot reliably predict whether perilesional or distant neural networks are able to compensate for any loss of function after the operation (i.e. there is a risk of false-positive results).

Despite practical and methodological issues, findings from cortical and subcortical mapping have greatly contributed to a new and more refined view on the functional representation of cognitive functions, in particular for language. This has not always been appreciated. In the previous century, starting with the work of Penfield, the results of cortical stimulation mapping were largely interpreted in support for the classic Broca–Wernicke model. Graphically, Penfield's maps of the speech areas closely resemble the maps that had already been deduced from post-mortem studies in language-disabled patients, and this may have contributed to the shallow interpretation of his clinical findings. As we have seen before, the conclusions that can be drawn from Penfield's studies are far more nuanced than this. In fact, Penfield himself did not attribute language functions to single brain areas but instead proposed a network of interconnected brain areas that was dynamic and able to compensate functionally in case of local damage of one of its areas. With the studies of Lüders [160] and Ojemann, and later those of several others, gradually a different and more complex view on language organization emerged. Language functions operate in parallel subsystems that show a large interindividual variability. Areas are scattered in a mosaic-like pattern and occupy a large part of the frontal, temporal and parietal lobes. More recently, attention shifted to the functional importance of white matter tracts. In particular Duffau contributed to newer clinical models whereby 'the language network is organized in parallel, segregated (even if interconnected) large-scale cortico-subcortical sub-networks underlying semantic, phonological and syntactic processing' [161]. It is of interest to see that this modern, and more anatomically based, view has a number of characteristics that were already described by other authors in the past but for various reasons never gained mainstream clinical attention. Some of these models and theories will be described in the next chapter.

References

1. Fritsch GT, Hitzig E. Über die elektrische Erregbarkeit des Grosshirns. Arch Anat Phys. 1870;37:300–32.
2. Carlson C, Devinsky O. The excitable cerebral cortex Fritsch G, Hitzig E. Uber die elektrische Erregbarkeit des Grosshirns. Arch Anat Physiol Wissen 1870;37:300–32. Epilepsy Behav. 2009;15:131–2.
3. Gross CG. The discovery of motor cortex and its background. J Hist Neurosci. 2007;16:320–31.
4. Boling W, Olivier A, Fabinyi G. Historical contributions to the modern understanding of function in the central area. Neurosurgery. 2002;50:1296–309; discussion 1309.
5. Aldini G. An account of the late improvements in galvanism with a series of curious and interesting experiments. London: Cuthell and Martin; 1803.
6. Ferrier D. The functions of the brain. New York: GP Putnam's Sons; 1886.
7. Withington ET. Hippocrates: on wounds in the head, vol. 3. Cambridge: Harvard University Press; 1927.
8. Finger S. Origins of neuroscience: a history of explorations into brain function. New York: Oxford University Press; 2001.
9. Young RM. Mind, brain and adaptation in the nineteenth century. New York: Oxford University Press; 1970.
10. Jefferson G. Selected papers. London: Pitman; 1960.
11. Jackson JH. Notes on the physiology and pathology of the nervous system. Med Times Gaz. 1868;2:696.
12. York GK, Steinberg DA. Hughlings Jackson's neurological ideas. Brain. 2011;134:3106–13.
13. Sherrington CS. Sir David Ferrier (1843-1928). London: Oxford University Press; 1937. p. 302.
14. Horsley V. The Linacre Lecture on the function of the so-called motor area of the brain: delivered to the Master and Fellows of St. John's College, Cambridge, May 6th, 1909. Br Med J. 1909;2:121.
15. Grünbaum ASF, Sherrington CS. Observations on the physiology of the cerebral cortex of some of the higher apes. Proc R Soc Lond. 1901;69:206–9.
16. Lemon RN. An enduring map of the motor cortex. Exp Physiol. 2008;93:798–802.
17. Archibald E. Surgical affections of the head. In: Bryant JD, Buck AH, editors. American practice of surgery. New York: William Wood and Co.; 1908. p. 3–379.
18. Phillips CG, Porter R. Corticospinal neurones. Their role in movement. London: Academic Press; 1977. p. 65.
19. Frohlich A, Sherrington CS. J Physiol. 1901;xxviii.
20. Leyton ASF, Sherrington CS. Observations on the excitable cortex of the chimpanzee, orang-utan, and gorilla. Exp Physiol. 1917;11:135–222.
21. Devinsky O, Beric A, Dogali M. Electrical and magnetic stimulation of the brain and spinal cord. New York: Raven Press; 1993.
22. Sanes JN, Schieber MH. Orderly somatotopy in primary motor cortex: does it exist? NeuroImage. 2001;13:968–74.
23. Brown GT. J Physiol. 1914;xlviii:xxix, xxx, xxxiii.
24. Brown GT, Sherrington CS. On the instability of a cortical point. Proc R Soc London, Ser B. 1912;85:250–77.
25. Krause F. Chirurgie des Gehirns und Rueckenmarks; 1911.
26. Vilensky JA, Gilman S. Horsley was the first to use electrical stimulation of the human cerebral cortex intraoperatively. Surg Neurol. 2002;58:425–6.
27. Pondal-Sordo M, Diosy D, Tellez-Zenteno JF, et al. Epilepsy surgery involving the sensory-motor cortex. Brain. 2006;129:3307–14.
28. Krause F, Schum H. Spezielle Chirurgie der Gehirnkrankheiten; 1931.
29. Foerster O, Penfield WP. The structural basis of traumatic epilepsy and results of radical operation. Brain. 1930;53:99–119.

30. Foerster O, Altenburger H. Elektrobiologische Vorgänge an der menschlichen Hirnrinde. J Neurol. 1935;135:277–88.
31. Luders HO, Comair YG. Epilepsy surgery. Philadelphia: Lippincott Williams and Wilkins; 2001.
32. Sarikcioglu L. Otfrid Foerster (1873-1941): one of the distinguished neuroscientists of his time. J Neurol Neurosurg Psychiatry. 2007;78:650.
33. Penfield WP, Boldrey E. Somatic motor and sensory representation in the cerebral cortex of man as studied by electrical stimulation. Brain. 1937;60:389–443.
34. Penfield WP, Rasmussen T. The cerebral cortex of man. New York: The Macmillan Company; 1957.
35. Feindel W. The physiologist and the neurosurgeon: the enduring influence of Charles Sherrington on the career of Wilder Penfield. Brain. 2007;130:2758–65.
36. Penfield WP, Roberts L. Speech and brain mechanisms. Princeton University Press: Princeton; 1959.
37. Catani M, Dell'acqua F, Vergani F, et al. Short frontal lobe connections of the human brain. Cortex. 2012;48:273–91.
38. Foerster O. Symptomatologie der Erkrankingen des Grosshirns: motorische Felder und Bahnen. In: Bumke O, Foerster O, editors. Handbuch der Neurologie. Berlin: Springer; 1936.
39. Luria AR. Higher cortical functions in man. 2nd ed. New York: Basic Books Inc.; 1980.
40. Schmahmann JD, Pandya DN. Fiber pathways of the brain. New York: OUP; 2006.
41. Sakamoto T, Porter LL, Asanuma H. Long-lasting potentiation of synaptic potentials in the motor cortex produced by stimulation of the sensory cortex in the cat: a basis of motor learning. Brain Res. 1987;413:360–4.
42. Ferezou I, Haiss F, Gentet LJ, et al. Spatiotemporal dynamics of cortical sensorimotor integration in behaving mice. Neuron. 2007;56:907–23.
43. Schott GD. Penfield's homunculus: a note on cerebral cartography. J Neurol Neurosurg Psychiatry. 1993;56:329–33.
44. Penfield WP. Ferrier lecture: some observations on the cerebral cortex of man. Proc R Soc London, Ser B. 1947;134:329–47.
45. Sanes JN, Donoghue JP. Plasticity and primary motor cortex. Annu Rev Neurosci. 2000;23:393–415.
46. Vargas-Irwin CE. Motor cortical control of naturalistic reachting and grasping actions [thesis]. Brown University; 2010.
47. Snyder PJ, Whitaker HA. Neurologic heuristics and artistic whimsy: the cerebral cartography of Wilder Penfield. J Hist Neurosci. 2013;22(3):277–91.
48. Bartholow R. Experimental investigations into the functions of the human brain. Am J Med Sci. 1874;67:305–13.
49. Harris LJ, Almerigi JB. Probing the human brain with stimulating electrodes: the story of Roberts Bartholow's (1874) experiment on Mary Rafferty. Brain Cogn. 2009;70:92–115.
50. Kim OJ. Experiment at bedside: Harvey Cushing's neurophysiological research. Korean J Med Hist. 2009;18(2):205–22.
51. Bliss M. Harvey Cushing: A life in surgery. New York: Oxford University Press; 2005.
52. Cushing H. The surgical aspects of major neuralgia of the trigeminal nerve. JAMA. 1905;44:860–5.
53. Miller JT, Rahimi SY, Lee M. History of infection control and its contributions to the development and success of brain tumor operations. Neurosurg Focus. 2005;18:e4.
54. Clark FC. A brief history of antiseptic surgery. Med Library Hist J. 1907;5:145–72.
55. Fodstad H, Kelly PJ, Buchfelder M. History of the Cushing reflex. Neurosurgery. 2006;59:1132–7; discussion 1137.
56. Jefferson G. Harvey Cushing, April 8, 1869-October 7, 1939. Surg Neurol. 1974;2:217–24.
57. Powell M. Sir Victor Horsley—an inspiration. BMJ. 2006;333:1317–9.
58. Cushing H. A note upon the faradic stimulation of the postcentral gyrus in conscious patients. Brain. 1909;32:44–53.
59. Pendleton C, Zaidi HA, Chaichana KL, et al. Harvey Cushing's contributions to motor mapping: 1902-1912. Cortex. 2012;48:7–14.

60. Horsley V. The Linacre lecture on the function of the so-called motor area of the brain. Br Med J. 1909;2:121.
61. Tan TC, Black PM. The contributions of Otfrid Foerster (1873-1941) to neurology and neurosurgery. Neurosurgery. 2001;49:1231–6.
62. Foerster O, Penfield WP. Der Narbenzug am und im Gehirn bei traumatischer Epilepsie in seiner Bedeutung für das Zustandekommen der Anfälle und für die therapeutische Bekämpfung derselben. Z ges Neurol Psychiat. 1930;125:475–572.
63. Preul MC, Feindel W. Origins of Wilder Penfield's surgical technique: the role of the "Cushing ritual" and influences from the European experience. J Neurosurg. 1991;75:812–20.
64. Penfield WP. Diencephalic autonomic epilepsy. Res Publ Assoc Nerv Ment Dis. 1930;9:645–63.
65. Nielsen JM. The possibility of pure motor aphasia. Bull Los Angel Neurol Soc. 1936;1:11–4.
66. Penfield WP, Jasper H. Epilepsy and the functional anatomy of the human brain. Boston: Little, Brown and Company; 1954.
67. Penfield W, Welch K. The supplementary motor area of the cerebral cortex: a clinical and experimental study. Arch Neurol Psychiatr. 1951;66:289.
68. Geschwind N. Selected papers on language and the brain. New York: Springer; 1974.
69. Schwab O. Über vorübergehenden aphasische Störungen nach Rindenexzision aus dem linken Stirnhirn bei Epileptikern. Dtsch Z Nervenheilk. 1927;94:117–84.
70. Bogen JE, Bogen GM. Wernicke's region—where is it? Ann N Y Acad Sci. 1976;280:834–43.
71. Prados M, Strowger B, Feindel W. Studies on cerebral edema: I. Reaction of the brain to air exposure; pathologic changes; II. Physiologic changes. Arch Neurol Psychiatry. 1945;54:163–74. 290
72. Crosson B. Subcortical functions in language and memory. New York: Guilford Press; 1992.
73. Ford AA, Triplett W, Sudhyadhom A, et al. Broca's area and its striatal and thalamic connections: a diffusion-MRI tractography study. Front Neuroanat. 2013;7:1–12.
74. Fedio P, Van Buren JM. Memory deficits during electrical stimulation of the speech cortex in conscious man. Brain Lang. 1974;1:29–42.
75. Ojemann GA. Cortical organization of language. J Neurosci. 1991;11:2281–7.
76. Lubrano V, Draper L, Roux FE. What makes surgical tumor resection feasible in Broca's area? Insights into intraoperative brain mapping. Neurosurgery. 2010;66:868–75; discussion 875.
77. Thiel A, Herholz K, Koyuncu A, et al. Plasticity of language networks in patients with brain tumors: a positron emission tomography activation study. Ann Neurol. 2001;50:629.
78. Stern Y. Cognitive reserve. Neuropsychologia. 2009;47:2015–28.
79. Prince M, Acosta D, Ferri CP, et al. Dementia incidence and mortality in middle-income countries, and associations with indicators of cognitive reserve: a 10/66 Dementia Research Group population-based cohort study. Lancet. 2012;380(9836):50–8.
80. Ojemann GA, Whitaker HA. Language localization and variability. Brain Lang. 1978;6:239–60.
81. Rutten GJ, van Rijen PC, van Veelen CW, Ramsey NF. Language area localization with three-dimensional functional magnetic resonance imaging matches intrasulcular electrostimulation in Broca's area. Ann Neurol. 1999;46:405–8.
82. Tomaiuolo F, MacDonald JD, Caramanos Z, et al. Morphology, morphometry and probability mapping of the pars opercularis of the inferior frontal gyrus: an in vivo MRI analysis. Eur J Neurosci. 1999;11:3033–46.
83. Benzagmout M, Gatignol P, Duffau H. Resection of World Health Organization Grade II gliomas involving Broca's area: methodological and functional considerations. Neurosurgery. 2007;61:741–52.
84. Ojemann GA, Ojemann JG, Lettich E, Berger MS. Cortical language localization in left, dominant hemisphere: an electrical stimulation mapping investigation in 117 patients. J Neurosurg. 1989;71:316–26.
85. Sanai N, Mirzadeh Z, Berger MS. Functional outcome after language mapping for glioma resection. N Engl J Med. 2008;358:18–27.
86. Ojemann GA. Some brain mechanisms for reading. In: von Euler C, Lundberg I, Lennerstrand G, editors. Brain and reading. New York: MacMillan; 1989. p. 47–59.

87. Roux FE, Lubrano V, Lauwers-Cances V, et al. Intra-operative mapping of cortical areas involved in reading in mono- and bilingual patients. Brain. 2004;127:1796–810.
88. Ojemann JG, Ojemann GA, Lettich E. Cortical stimulation mapping of language cortex by using a verb generation task: effects of learning and comparison to mapping based on object naming. J Neurosurg. 2002;97:33–8.
89. Lubrano V, Roux FE, Demonet JF. Writing-specific sites in frontal areas: a cortical stimulation study. J Neurosurg. 2004;101:787–98.
90. Hamberger MJ, Goodman RR, Perrine K, Tamny T. Anatomic dissociation of auditory and visual naming in the lateral temporal cortex. Neurology. 2001;56:56–61.
91. Ojemann G, Mateer C. Human language cortex: localization of memory, syntax, and sequential motor-phoneme identification systems. Science. 1979;205:1401–3.
92. Schaffler L, Luders HO, Dinner DS, et al. Comprehension deficits elicited by electrical stimulation of Broca's area. Brain. 1993;116:695–715.
93. Rapport RL, Tan CT, Whitaker HA. Language function and dysfunction among Chinese- and English-speaking polyglots: cortical stimulation, Wada testing, and clinical studies. Brain Lang. 1983;18:342–66.
94. Roux FE, Tremoulet M. Organization of language areas in bilingual patients: a cortical stimulation study. J Neurosurg. 2002;97:857–64.
95. Lucas TH, McKhann GM, Ojemann GA. Functional separation of languages in the bilingual brain: a comparison of electrical stimulation language mapping in 25 bilingual patients and 117 monolingual control patients. J Neurosurg. 2004;101:449–57.
96. Boatman D. Cortical bases of speech perception: evidence from functional lesion studies. Cognition. 2004;92:47–65.
97. Ojemann GA. Organization of short-term verbal memory in language areas of human cortex: evidence from electrical stimulation. Brain Lang. 1978;5:331–40.
98. Ojemann GA. Models of the brain organization for higher integrative functions derived with electrical stimulation techniques. Hum Neurobiol. 1982;1:243–9.
99. Fried I, Mateer C, Ojemann G, et al. Organization of visuospatial functions in human cortex. Evidence from electrical stimulation. Brain. 1982;105:349–71.
100. Kho KH, Rutten GJ, Leijten FS, et al. Working memory deficits after resection of the dorsolateral prefrontal cortex predicted by functional magnetic resonance imaging and electrocortical stimulation mapping. Case report. J Neurosurg. 2007;106:501–5.
101. Perrine K, Uysal S, Dogali M, et al. Functional mapping of memory and other nonlinguistic cognitive abilities in adults. Adv Neurol. 1993;63:165–77.
102. Thiebaut de Schotten M, Urbanski M, Duffau H, et al. Direct evidence for a parietal-frontal pathway subserving spatial awareness in humans. Science. 2005;309:2226–8.
103. Roux FE, Boetto S, Sacko O, et al. Writing, calculating, and finger recognition in the region of the angular region: a cortical stimulation study of Gerstmann syndrome. J Neurosurg. 2003;99:716–27.
104. Boatman D, Gordon B, Hart J, et al. Transcortical sensory aphasia: revisited and revised. Brain. 2000;123:1634–42.
105. Martin A. The representation of object concepts in the brain. Annu Rev Psychol. 2007;58:25–45.
106. Lambon Ralph MA. Neural basis of memory. Brain mapping. Wien: Springer; 2012. p. 145–54.
107. Alexander GE, DeLong MR, Strick PL. Parallel organization of functionally segregated circuits linking basal ganglia and cortex. Annu Rev Neurosci. 1986;9:357–81.
108. Sanai N, Berger MS. Operative techniques for gliomas and the value of extent of resection. Neurotherapeutics. 2009;6:478–86.
109. Skirboll SS, Ojemann GA, Berger MS, et al. Functional cortex and subcortical white matter located within gliomas. Neurosurgery. 1996;38:678–84.
110. Duffau H. The "frontal syndrome" revisited: lessons from electrostimulation mapping studies. Cortex. 2012;48:120–31.
111. Mandonnet E, Delattre JY, Tanguy ML, et al. Continuous growth of mean tumor diameter in a subset of grade II gliomas. Ann Neurol. 2003;53:524–8.

112. Schomas DA, Laack NN, Rao RD, et al. Intracranial low-grade gliomas in adults: 30-year experience with long-term follow-up at Mayo Clinic. Neuro-Oncology. 2009;11: 437–45.
113. van den Bent MJ. Practice changing mature results of RTOG study 9802: another positive PCV trial makes adjuvant chemotherapy part of standard of care in low-grade glioma. Neuro-Oncology. 2014;16:1570–4.
114. Smith JS, Chang EF, Lamborn KR, et al. Role of extent of resection in the long-term outcome of low-grade hemispheric gliomas. J Clin Oncol. 2008;26:1338–45.
115. Sanai N, Berger MS. Glioma extent of resection and its impact on patient outcome. Neurosurgery. 2008;62:753–64.
116. Teunissen F, Verheul HB, Rutten GJ. Functionality of glioma-infiltrated precentral gyrus: experience from fourteen patients. J Neurosurg Sci. 2017;61(2):140–50.
117. Duffau H. Lessons from brain mapping in surgery for low-grade glioma: insights into associations between tumour and brain plasticity. Lancet Neurol. 2005;4:476–86.
118. Plaza M, Gatignol P, Leroy M, Duffau H. Speaking without Broca's area after tumor resection. Neurocase. 2009;15:294–310.
119. Sarubbo S, Le Bars E, Moritz-Gasser S, Duffau H. Complete recovery after surgical resection of left Wernicke's area in awake patient: a brain stimulation and functional MRI study. Neurosurg Rev. 2012;35:287–92; discussion 292.
120. Ungerleider LG, Haxby JV. 'What' and 'where' in the human brain. Curr Opin Neurobiol. 1994;4:157–65.
121. Duffau H, Thiebaut de Schotten M, Mandonnet E. White matter functional connectivity as an additional landmark for dominant temporal lobectomy. J Neurol Neurosurg Psychiatry. 2008;79:492–5.
122. Duffau H. The anatomo-functional connectivity of language revisited. New insights provided by electrostimulation and tractography. Neuropsychologia. 2008;46:927–34.
123. Sarubbo S, De Benedictis A, Maldonado IL, et al. Frontal terminations for the inferior fronto-occipital fascicle: anatomical dissection, DTI study and functional considerations on a multi-component bundle. Brain Struct Funct. 2011;218(1):21–37.
124. Martino J, Brogna C, Robles SG, et al. Anatomic dissection of the inferior fronto-occipital fasciculus revisited in the lights of brain stimulation data. Cortex. 2010;46(5):691–9.
125. Duffau H. Brain mapping: from neural basis of cognition to surgical applications. Wien: Springer; 2011.
126. Duffau H. Brain plasticity and tumors. Adv Tech Stand Neurosurg. 2008;33:3–33.
127. Sanai N, Polley MY, Berger MS. Insular glioma resection: assessment of patient morbidity, survival, and tumor progression. J Neurosurg. 2010;112:1–9.
128. Hebb AO, Yang T, Silbergeld DL. The sub-pial resection technique for intrinsic tumor surgery. Surg Neurol Int. 2011;2:180.
129. Dewarrat GM, Annoni JM, Fornari E, et al. Acute aphasia after right hemisphere stroke. J Neurol. 2009;256:1461–7.
130. Alexander MP, Fischette MR, Fischer RS. Crossed aphasias can be mirror image or anomalous. Case reports, review and hypothesis. Brain. 1989;112:953–73.
131. Zangwill OL. Speech and the minor hemisphere. Acta Neurol Psychiatr Belg. 1967;67:1013–20.
132. Gloning I, Gloning K, Haub G, Quatember R. Comparison of verbal behavior in right-handed and non right-handed patients with anatomically verified lesion of one hemisphere. Cortex. 1969;5:43–52.
133. Pedersen PM, Jorgensen HS, Nakayama H, et al. Aphasia in acute stroke: incidence, determinants, and recovery. Ann Neurol. 1995;38:659–66.
134. Wade DT, Hewer RL, David RM, Enderby PM. Aphasia after stroke: natural history and associated deficits. J Neurol Neurosurg Psychiatry. 1986;49:11–6.
135. Ross ED. Affective prosody and the aprosodias. In: Mesulam MM, editor. Principles of behavioral and cognitive neurology. 2nd ed. New York: Oxford University Press; 2000. p. 316–31.

136. Gazzaniga MS, Sperry RW. Language after section of the cerebral commissures. Brain. 1967;90:131–48.
137. Wada J. A new method for the determination of the side of cerebral speech dominance: a preliminary report on the intracarotid injection of sodium amytal in man. Igaku Seibutsugaku. 1949;14:221–2.
138. Loring DW, Meador KJ, Lee GP, King DW. Amobarbital effects and lateralized brain function: the Wada test. New York: Springer; 2012.
139. Hart Jr J, Lesser RP, Fisher RS, et al. Dominant-side intracarotid amobarbital spares comprehension of word meaning. Arch Neurol. 1991;48:55–8.
140. de Paola L, Mader MJ, Germiniani FM, et al. Bizarre behavior during intracarotid sodium amytal testing (Wada test): are they predictable? Arq Neuropsiquiatr. 2004;62:444–8.
141. Yetkin FZ, Swanson S, Fischer M, et al. Functional MR of frontal lobe activation: comparison with Wada language results. Am J Neuroradiol. 1998;19:1095–8.
142. Rasmussen T, Milner B. The role of early left-brain injury in determining lateralization of cerebral speech functions. Ann N Y Acad Sci. 1977;299:355–69.
143. Risse GL, Gates JR, Fangman MC. A reconsideration of bilateral language representation based on the intracarotid amobarbital procedure. Brain Cogn. 1997;33:118–32.
144. Kurthen M, Helmstaedter C, Linke DB, et al. Quantitative and qualitative evaluation of patterns of cerebral language dominance. An amobarbital study. Brain Lang. 1994;46:536–64.
145. Rutten GJ, Ramsey NF, van Rijen PC, et al. fMRI-determined language lateralization in patients with unilateral or bilateral language dominance according to the Wada test. NeuroImage. 2002;17:447–60.
146. Wyllie E, Luders H, Murphy D, et al. Intracarotid amobarbital (Wada) test for language dominance: correlation with results of cortical stimulation. Epilepsia. 1990;31:156–61.
147. Kho KH, Leijten FS, Rutten GJ, et al. Discrepant findings for Wada test and functional magnetic resonance imaging with regard to language function: use of electrocortical stimulation mapping to confirm results. Case report. J Neurosurg. 2005;102:169–73.
148. Duffau H, Lopes M, Arthuis F, et al. Contribution of intraoperative electrical stimulations in surgery of low grade gliomas: a comparative study between two series without (1985-96) and with (1996-2003) functional mapping in the same institution. J Neurol Neurosurg Psychiatry. 2005;76:845–51.
149. Haglund MM, Berger MS, Shamseldin M, et al. Cortical localization of temporal lobe language sites in patients with gliomas. Neurosurgery. 1994;34:567–76.
150. Borchers S, Himmelbach M, Logothetis N, Karnath HO. Direct electrical stimulation of human cortex—the gold standard for mapping brain functions? Nat Rev Neurosci. 2012;13:63–70.
151. Rutten GJ, Ramsey NF. The role of functional magnetic resonance imaging in brain surgery. Neurosurg Focus. 2010;28:E4.
152. Duffau H, Lopes M, Denvil D, Capelle L. Delayed onset of the supplementary motor area syndrome after surgical resection of the mesial frontal lobe: a time course study using intraoperative mapping in an awake patient. Stereotact Funct Neurosurg. 2001;76:74–82.
153. Krainik A, Duffau H, Capelle L, et al. Role of the healthy hemisphere in recovery after resection of the supplementary motor area. Neurology. 2004;62:1323–32.
154. Luders H, Lesser RP, Hahn J, et al. Basal temporal language area. Brain. 1991;114:743–54.
155. Ishitobi M, Nakasato N, Suzuki K, et al. Remote discharges in the posterior language area during basal temporal stimulation. Neuroreport. 2000;11:2997–3000.
156. Orwell G. Animal farm. 1st World Library—Literary Society; 2005.
157. Nathan SS, Sinha SR, Gordon B, et al. Determination of current density distributions generated by electrical stimulation of the human cerebral cortex. Electroencephalogr Clin Neurophysiol. 1993;86:183–92.
158. Mandonnet E, Winkler PA, Duffau H. Direct electrical stimulation as an input gate into brain functional networks: principles, advantages and limitations. Acta Neurochir. 2010;152(2):185–93.

159. Boisgueheneuc G, Levy R, Volle E, et al. Functions of the left superior frontal gyrus in humans: a lesion study. Brain. 2007;129:3315–28.
160. Luders H, Lesser RP, Dinner DS, et al. Localization of cortical function: new information from extraoperative monitoring of patients with epilepsy [published erratum appears in Epilepsia 1988 Nov-Dec;29(6):828]. Epilepsia. 1988;29(Suppl 2):S56–65.
161. Duffau H, Moritz-Gasser S, Mandonnet E. A re-examination of neural basis of language processing: proposal of a dynamic hodotopical model from data provided by brain stimulation mapping during picture naming. Brain Lang. 2013.
162. Luciani L, Tamburini A. Ricerche sperimentali sui centri psico-motori corticali; 1878.
163. Mills CK. Cerebral localization in its practical relations; 1888.
164. Krause F, Heymann E. Chirurgische Operationslehre des Kopfes. 1912;2.
165. Catani M, Stuss DT. At the forefront of clinical neuroscience. Cortex. 2012;48:1–6.
166. Rahm Jr WE, Scarff JE. Electrical stimulation of the cerebral cortex: description of a new stimulator. Arch Neurol Psychiatr. 1943;50:183–9.
167. Gill AS, Binder DK. Wilder Penfield, Pio del Rio-Hortega, and the discovery of oligodendroglia. Neurosurgery. 2007;60:940–8; discussion 940.
168. Penfield W. Cytology and cellular pathology of the nervous system. New York: Paul B Hoeber; 1932.
169. Sherman SM, Guillery RW. Functional connections of cortical areas: a new view from the thalamus. booksgooglecom; 2013.
170. Jelgersma G. Atlas anatomicum cerebri humani: 168 sections of the human brain. Amsterdam: Scheltema & Holkema; 1931.
171. Alexander M, Naeser M, Palumbo C. Correlations of subcortical CT lesion sites and aphasia profiles. Brain. 1987;110:961–88.
172. Murdoch BE. Speech and language disorders associated with subcortical pathology. Hoboken: Wiley; 2009.
173. Bates E. Plasticity, localization, and language development. The changing nervous system. Neurobehavioral consequences of early brain disorders. New York: Oxford University Press; 1999. p. 214–53.
174. Boatman DF, Miglioretti DL. Cortical sites critical for speech discrimination in normal and impaired listeners. J Neurosci. 2005;25:5475–80.

7 Neo-connectionism, Neurodynamics and Large-Scale Networks

After World War II, the centre of gravity in aphasiology research shifted from Europe to North America [1]. In this period, interest in the German and French localist theories waned to virtual non-existence in clinical practice. The localist view was replaced by ideas with a more holistic character. A factor that likely contributed to this transition was the many war casualties, whose complex disturbances and potential for recovery were not very well explained by the contemporary language theories. These observations triggered basic research on aphasia, as well as efforts to rehabilitate patients [2]. Somehow, then, interest renewed to a point where localism again became the dominant clinical view that it remains today.

7.1 Geschwind

One of the investigators who contributed much to this renaissance was Norman Geschwind (1926–1984), who summarized his own view on historical findings as follows (1974):

> It is worth recalling what the state of the art was in 1961. Until the First World War the accomplishments in the field of understanding the relationships of the brain to language had been rightly regarded as among the brightest treasures of the discipline of clinical neurology, and the list of major contributers to it would have included a majority of the creators of the discipline—Broca, Meynert, Flechsig, Jackson, Wernicke, Liepmann, Lichtheim, Dejerine, to mention only a few.
>
> Yet by the 1950s interest in the area had nearly vanished from the field of clinical neurology. There were a few exceptions, such as Luria in the Soviet Union, but much of his work has been published only recently. In all of the French, German, and English-speaking worlds there were probably not a dozen neurologists for whom the higher functions of the brain represented a major interest [3].

Geschwind's training as a neurologist imbued him with 'an overwhelming skepticism toward the view that there were highly characteristic aphasic syndromes associated with different lesions of the brain' [3]. The dominant theories of that time were strongly holistic. As he himself explained, he had 'forcefully (…) accepted the

G.-J. Rutten, *The Broca-Wernicke Doctrine*, DOI 10.1007/978-3-319-54633-9_7

view that any attempt at "explaining" the syndromes on the basis of anatomy was a futile endeavor' [3]. At that time Geschwind also regarded himself a believer of the ideas of more holistic-orientated neurologists such as Jackson, Goldstein and Head. Triggered by some of his teachers and the recent work of Myers and Sperry on callosum-sectioned animals, he began doubting these ideas and decided to study the classic localizationist school by reading the original papers 'rather than by reading the interpretations of later hostile authors' [3–4]. As he recalls in his *Selected Papers on Language and the Brain* (1979):

> Somewhere about 1960 I awoke, perhaps belatedly, to my own profound confusion. (…) I was persistently troubled by the fact that people who had left their mark so indelibly in many areas of neurology, such as Wernicke, Bastian, Dejerine, Charcot, and many others, could apparently have shown what was asserted to be the sheerest naiveté and incompetence in the area of higher functions. It seemed difficult to accept the view that men who had established long-honored clinical pictures should have apparently been so incapable of examining an aphasic, or that scholars who had made fundamental anatomical investigations of permanent worth should have been so perfunctory and sloppy in their descriptions of the brains of aphasics [3].

The first paper he read was that of Dejerine on his case of pure alexia without agraphia (the case was briefly described in Fig. 4.5 in Chap. 4):

> The impact of the paper was multiple. In the first place the description was so lucid that it was immediately clear that this 'pure' syndrome must exist, despite the insistence of some modern writers that these selective syndromes 'could no longer be seen' (a statement implying that the earlier descriptions had been grossly in error). Indeed it was obvious that Dejerine's standard of examination was superior to that of most modern students of aphasia. Furthermore it was a shock, but a salutary one, to discover that even so masterful a paper had been neglected by later writers, or grossly misquoted. Henry Head (1926), the often cited critic of classical approaches, did not even list it in his biography. This paper was therefore instrumental in making me aware of the inaccuracy of most of the histories of aphasia in English. [3]

Within a few weeks after reading this paper, Geschwind observed a similar case himself. Not much later he came across another callosal case. This patient had initially been seen by a colleague, Edith Kaplan. The patient had undergone surgery for a left frontal glioblastoma. After surgery, Kaplan found out that the man could write normally with his right hand, but—astonishingly—wrote aphasically with his left hand. In a book that is a tribute to Geschwind, Devinsky (1997) writes about this case:

> She [Kaplan] could have told a thousand neurologists about this case, and it would have remained one of those curiosities that doesn't seem to fit into the comfortable pigeon-holes of knowledge, but she told Norman Geschwind. (…) The timing was fortuitous, but Pasteur's dictum applied - chance favors the prepared mind [5].

The 41-year-old patient presented with headache, nausea and vomiting. Members of the family had noticed increasing behavioural changes over the previous months. The patient was right-handed and, except for a few paraphasias and some hesitancy in naming objects, had normal speech. Further neurological examination revealed an alert and cooperative patient with considerable frontal lobe dysfunction:

> He exhibited inappropriate jocularity against a background of general apathy. He had no insight into his illness and appeared unconcerned about it. He kept repeating questions as if he did not quite understand them. He could remember only 1 of 3 items after five minutes. He repeated 6 digits forward and 4 backward. He made many errors in subtracting sevens serially from 100. His proverb interpretations were very poor, being little more than restatements without interpretation. He did simple written arithmetic correctly but failed on more complex material such as multiplying 214 by 35. When asked to draw a clock face at a certain hour, he frequently reversed the large and small hands [5].

After surgery (where a partial left frontal lobectomy had been performed), there was initially a dense right hemiplegia and a marked aphasia. This partially improved with time, in particular the language abilities. One and half months after surgery, the patient's speech showed 'at most a few paraphasic errors', although he 'followed complex commands poorly'. Neuropsychological examination showed no essential differences as compared with the preoperative test results. Around that time it was discovered that the patient had writing difficulties with his left (!) hand. Upon subsequent testing, several other unusual findings were found, based on an aphasic disturbance in naming:

> He named objects placed in the left hand (concealed from vision) incorrectly; he could select them afterwards with his left hand by touch or pointing; and he could draw the object afterwards with his left hand. Even while giving an incorrect verbal description, he could demonstrate correctly the use of the object being held in the left hand. If an object was placed in one hand (concealed from vision), he could not select it from a group or draw it with the other hand. He frequently performed verbal commands incorrectly with his left hand.
>
> The authors feel that the simplest explanation of the phenomena is that the patient behaved as if his two cerebral hemispheres were disconnected and that the probable cause of this was a lesion in the corpus callosum [5].

Kaplan and Geschwind reasoned that as a consequence of surgery, the anterior portion of the corpus callosum had been lesioned, thereby disconnecting left hemisphere language areas from the right hemisphere areas that controlled the left hand (this was later confirmed in post-mortem examination) [6]. They presented this case of tactile aphasia, as well as the one with alexia without agraphia, at a conference in 1961. A brief report of this meeting was published in *The New England Journal of Medicine* [7] and a full paper somewhat later in *Neurology* [8]. From that period on, Geschwind and collaborators further elaborated on the behavioural syndromes that resulted from localized brain lesions and in particular on the various 'syndromes of disconnection'. Geschwind is not only to be credited for these efforts, and as a founder and teacher of the new discipline of behavioural neurology, but also for 'resurrecting the German and French literature from the late nineteenth and early twentieth centuries that had been buried by neglect and misinterpretation' [5]. It is indeed striking to read in his work the many references to the historical literature and his tributes to its many pioneering authors. A genuine surprise can often be felt with him when he describes in his papers that many of his cases and syndromes that he initially thought were original and new had been so eloquently documented in the past:

> What was astonishing [Geschwind is now referring to the works of Dejerine and Liepmann on callosal disorders] was the fact that this work had been so grossly neglected. It was published in widely-read journals and received wide acclaim in its day. Indeed, it was discussed fully in Lange's (1936) article contained in the standard German neurological reference, the Bumke-Foerster Handbuch. Furthermore, Liepmann's results had been reported with Otto Maas, who was still alive in the 1950s and had been confirmed by a whole array of authors such as Kurt Goldstein, who himself described a callosal syndrome in 1908. Despite this fact, the students of Kurt Goldstein whom I met were generally unaware of this fact.
>
> I was again made aware not merely of how inaccurate most of the histories of the higher functions were, but also that important confirmed scientific observations could almost be expunged from the knowledge of contemporary scientists. My presumption is that this must occur in other fields as well. The reasons for this phenomenon were fairly standard: neglect of work written in a foreign language, neglect of work done by someone in a different field, excessive reliance on the authority of certain towering individual figures. [3]

Geschwind wrote several papers that directly aimed to correct some of the 'grossly incorrect views' on historical events and some of its key figures. For instance, in a lengthy paper, he pays tribute to Meynert and Wernicke, re-analysing their work and restating the importance of their contributions for modern neurology and neuroscience [9]. Geschwind was also keen to acknowledge some errors in Wernicke's earliest work, in particular regarding the concept of his conjunction aphasia [Leitungsaphasie]:

> There was, however, an even more important source of error which resulted from Wernicke having failed to analyze his own diagram correctly; he had omitted the deduction that there should be a disturbance of repetition in conduction aphasia. Lichtheim in 1885 correctly added the deduction and cited a patient showing this condition [10]. Freud returned to this problem in 1891 and, although he cited Lichtheim extensively, he seemed to have missed in part some of Lichtheim's extension of Wernicke's theory [11]. Freud argued, as Lichtheim had, that the lesion disconnecting the motor from the sensory speech area should produce a loss of repetition in the face of intact comprehension, and went on to remark that this situation is highly unlikely. Yet within the next twenty years the triad of paraphasia, intact comprehension and impaired repetition was to become well known. In 1904 Karl Kleist, then an assistant of Wernicke's at Halle, demonstrated a case which convinced Wernicke of the existence of this entity. [3]

Geschwind refined and expanded the older connectionist view into what some have called neo-associationism [12]. In 1961, Geschwind met Zangwill, who was at the time an associate editor of the journal *Brain*. Zangwill persuaded him to expand his ideas on callosal disconnections, and this led to Geschwind's landmark paper, 'Disconnexion syndromes in animal and man', which was published in *Brain* in 1965. This paper, which was essentially a monograph, spanned 116 unillustrated pages and was published in two parts for editorial reasons [13, 14]. The work starts with anatomical considerations on connecting pathways in the brain and in particular with Geschwind's description and 'rediscovery' of Flechsig's rule (1901). This rule states that 'primary receptive areas (…) have no direct neocortical connexions except with immediate adjacent, "parasensory" areas' [13, 15]. Thus, according to Flechsig, all cortical connections between primary sensory areas pass through surrounding areas of associative cortex, and there are no direct intra-hemispheric connections between these areas (there are—of course—direct connections to and from subcortical areas, e.g. via the thalamus). Geschwind generalized the rule to include also primary motor cortex and inter-hemispheric connections [12, 13].

Flechsig (1847–1929) was a German neuroanatomist who is remembered best for his work on myelination of the central nervous system. He had noted that different areas myelinated at different times during embryonic and postnatal development of the brain and 'introduced the 'fundamental law of myelinogenesis' that the sequence of myelination during individual development repeats their phylogenetic appearance' [16]. Flechsig categorized cerebral areas into three large groups, depending on their state of myelination. The 'primordial regions' were those areas that were already myelinated before birth; these areas largely correspond to the 'primary sensorimotor areas' (see also Fig. 7.1). The second group consisted of 'intermediate areas' that were myelinated approximately 1 month after birth. Finally, there were 'terminal areas' that myelinated even later than that. Flechsig was led by the idea that evolutionarily younger systems myelinated later than the phylogenetically older and more primitive systems. He linked this temporal differentiation in myelination to differences in brain functions. The slower the development of an area, the more 'higher' intellectual brain functions were performed by that area. Flechsig named these areas 'associative centres':

> Flechsig (1905) thought his posterior association center [located in the temporoparietal region] was the most important part of the brain for intellectual functions. He wrote that the posterior association center was responsible for connecting words with their content, understanding notations, forming intellectual conceptions of the external world, and grasping complex situations. In contrast, Flechsig emphasized the role of the frontal cortex in emotion and consciousness, but concluded that the frontal lobes were not the centers for abstract thinking, as Eduard Hitzig had proposed. [17]

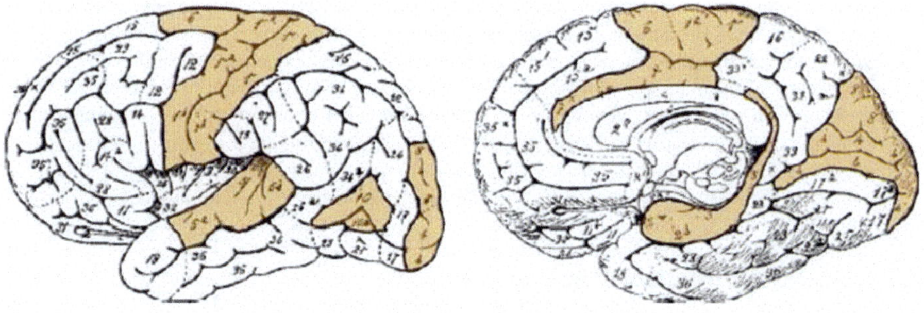

Fig. 7.1 Flechsig's myelogenetic map of the human cortex (1901). Flechsig constructed a brain map that initially consisted of 40 areas; later he reduced the total to 36. The coloured areas constitute the so-called primordial zones that are already myelinated at birth. Myelinization of the other areas is completed in later stages. Numbers reflect the chronological order of myelinization (Figure taken from Catani and ffytche, 2005 [12])

For Geschwind, Flechsig's rule and his forgotten work was an important part of his theories, whereby lesions to association cortex itself could lead to disconnection syndromes:

> These anatomical facts imply that a large lesion of the association areas around a primary sensory area will act to disconnect it from other parts of the neocortex. Thus, a 'disconnexion lesion' will be a large lesion either of association cortex or of the white matter leading from this association cortex. The specification of the association areas as way-stations between different parts of the neocortex is certainly too narrow, but it is at least not incorrect. This view, as we shall see, simplifies considerably the analysis of effects of lesions of these regions. Since a primary sensory region has no callosal connections, a lesion of association cortex may serve both to disconnect such an area from other regions in the same hemisphere and also to act in effect as a lesion of the callosal pathway from this primary sensory area. [13]

Geschwind's theories were also shaped by other evolutionary facts (see also Catani's excellent review *The rises and falls of disconnection syndromes* [12]). In subprimate animals, Flechsig's rule does not hold, and the primary sensory cortices of different modalities are connected either directly or via the limbic system [12]. As organisms ascend the evolutionary ladder, associative areas arise and these 'associative areas become separated to a great extent from the receptive'. Primary sensory areas become surrounded by associative cortical regions. In the monkey, associative cortices are connected to one another via the limbic system, and all primary cortices are thus indirectly connected via associative cortex and the limbic system. In man, according to Geschwind, a higher-order associative area developed in the inferior parietal lobe, and this area was able to integrate information from multiple modalities independent of the limbic system. It formed at the crossroads of visual, auditory and sensory areas. Geschwind calls this area the 'association area of association areas' (i.e. a secondary association area). He speculated that this area freed humans from the dominant pattern of sensory-limbic associations, and postulated that the evolutionary development of language was dependent on the emergence of this parietal association area (1965):

> The situation (...) is not simply a slightly more complex version of the situation in the higher primates but depends on the introduction of a new anatomical structure, the human inferior parietal lobule, which includes the angular and supra marginal gyri, to a rough approximation areas 39 and 40 of Brodmann. In keeping with the views of many anatomists Crosby et al. (1962) comment that these areas have not been recognized in the macaque. Critchley (1953), in his review of the anatomy of this region, says that even in the higher apes these areas are present only in rudimentary form. (...) In addition this area is one of the late myelinating regions or 'terminal zones' as Flechsig termed them. In fact, this region was, in Flechsig's map, one of the last three to myelinate. (...) Yakovlev (personal communication) has pointed out that this region matures cytoarchitectonically very late, often in late childhood. [13]

7.1.1 Neo-connectionism

Geschwind's language model was essentially a modification of that of Wernicke and Lichtheim; hence, it is now usually referred to as the Wernicke–(Lichtheim–) Geschwind model of language [18, 19]. Geschwind built on the traditional approach whereby words are transferred from one region to another, the regions being interconnected as functionally distinct modules. It is of note that the only connection that is anatomically specified is the arcuate fasciculus, although even of this pathway, the exact course and terminations are not given. Geschwind agreed with Wernicke—and later Konorski (1961), who revived interest in conduction aphasia—that conduction aphasia was best explained by a lesion of the arcuate fasciculus [13]. Geschwind added the angular region as a centre for 'visual word memory'—or 'more correctly'—the region of the temporoparietal-occipital junction. Within this region, written language is turned into spoken language and vice versa. It contains the 'rules for associating stimuli in two modalities, i.e. visual and auditory' [3]:

> Thus a visual stimulation can evoke an auditory association by means of the pathway: visual cortex—visual association cortex—angular gyrus—auditory association cortex. The auditory association cortex shown here [Fig. 7.2, top left] is the classical Wernicke's area.

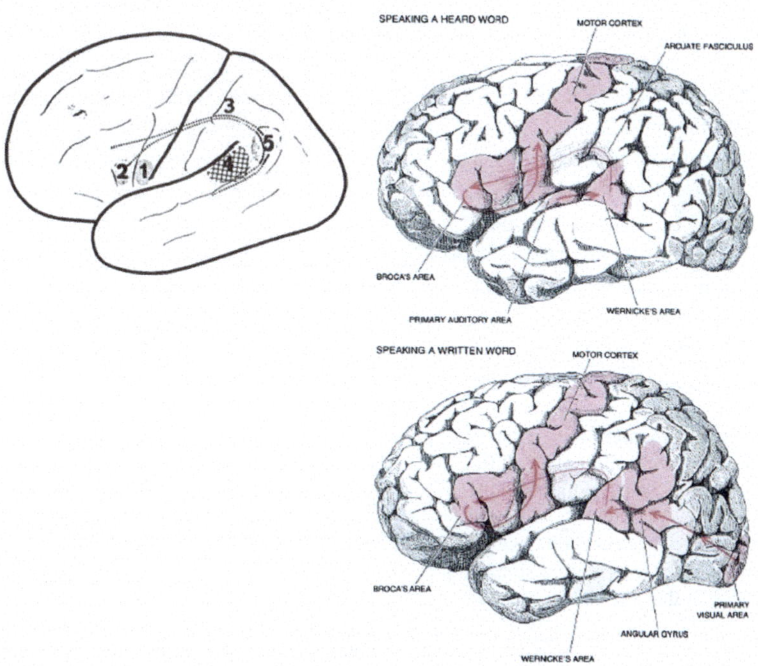

Fig. 7.2 Some of the schemes that Geschwind (1969) used to illustrate his ideas of the organization of auditory and visual language functions in the brain [3]. (*Left*) (*1*) face area of motor cortex, (*2*) Broca's area, (*3*) lesion involving arcuate fasciculus (*dotted lines*), (*4*) Wernicke's area and (*5*) angular gyrus. (Figure taken from Geschwind, 1974 [3]). (*Right*) 'Linguistic competence requires the cooperation of several areas of the cortex. When a word is heard (*upper diagram*), the sensation from the ears is received by the primary auditory cortex, but the word cannot be understood until the signal has been processed in Wernicke's area nearby. If the word is to be spoken, some representations of it is thought to be transmitted from Wernicke's to Broca's area, through a bundle of nerve fibers called the arcuate fasciculus. In Broca's area the word evokes a detailed program of articulation, which is supplied to the face area of the motor cortex. The motor cortex in turn drives the muscles of the lips, the tongue, the larynx and so on. When a written word is read (*lower diagram*), the sensation is first registered by the primary visual cortex. It is then thought to be relayed to the angular gyrus, which associates the visual form of the word with the corresponding auditory patterns in Wernicke's area. Speaking the word then draws on the same systems of neurons as before' (Figure and text taken from Geschwind, 1979 [128])

> Broca's region is a region of motor association cortex lying anterior to the face region of the classical motor cortex. It may be thought of as containing the learned rules for translating a particular heard sound pattern into a motor sequence. [3]

The model permits certain new predictions, for instance, that lesions in Wernicke's area will also lead to problems with written language, as seen words cannot longer arouse their auditory forms. However, this was, and still is, a matter of debate:

> Some of the classical authors were inclined to feel that the sight of an object could directly arouse the written production. (…) But what makes this supposition unlikely is that writing is invariably abnormal in patients with the speech pattern of Wernicke's aphasia, while one might expect it to be spared at least occasionally if this were the case. Hence it appears that to write the word, the spoken form must be aroused first. [3]

Geschwind realized that his model had limitations and that not all aphasic phenomena could be adequately explained by it. This is extensively discussed in his work, and—when appropriate—possible solutions are suggested. For instance, in case of writing disturbances, Geschwind hypothesized a separate pathway from the angular gyrus to Broca's area that 'runs forward and mixes in the lower parietal lobe with the fibers from Wernicke's area' [3]. Despite the shortcomings, Geschwind considered the anatomical approach the best way to explain the known data, as well as a good starting point for experimental testing of new hypotheses. Of course, Geschwind knew that reality was far more complex than his reductionist model, a fact he kept repeating in his work:

> By far the most common types of criticism of the lesion method have been based on (a) the inappropriate attribution of a lost function to the locus of a lesion, and (b) exceptions to and inconsistencies of anatomical localization. Only an uninformed or naive thinker can believe that Wernicke regarded the area which came to carry his name as the site where the full process of auditory comprehension took place; yet, Wernicke and many of his most important followers have been accused of precisely that belief. No effort should be spared to make it clear that the effects of a circumscribed locus of damage can be understood only by taking into account that the healthy brain tissue that has been destroyed was a component of a neural network, the activity of which mediates the normal function; and furthermore, that to understand post-lesional alterations in function one must take into account changes in other parts of that network.
>
> The argument regarding exceptions and inconsistencies in localization depends, as usually stated, on invalid assumptions. No one can disagree with the assertion that exceptions or inconsistencies demand attention. In some instances they lead to the realization that the asserted localization is incorrect, and thus serve the same function that discrepant data serve in every branch of science. All to often, however, the argument is advanced that the existence of discrepancies shows that the very concept of localization of function is untenable. Implicit in this argument is the belief that a particular localization is acceptable only if it is universally valid. (…) It is clear that, in order to establish a valid correlation between disordered behavior and the site of anatomical brain damage related to it, it is necessary to take into account a large number of factors, the principal of which are: (a) the nature of the pathological process, (b) the size of the lesion, (c) the speed of the pathological process, (d) the timing of anatomical and behavioral observations, (e) individual variations in neurological organization, (f) the age, level of education, sex, and premorbid psychological and social factors of the individual under study. [20]

Geschwind and Wernicke shared a common opinion on many issues. Geschwind regarded the loss of Wernicke's area as the destruction of a memory store, 'as it was in fact regarded classically' [13]. He named Wernicke's area 'the storehouse of auditory associations' [13]. When a name passes through Wernicke's area and the angular gyrus, it will arouse associations in the other parts of the brain. 'It is probably thus that Wernicke's area attains its essential importance in "comprehension", i.e. the arousal of associations'. Geschwind and Wernicke also shared a similar critical view on agnosia and its interrelationship with aphasic disturbances, stressing that these are not two distinct physiological phenomena (see also Chap. 3). Geschwind held the opinion that the contemporary criteria that had been formulated to classify abnormal behaviour into one or the other category were inadequate:

> The fundamental difficulty has been in the acceptance of a special class of deficits of 'recognition', lying somewhere between defects of 'perception' and 'naming'. What indeed are the criteria for 'recognition' and is it a single function? I believe in fact that there is no single faculty of 'recognition' but that the term covers the totality of all the associations aroused by any object. (…) this view abolishes the notion of a unitary step of 'recognition'; instead, there are multiple parallel processes of appropriate response to a stimulus. [14]

Geschwind's neoclassical approach in the 1960s became internationally known, and Wernicke's classification was to some extent 'repackaged as the Boston classification' [1]. The Boston School became renowned for its research on aphasia and for a test battery that was developed by Goodglass and Kaplan, the Boston Diagnostic Aphasia Examination [21]. This battery, developed at a time when brain imaging was still in its infancy, aimed to localize structural lesion from impaired functions [1]. Until today, Geschwind's legacy remains a prominent part of our clinical approach to patients with aphasia.

7.2 Luria

Alexandr Romanovitsj Luria (1902–1977) was a Soviet psychologist whose work had a great influence on (Western) modern neuropsychology and cognitive neuroscience. He was one of the leading psychologists of the twentieth century and is considered one of the founders of modern neuropsychology and cognitive neuroscience. His work covered many different topics and has been characterized by adjectives such as *dynamic*, *functional* and *evolutionary* [22].

Luria graduated from university in 1921, aged just 19. While still a student, he established the Kazan Psychoanalytic Association and planned a career as a psychologist [23]. Initially, he worked on methods that could objectively measure abnormal mental processes, based on theories of Jung and Freud. At the time, Luria's intellectual life was heavily influenced by German writings. German was the family's second language, and Luria had learned it at an early age. He was therefore able to read German literature long before it would be available in Russian translation [23].

In 1924, Luria met Lev Vygotsky (1896–1934), who would become his lifelong teacher and friend. In Vygotsky's view, higher mental processes are predominantly the result of a person's complex social–historical development. 'They are formed under the influence of people's concrete activity in the process of their communication with each other and in fact always represent complex functional systems based on jointly working zones in the brain' [24]. Luria based his work on this more philosophical approach of Vygotsky, attributing a large role to a person's environment as a determinant of his or her individual behaviour. He thereby maintained a more comprehensive view than would have been possible if he had endorsed either the psychic or physical reductionist positions [25]. Later in life, Luria also embarked on a career in neurology and was able to link theories of higher mental functions and language to anatomical and clinical evidence of brain dysfunction. Throughout his whole career, he tried to fuse elements of the biological and the environmental approach, as for him the study of the individual brain alone was not sufficient to reveal the organization of complex behaviour [26]. In addition to that, Luria stated in his autobiography:

> The time was long past to consider psychological processes as the results of either strictly localized brain activities or the 'mass action' of the brain in which all of its parts were 'equipotential'. It was time for us to begin the next step in our work: to explain the neurophysiological or, to use a Russian phrase, 'neurodynamic' mechanisms underlying the activity of brain loci implicated in specific syndromes. (…) Progress depended upon advances in both of the areas that had concerned me all my life. On the one hand, I had to move from brain structures to a deeper understanding of the neurophysiological mechanisms that were operating in these structures. On the other hand, our psychological analysis of higher cortical functions was by no means complete, and we needed improved psychological analyses as well. To signify the combination of these two enterprises, the 'neurological' and the 'psychological', the term neuropsychology was coined. Developing this field of science has taken a long time and the help of many people [26].

For a large part of his professional life, Luria worked in Moscow's Burdenko Institute of Neurosurgery. Intracranial cases from all over Russia were referred to this large (300-bed) institute. After the outbreak of World War II, it became in fact an enormous testing ground for new neuropsychological approaches, as Luria organized a rehabilitation hospital for the massive numbers of brain-injured patients. This was fertile ground to develop new techniques to diagnose and localize brain lesions and to restore lost neurological functions (Fig. 7.3) [26]. Besides the huge number of cases available, it was of great advantage that the location of the lesion was often known because of surgical findings or post-mortem examinations. One of Luria's many books, *Traumatic Aphasia: Its Syndromes, Psychology and Treatment*, resulted from these experiences and was published in 1947 [27]. Unfortunately, it did not appear in English translation until as late as 1970 [28]. Geschwind, in a review of this book, acknowledged its great importance and called Luria 'one of the modern giants in the field':

> Indeed, some of Luria's original contributions, as expressed in his other writings over the past twenty years, have become such integral components of thought in this area, and have appeared so frequently in the writings of others—who often fail to realize their origin—that we are likely to underestimate the statue of this book. At the time of its publication in 1947, it was undoubtedly the most important work on aphasia since the close of the classical period in the 1920s. Had it been translated into English at the time, it might have hastened

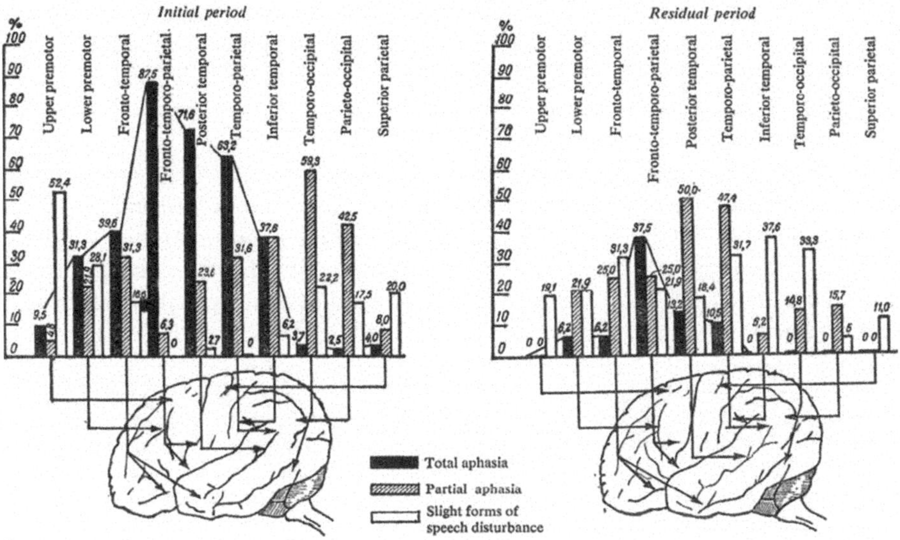

Fig. 7.3 Luria (1947) systematically documented traumatic aphasia in war casualties [27]. The diagrams show the distribution of speech disturbances from gunshot wounds to different parts of the brain in approximately 800 patients. The vertical axis indicates the percentage of speech disorders (of various degrees of severity) as a function of the particular area of the left hemisphere that was damaged. Note that Luria studied patients for some time after the initial brain injury in order to capture the effects of recovery (Figure and text (adapted) from Luria (1970) [28])

the modern revival of interest in this area in France, Germany and the English-speaking countries, where knowledge of Russian and even sensitivity to the possibility of important Soviet contributions were unfortunately generally lacking [22].

7.2.1 Functional Systems

Luria regarded 'functions' as complex functional systems with dynamic levels of localization in the brain. He defined a function as a complex form of goal-directed adaptive behaviour and characterized functions by the variability of the participating systems. Take for instance locomotion, which can be achieved by different sets of muscles; or writing, which can be achieved both by mouth or foot if required by circumstances [29]. Initial ideas on functional systems in Russia had been formulated by Anokhin [30] and Bernstein [31] and were used by Luria (1973):

> Of utmost importance is the fact that all higher behavioral processes (or 'higher cortical functions') are really such complex functional systems, which are based on the coordinated functions (or 'constellations') of cerebral zones constructed in such a way that separate links of this system can be interchanged and that such a change does not affect the whole functional system. This approach is associated with a radical revision of the whole problem of 'cerebral localization of functions'. We do not start with any attempts to 'localize' a complex function in a limited part of the brain; rather we try to discover how a 'functional system' is distributed in different parts of the brain, and the role that every part of the brain plays in the realization of the whole 'functional system'. [29]

Luria thus never wholly associated functions with particular, isolated areas of the cerebral cortex, a view comparable to that of Wernicke, although he accepted the fact that different regions participated in different functions (he was therefore not a *holist*). Luria assumed that restoration of functions led to reorganization of a new and widely dispersed dynamic system and that it did not take place solely by transfer of function to equipotential brain areas. For details of his work and his many ideas and theories, the reader is referred to Luria's autobiography [26] and the series of books he published, in particular *Higher Cortical Functions in Man* [32]. Luria had an enormous capacity for work and left hundreds of publications and dozens of books; he often had several jobs at once and was typically engaged in multiple scientific projects. In Luria's obituary, Critchley (1978) described his visit to the Burdenko Institute:

> To work as a visitor in his Burdenko Institute was a heady but exasperating experience. Oblivious to all sense of time, Luria seemed immune to such frailties as hunger and fatigue. He would move on from patient to patient, coping with one brash interruption after another. Discipline orderliness and regimentation were not there. His working conditions, let us face it, were astounding. Among the plethora of patients with brain trauma, Luria's researches were carried out in an office which was more like a busy airport than the sanctum of a reflective scholar. Doors to the right and left of him opened and shut. Bells rang. All and sundry—nurses, technicians, porters, assistants—seemed to use his study like a public highway. Privacy there was none. Luria's careful case studies were impervious to noise and distraction. [33]

In addition to his scientific, classical writings, Luria was also committed to what he himself called 'romantic science'. This culminated in two books which Luria termed 'biographical studies'—in fact, extensive case histories of men with 'abnormal' brain functions. One, *the Mnemonist*, is about Shereshevsky, a man who had an innate talent of remembering almost anything that he had experienced. He was able to recall near-infinite lists of words or sounds and repeat them years later. His synaesthetic memory (he described his recall as a journey through a remembered landscape) seemed to have no limits to its capacity nor durability. Characteristic of Luria is that he not only (thoroughly) researched Shereshevsky in a classical psychological manner, by measuring and testing his behaviour, but that he was also just as interested in how Shereshevsky's strange condition affected his personality and his life. His brain seemed constantly involved in associations; those of new events and of those that he had earlier picked up in the outside world. These cerebral processes were virtually unstoppable and interfered with other brain functions, creating a downside to his incredible mnemonic talents. Shereshevsky had problems with more abstract concepts and personal relationships, and in the end, his 'hypertrophied' memory was probably more a curse than a gift. Remarkably, his memory also affected his behaviour in a more physical way:

> He was able to control his involuntary processes, such as his heart rate and the temperature of his body, in the same way that a yogi does. A clear image of himself running fast increased his pulse rate. An image of a piece of ice on his hand decreased the temperature

> of his hand. And an image of his hand holding a glass of hot water increased his skin temperature. By this process he could increase or decrease the temperature of his hands by 5 degrees. [26]

The other book, *The Man with the Shattered World*, is about Zazetski, who as a young soldier was struck by a bullet in the left parieto-occipital region and was left with a myriad of neurological and psychological problems. Despite his visual disturbances, his amnesia and aphasia, Zazetski kept writing whenever he could, struggling for every word and sentence. He did so in an attempt to reconstruct his lost life and to recover the use of his damaged brain. Over a period of 20 years, Zazetski produced 3000 pages and this activity gave him, according to Luria, a reason to live. Luria edited these writings and added observations during the period of almost three decades in which he followed Zazetski in his clinic:

> This book describes the damage done to a man's life by a bullet that penetrated his brain. Although he made every conceivable effort to recover his past, and thereby have some chance of a future, the odds were overwhelmingly against him. Yet I think there is a sense in which he may be said to have triumphed. It is not false modesty on my part to wish no credit for this book. The real author is its hero.[a]

In letters to Oliver Sacks, who was a great admirer of his work and himself a master of 'neurological novels', Luria wrote (1973):

> Frankly said, I myself like very much the type of 'biographical' study, such as Shereshevsky (the Mnemonist) and Zazetski … firstly because it is a kind of 'Romantic Science' which I wanted to introduce, partly because I am strongly against a formal statistical approach and for a quantitative study of personality, for every attempt to find factors underlying the structure of personality … only the style of these two books is different from the others; the principle remains the same. (…) I was ever conscious and sure that a good clinical description of cases plays a leading role in medicine, especially in Neurology and Psychiatry. Unfortunately, the ability to describe, which was so common to the great Neurologists and Psychiatrists of the 19th century … is almost lost now. [34]

For Luria, these case histories were as important as his scientific work. It shows his conviction that normal brain function is impossible without the integrity of the whole brain and the whole 'personality'. To study and describe the nature of in particular the higher cognitive functions, Luria committed himself to two strategies:

> The first was to trace their development; the second was to follow the course of their dissolution under conditions of local brain damage. In the mid-1920s Vigotsky first suggested that an investigation of localized brain damage could provide a way to analyze the cerebral structure and development of higher psychological processes. At that time, neither the structure of the higher psychological processes themselves nor the functional organization of the brain was clear. [26]

---

[a] Foreword by Luria to *The Man with a Shattered World* [34]

7.2.2 Aphasia

Around 1937, Luria turned his attention to 'the tangled knot of disorders that were and still are referred to under the general rubric "aphasia"' [26]. He began studying the three classes of aphasia that were at the time generally recognized: sensory, motor and semantic (or amnestic) aphasia. Luria used Pavlov's fundamental idea of 'nuclear zones' and 'systems of cortical analysers' (see Fig. 7.4). Pavlov defined the nuclear zone as that part of the cortex that deals with the most precise differentiation and the most complex integration of special stimuli [26]. Within the centre of a zone lies the primary analyser. Examples are BA 17 in the visual zone, BA 21 in the auditory zone and BA 3 in the cutaneous–kinaesthetic zone. Primary analysers have a well-defined somatotopic projection for information that enters the cortex from the peripheral receptors, that is, proportional to their physiological importance (not their size). It corresponds to our current concepts of a primary cortical area. Luria noted that even

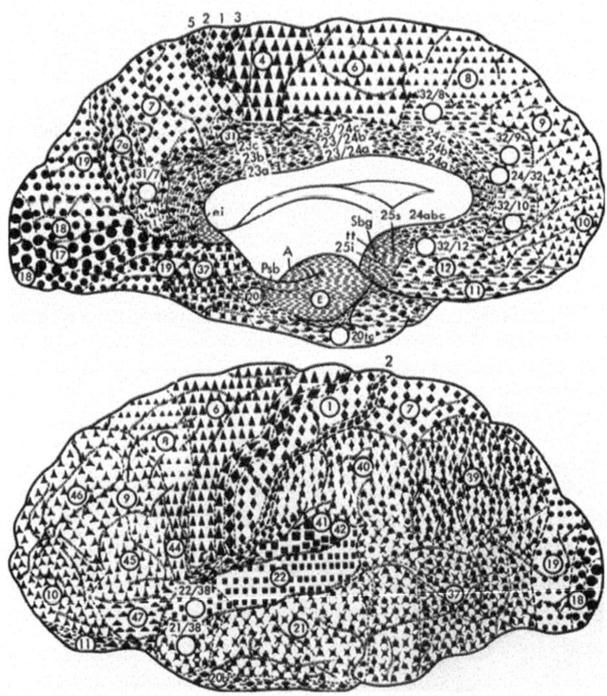

Fig. 7.4 Diagram showing the interrelationship among fields of nuclear zones and analyser systems [129]. The nuclear zones (after Pavlov) consist of primary, secondary and tertiary analysers that deal with information that is subsequently more complex and multimodal [32]. These zones are denoted by *circles* (visual zone), *squares* (auditory zone), *rhombi* (general sensory zone) and *triangles* (motor zone); central fields within a nuclear zone are demarcated by larger symbols (note that numbers correspond to Brodmann areas). 'The speech-motor (fields 44 and 45) and speech-auditory (back part of field 22) sections are identified by a slightly different shape of signs selected for the corresponding nuclear zones' [129] (Figures and text from *Neuron Structure of the Brain* by Poliakov (1961, 1974) [32])

these primary cortical areas already have a dynamic and not a static organization. He referred to the earlier investigations of Foerster and of Sherrington, whereby electrical stimulation of the same cortical point could lead to different functional responses. Luria also stressed that primary analysers already participated in the analysis of information and were not simply a representation of the peripheral stimuli:

> From the very beginning the sensory cortical divisions participate in the analysis and integration of complex, not elementary, signals. The units of any sensory process (including hearing) are not only acts of reception of individual signals, measurable in terms of thresholds of sensation, but also acts of complex analysis and integration of signals, measurable in units of comparison and discrimination. The sensory divisions of the cortex are the apparatuses responsible for this analysis, and indications of a lesion of these apparatuses are to be found, not so much in a lowering of the acuity of the sensations, as in a disturbance of the analytic-synthetic function. [32]

Secondary and tertiary zones were considered by Luria as 'specifically human parts of hemispheres': lesions within these areas do not generally result 'in any elementary sensory or motor defects and remain inaccessible for classical neurological examination' [29]. Secondary analysers occupy the periphery of the nuclear zones (e.g. the secondary auditory analyser is located in BAs 41, 42 and 22). Their cellular formation in the second and third layers of the cortex is more complex, allowing for more complicated operations [25]. Fibres from the thalamus also arrive in the secondary auditory zones, belonging to 'the internal portion of the vertical connections transmitting impulses that have already been analysed and integrated' [32]. Tertiary analysers are less modality specific and integrate information from different analysers.

To this cytoarchitectonic background, Luria added other anatomical studies, for instance, that of neuronographic investigations that showed connections of the secondary auditory zones in the temporal lobe to the inferior premotor cortex and frontal regions (Fig. 7.5). These connections established a functional network of

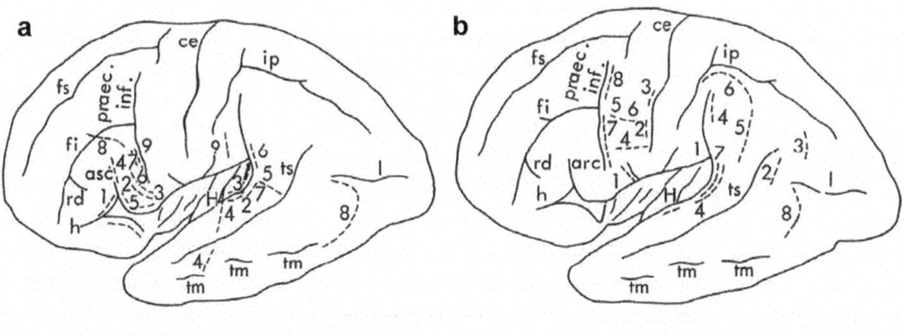

Fig. 7.5 Schematic drawings of the brain showing connections between temporal and inferior frontal regions. Connections of each bundle of fibres with the cortex are denoted by identical numbers on the surface of the corresponding parts of the brain: (**a**) connections between the temporal lobe and inferior frontal gyrus and (**b**) connections between the temporal lobe and anterocentral gyrus (Luria reproduced this figure from Blinkov (1955) [130] and used it in his book *Higher Cortical Functions in Man* [32])

anatomically distinct areas. 'In this way, conditions are created for the auditory and motor analysers, especially those parts of the latter concerned with the innervation of the vocal organs, to work together' [32].

Luria once defined language as 'the culturally determined syntactic systematization of signs and/or symbols' [25]. His ideas on speech and language originated from several different scientific backgrounds. First, there was the obvious—but often forgotten—notion that language is important in communication and social interaction. This was the Vygotskian influence. In addition to that, Luria based his theories on neuroanatomical information and his own work in brain-damaged patients. A last important influence came from linguistics, which he studied at length in order to understand better the neural basis of language-related behaviour:

> The two years I spent in the study of linguistics early in my career stood me in good stead when I began to work seriously on the problem of semantic aphasia because I could understand more fully the different mental requirements that seemingly similar linguistic acts placed on people. I was thus in a better position to carry out differential diagnosis of pathological symptoms which previously had been lumped together in the neurological literature. As my work continued to involve me in attempts to understand the brain basis of language-related behavior, I found it necessary to continue to study the psychology of language at the same time that I searched for its neurological bases. And just as advances in neurology and neurophysiology were instrumental to our study of brain mechanisms, advances in the study of linguistics were crucial to advancing our understanding of those phenomena of speech which brain pathology was interrupting; the two enterprises are inextricably bound together. Time and time again I found myself returning to old data, armed with new insights from advances in linguistics. [26]

Luria divided aphasia into general categories according to the subsystem of spoken language that is disturbed: phonetic, articulatory, semantic or dynamic. Vocate, in her book on Luria's language research, remarks that these distinctions are predominantly for academic convenience as in practice they are much more difficult to distinguish because of 'interdependent components of the functional systems of speech' [25]. Both phonemic and articulatory aphasia are each subdivided into two types of aphasia, listing a total of six different syndromes. Luria related each of these aphasias to different regions of the brain and considered them neurologically different (Fig. 7.7). These different types of aphasia will be reviewed briefly below.

7.2.2.1 Phonemic (Sensory) Aphasia

For Luria, the basic difficulty underlying the symptoms of patients with classic sensory aphasia was 'the inability to discriminate the distinctive features of phonemes, the basic units of word sounds' [26]. Thus, in patients with a temporal lobe lesion, the difficulties in understanding words, naming objects, retrieving words or writing were secondary consequences of the primary defect in phonemic hearing [26]. Luria interpreted this disturbance of discriminative hearing as a 'disturbance in the analytic-synthetic activity of the auditory cortex' and regarded it as a fundamental symptom of a lesion in the posterior third of the left superior temporal gyrus (Fig. 7.6). He considered the resulting acoustic agnosia as the fundamental source of the speech disturbance, whereby disturbances could be divided into phonemic errors and conceptual errors [32]. Clinically, a patient with temporal aphasia

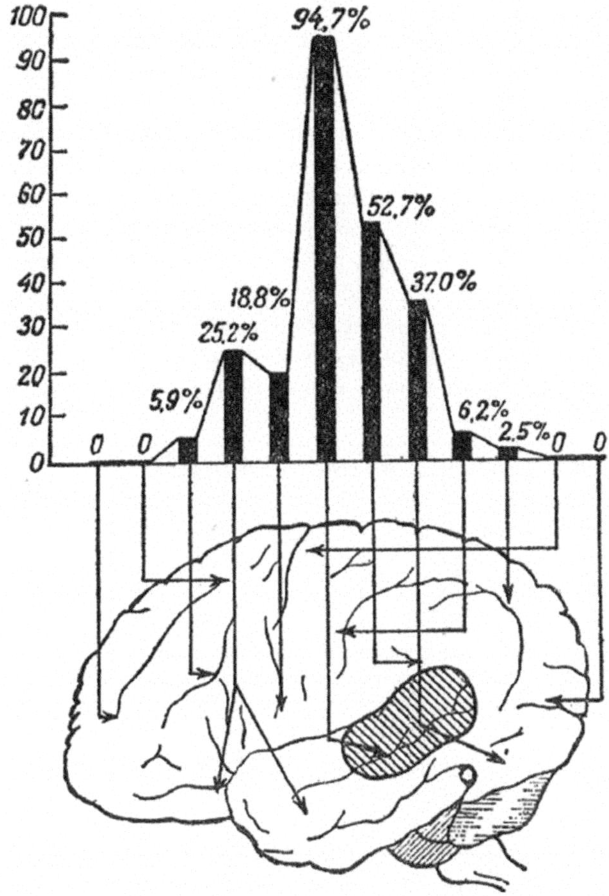

Fig. 7.6 Schematic diagram of the distribution of cases (percentage) with impaired phonemic hearing as a result of traumatic lesions in different areas of the left hemispheres (Figure from Luria (1970) [28])

typically has difficulty pronouncing a word that is heard, resulting in paraphasias (although paraphasias are not specific for the disorder). Disturbances of the motor aspect of speech may also be present. As the phonemic (sound) structure of speech is disturbed, it is 'quite natural', according to Luria, that the system of word meaning must also be disordered [32]. This was the second characteristic of sensory aphasia. 'Often, only diffuse conceptual associations of the original meaning of the word remain, arising from the individual fragments of the sound complex' [32]. Consequently, verbal comprehension will be impaired, the classic hallmark of sensory aphasia. The third essential part of temporal aphasia is disturbance in speech memory, invariably accompanying lesions in the superior temporal region:

> A disturbance of simple differentiation of audio verbal complexes, which arises as an inevitable result of temporal aphasia, causes words or groups of sounds similar in their acoustic composition to be no longer clearly distinguished, so that a person who normally could easily find the required name can no longer do so. Irrelevant acoustic complexes, which under normal conditions would have a very small chance of appearing during the perception of that particular object or the appearance of a given pattern, now appear just as

probably as the appropriate words; the number of different acoustic complexes springing up in such a patient may often be so great that he is unable to pick out the required name, and the selectivity of the process of word finding is lost. [32]

Luria distinguished another type of temporal aphasia, namely, acoustic–amnestic aphasia. Such disturbances are the result of a lesion of the middle segment of the temporal lobe (i.e. the extranuclear divisions of the auditory analyser, BA 21 and 37). In contrast to the abovementioned acoustic–gnostic aphasia, in these cases phonemic hearing is largely preserved. Patients exhibit marked disturbances when they attempt to remember and reproduce words given to them orally (i.e. disturbances in verbal memory). They also cannot reproduce them in the proper order. To Luria, this 'disturbance of the order of verbal traces is evidence of weakness of the tracer function of the cortex in this region' [32].

7.2.2.2 Articulatory (Motor) Aphasia

A hallmark of motor aphasia is a disturbance in articulation. Again, this form of aphasia was not seen as a single syndrome by Luria, and he differentiated two main types. Luria referred to Bernstein, whose work demonstrated that correct performance of any motor act requires both efferent and afferent inputs:

By distinguishing between two components essential for the performance of a motor act—its kinesthetic basis, providing the differential composition of complex movements, and its kinetic structure, responsible for the formation of smooth skilled movements in easy consecutive order—a new approach in the analysis of the motor aspect of speech can be adopted and two different components (or two different forms) of motor aphasia can be described. [32]

The first type is afferent or kinaesthetic motor aphasia. This can be found in patients 'with lesions in the lower parts of the posterior (kinaesthetic) parts of the left hemisphere' [29]. In his book *Language and Cognition*, Luria specified this further to 'the postcentral region of the speech region (Operculum Rolandi)' [35]. The principle component is mispronunciation of individual speech sounds, which Luria called 'articulemes'. The reason for this is that patients do not register feedback from the movements that are made to produce the articulemes. Therefore, movements lose selectivity and patients cannot assume the correct positions of the tongue and lips. These clinical features, Luria remarked, closely resemble a condition that had been described by German and French neurologists [29, 36, 37]. The second type of motor aphasia is efferent or kinetic motor aphasia. In these patients, articulation may be intact, but there are problems with the transition from one articuleme to another. Their 'kinetic melody' is disturbed. These patients have lesions within the lower part of the left premotor cortex, the classic Broca's area.

7.2.2.3 Semantic (Amnestic) Aphasia

In this form of aphasia, patients have a good understanding of the meaning of individual words, but they cannot grasp the meaning of more complex ideas. Characteristic is disturbances of logico-grammatical relations, whereas there are rarely signs of disturbed articulation or phonemic hearing (i.e. no sensory or motor deficits). 'Such patients find it almost impossible to understand phrases and words which denote relative position and cannot carry out simple instructions such as "draw a triangle above a circle"'. Head had described in 1926 that there were accompanying difficulties in the perception of complex spatial relations. The syndrome is caused by lesions that extend within the temporoparietal-occipital ('tertiary') area. That is why, according to Luria, 'the syndrome of "semantic aphasia" includes, as a rule, deterioration of orientation in space, constructive apraxia, and defects in computation' [29].

7.2.2.4 Dynamic Aphasia

In these patients spontaneous and original speech is severely reduced, while naming, repetition of words or even simple sentences is preserved. When required to tell a story, patients complain of an 'emptiness in the head' [28]. There is doubt to what extent syntactic and grammatical processes are affected, as Luria only provided qualitative descriptions of patients with dynamic aphasia [38]. Luria referred to pioneering investigators who had described this syndrome, notable Lichtheim's transcortical motor aphasia [29]. The lesion is located anterior to Broca's area, in the 'tertiary zone'. Current opinion does not agree on any specific frontal region nor on its underlying mechanism [38, 39].

Figure 7.7 shows an abstraction of the localization of the different types of aphasia that were considered by Luria. Graphically, at least, the model has similarities to the Wernicke–Lichtheim model. However, there are crucial differences, as formulated by Tesak and Code (2008):

> First, he [Luria] emphasizes the individual processes (analysis, synthesis, integration) engaged in language; his is a process model. Second, the possibility of aphasic symptoms being connected at different linguistic levels on the basis of abstract principles is implied in Luria's work. The disturbance of the linear scheme, which shows itself in sound production, sentence production and in writing is an example. Third, Luria's process model provides routes for the formulation of strategies for rehabilitation, because the model is flexible and dynamic in contrast to the static classical model and because the brain is conceptualized overall as an interactive system. [1]

Luria certainly was ahead of his time by anticipating that clinical, cognitive and anatomical studies need to be integrated to enhance our understanding of

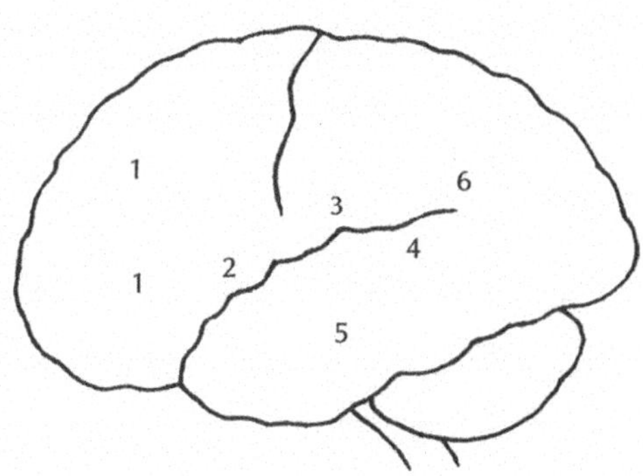

Fig. 7.7 Luria's classification of the location of lesions that produce different types of aphasias. (*1*) Dynamic aphasia, (*2*) efferent motor aphasia, (*3*) afferent motor aphasia, (*4*) sensory aphasia, (*5*) acoustic–amnestic aphasia and (*6*) semantic aphasia (Figure from Kagan and Saling [131])

brain–behaviour relationships. Even now, his work is of much inspiration and importance [40]. Luria's theoretical concepts of distributed and dynamic functional systems, for instance, remain highly relevant in our era of modern neuroimaging. Although our new techniques are powerful and provide us with detailed maps of structural and functional connectivity within individuals, their use often seems more data-driven than guided by predictive theoretical models.

7.3 Computational Models and Parallel Processing

At the end of the nineteenth century, Cajal had postulated that information could be dynamically stored in the brain by anatomical changes in connections between neurons. For a long time, this hypothesis could not be tested, if it was ever taken seriously by scientists. However, this changed in the late 1940s with the introduction of intracellular micro-electrode recordings and electron microscopy. Synaptic activity from individual neurons could now be measured and the structure of synapses visualized. This led to the discovery of synaptic plasticity and the formulation of a number of rules whereby synaptic connections could be modified [41]. The most well known is the Hebbian learning rule, named after Donald Hebb (1904–1985), who proposed that the strength of a connection between two neurons increases when presynaptic and postsynaptic neuronal firing is synchronized in time. His rule is usually more popularly phrased as 'neurons that fire together, wire together' [41]. The first instances of this rule were discovered in kindling (1969) [42] and long-term potentiation (1973) [43]. Later, with the development of the computer, it became the most basic learning algorithm for adjusting connection weights in artificial neural network models (see Fig. 7.8 for an example). Hebb's work had a great influence on the development of neuropsychology and neuroscience. He collaborated, among others, with Lasley and Penfield and wrote an influential book, *The Organization of Behavior* (1949), where 'he outlined an entirely new way of relating brain and behaviour, based on the conviction, not then prevalent among psychologists, that the only scientific way to explore behaviour was in terms of brain function' [44].

With the introduction of computing and simulations of neural functioning, the question arose why people are so much smarter than machines. Computers are faster and more precise and also have more processing power. Here is a quote from one of the earlier books on computer algorithms and cognition, by McClelland and Rumelhard (1986):

> Yet people are far better at perceiving objects in natural scenes and noting their relations, at understanding language and retrieving contextually appropriate information from memory, at making plans and carrying out contextually appropriate actions, and at a wide range of other natural cognitive tasks. People are also far better at learning to do these things more accurately and fluently through processing experience. [45]

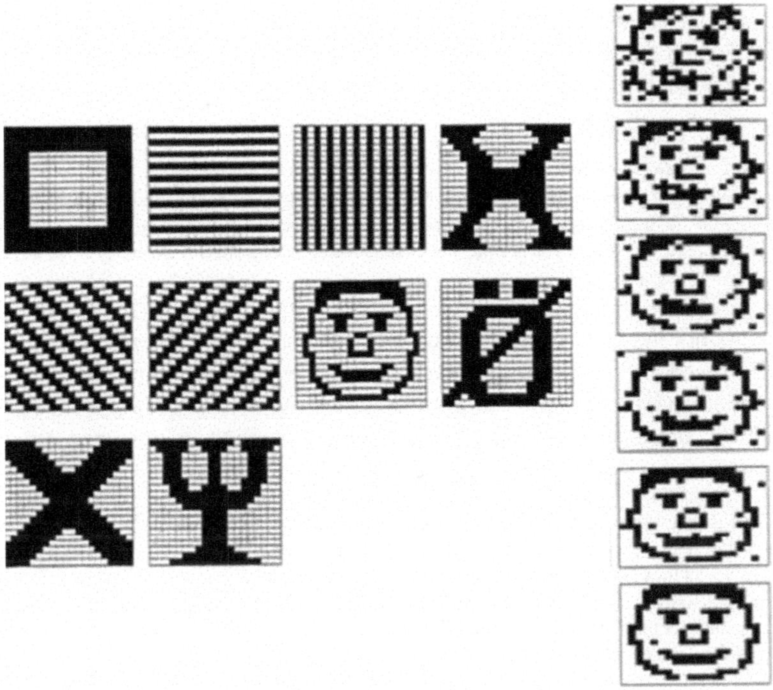

Fig. 7.8 Hebb was one of the first researchers to propose that information in the brain can be modified via synaptic transmissions in neural networks, although quite similar ideas on this mechanism for plasticity had been formulated much earlier by Wernicke, James and Cajal [132–134]. Dynamic neural networks have several interesting properties that are also thought to be fundamental underlying principles for brain function. Their continuous modifications allow for storage, optimization and automatization of information processing and for adaptation to structural changes (e.g. normal ageing or a brain tumour). An important property is that information is stored diffusely and non-locally, which makes these networks very robust. Also, neural networks process information in a reciprocal and parallel manner, whereby some networks are able to learn without the need for explicit rules (unsupervised learning). The mathematical Hopfield model for associative memory can be used as a simple metaphor to illustrate some of the properties of neural networks [135]. In this particular example, there are 400 binary neurons that are all interconnected (i.e. 400 by 399 connections) [48]. Each neuron is either in an 'active' state or 'in rest' (visualized, respectively, with a white or a black voxel). The state of every neuron is continuously updated and is determined by its local input, that is, the mathematical sum of all other neurons modulated by the strength of the (synaptic) connection. (*Left figure*) A learning rule is used to store information in the network. Hopfield used Hebb's rule, where the strength of the connection between neurons increases when the two neurons are in the same state and decreases when they are in different states [132]. In this example, the network first 'learned' ten different patterns. (*Right figure*) In a simulation experiment, one of the patterns was then mutilated by randomly changing 20% of the state of the neurons. By randomly updating the neurons, the pattern is gradually recovered, illustrating the robustness of this 'holistic' form of information storage (Figure from Rutten and Ramsey (2011) [136])

The authors, in a trendsetting book on parallel distributed processing (PDP) from 1986, explained that this was because the brain's computational processes work in parallel [45, 46]. Serial processing is simply too slow to be able to account for human thought processes:

> And the time limitation only gets worse, not better, when sequential mechanisms try to take large numbers of constraints into account. Each additional constraint requires more time in a sequential machine, and, if the constraints are imprecise, the constraints can lead to a computational explosion. Yet people get faster, not slower, when they are able to exploit additional constraints. [47]

Speed is not the most critical issue here. The frequency of neural firing usually does not exceed 200 Hz; compare that with modern personal computers with clock times around 2 GHz (i.e. ten million times faster). What is crucially different is the fact that the brain analyses many pieces of information at the same time. The approach taken by PDP is based on simple and local rules:

> First, we do not assume that the goal of learning is the formulation of explicit rules. Rather, we assume it is the acquisition of connection strengths which allow a network of simple units to act as though it knew the rules. Second, we do not attribute powerful computational capabilities to the learning mechanism. Rather, we assume very simple connection strength modulation mechanisms which adjust the strength of connections between units based on information locally available at the connection. [47]

Problems are solved by iteratively seeking a solution that will return the system to 'a state of least conflict' or mathematically speaking, of least energy [46, 48]. One of the remarkable properties of PDP is that, when constraints are added to the computational problem, the system gets faster, not slower. Even now, 30 years after McClelland and Rumelhard's book, and after many revolutionary improvements in software and hardware, there are still no algorithms that can rival the human capacity to perform natural tasks such as motor control or speech recognition.

It is interesting to read that McClelland and Rumelhard were inspired by several other scientists. Among those that they mentioned in their historical review are Hughlings–Jackson and Luria. 'Neither Hughlings-Jackson nor Luria is noted for the clarity of his views, but we have seen in their ideas a rough characterization of the kind of parallel distributed processing system we envision' [47].

The key to the human brain's high level of performance thus needs to be sought in its architecture and not in the number or the speed of its processing units. Otherwise, whales and elephants should be a lot smarter, as their brains weigh significantly more than that of a human brain (7.8 kg and 4.8 kg versus 1.4 kg, respectively). Some dolphin species even have a larger cortical surface area than that of humans and a superior number of cortical foldings [49]. Still, 'higher cortical functions' did not develop in these species and the evolutionary ground for human skills such as language and social behaviour must somehow relate to the wiring of the human brain [50].

7.4 Language and Evolution

It is beyond scientific dispute that the anatomical and functional organization of the brain is a product of evolutionary history. Over time, phylogenetically older structures have been modified and, to a variable extent, included in newer systems [51]:

> The mammalian brain did not lose the reptilian complex. Mammals added the cingulate cortex, which adds some other behaviors and enhances motor control, but the older part of the brain was not unplugged and replaced by a new 'module'. It continued to function in concert with the newer part. Nor did primates trade in the cingular cortex for the new improved neocortex.
>
> The older parts of the brain also have evolved and changed in more advanced animals in response to selective pressures from new functions and the newer parts of the brain. Although most comparative anatomists stress the enlargement of the newest part, the neocortex, a similar change occurred in the human basal ganglia: the caudate nucleus and putamen (…) are fourteen times larger in humans than they would be if we had the brain of an insectivore, a primitive primate with the same body weight. The basal ganglia have also become more complex and differentiated. In rodents, who are similar in many ways to the first mammals, the putamen and caudate nucleus form a single entity. In higher primates such as squirrel monkeys and human beings they have differentiated and play a part in different aspects of behavior. [52]

When the origin of language is discussed, most scholars believe in 'continuity theories' that assume an evolutionary onset of language abilities. Language is hereby thought to have evolved in a very complex and slow Darwinian manner from prelinguistic systems among our ancestors [53]. Parallels are frequently sought between human speech and vocalization in animals because of obvious similarities in mode and means of communication. However, animals use many other forms of communication, notably bodily expression and behaviour. Think of an animal that hisses or growls when it is angry but also always uses its face and body to express its emotions. Innumerable complex forms of behaviour are found in nature, all meant for one animal to 'convince' another animal of its intentions (i.e. to induce a behavioural response in the other animal).[b] All primates will regularly use vocalization and body postures to communicate with other animals. Facial and brachial gestures, however, seem largely limited to apes and humans. Apes can use these gestures in an intentional manner, whereby the meaning of a gesture is often determined by its context and not only by the gesture itself. Gestures are therefore seen by some scientists as the starting point of human language evolution (Fig. 7.9) and even hypothesized to have been the first linguistic utterances in our ancestors. Pollick and de Waal (2007) studied facial and brachial gestures in different species of apes and conclude that:

> Our closest primate relatives use brachiomanual gestures more flexibly across contexts than they do facial expressions and vocalizations. Gestures seem less closely tied to particular emotions, such as aggression or affiliation, hence possess a more adaptable function. Gestures are also evolutionarily younger, as shown by their presence in apes but not monkeys, and likely under greater cortical control than facial/vocal signals (…). This observation makes gesture a serious candidate modality to have acquired symbolic meaning in early hominins. [54]

---

[b] See for a colourful example the Australian bowerbird, https://www.youtube.com/watch?v=GPbWJPsBPdA.

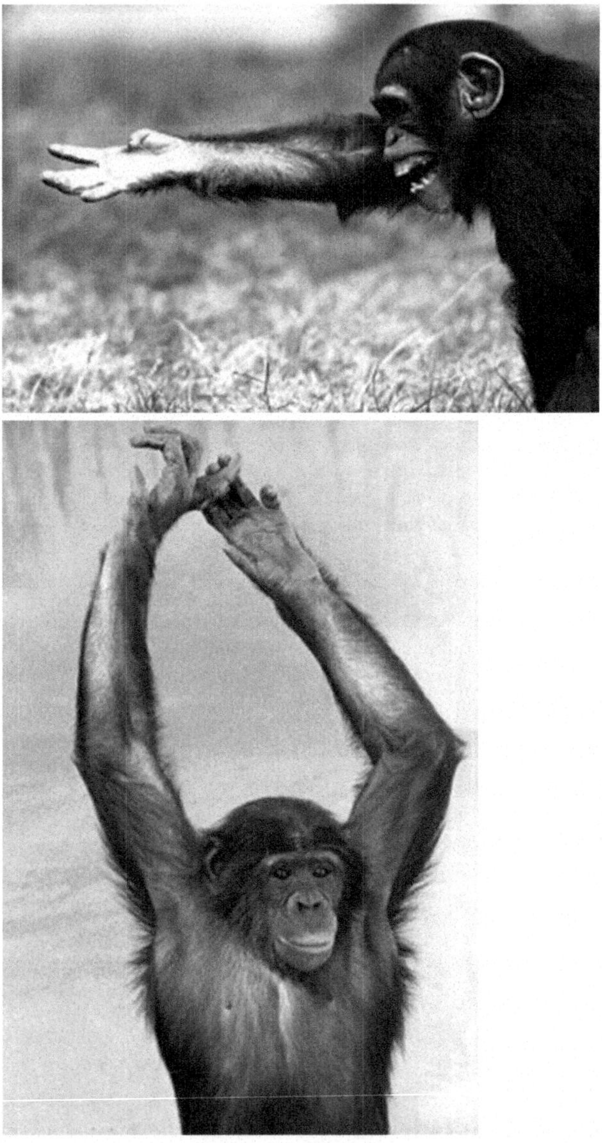

Fig. 7.9 As stated by Pollick and de Waal (2007), 'Meaning often needs to be extracted from the specific context in which a gesture is being used'. All primates regularly communicate by means of vocalizations, orofacial movements, body postures and locomotion patterns. However, free brachiomanual gestures (i.e. manual communication without touching another individual) are typical of humans and apes. (*Top*) A juvenile chimpanzee tries to reclaim food that a dominant has taken away by combining the *reach out up* begging gesture with a *scream* vocalization. (*Bottom*) An adolescent bonobo male making sexual advances to a female adds the *arm raise* gesture (Figure and text (adapted) from Pollick and de Waal (2007) [54])

There are other arguments in favour of this gestural hypothesis of language, for instance:

> the appearance of gestural communication in human infants before speech, and the right-hand (hence left-brain) bias of both ape and human gestures. The ape homologue of Broca's area (i.e., Brodmann's area 44) is enlarged in the left hemisphere. In monkeys, this area is activated during both the production and perception of gestures but not vocalizations. It has been speculated, therefore, that the neural structures underlying manual movements in the great apes, perhaps also including tool use, are homologue with the lateralized language areas in the human brain. [54]

But do animals or primates really have language capabilities, or is this phenomenon exclusively restricted to humans? The answer likely depends on how we define 'language'. As Lieberman writes (1993):

> Until the 1960s linguists and philosophers believed that the ability to use words was the key to human language. For example, in 1964 Norman Geschwind claimed that other animals lacked neocortical brain circuits that supposedly were necessary for learning the meaning of words. [52]

If we take the ability to 'use words' as a criterium for language, then there are many animals that possess at least some language capabilities. It is well known that animals react to verbal instructions. Dogs, sea lions and apes, for example, can respond to a set of different commands and can even learn the meaning of words. Trained dogs have a remarkable vocabulary [55]. When chimpanzees and gorillas are raised as children with American sign language, they can master up to a maximum of about 200 words [52]. In this respect, differences between non-human and human primates seem at first glance merely quantitative. However, there is an important and more qualitative difference, and that is the superior ability of humans to create expressions and sentences from a lexicon of words. This allows the expression of ideas in a fast and creative manner and description of new situations and experiences. Even the most highly trained apes can master only the most simple syntactic rules and are surpassed in this respect by human 3-year-olds. One or two years later, these humans are able to produce an infinite number of new sentences [52]. Chomsky, who was one of the most influential linguists of his time, considered 'syntax' an exclusive human capability that could not have evolved via natural selection. He was a proponent of a 'discontinuity' theory that states that language emerged almost instantly, presumably due to favourable genetic mutations that occurred some 100,000 years ago [56]. Around this time, the modern human vocal tract also evolved, contributing significantly to a more efficient communication. For comparison, non-human primates are only able to produce a subset of human speech sounds due to restrictions of the anatomy of their vocal organs, although a recent study suggests that their vocal capabilities may have been underestimated [57].

From an anatomical perspective, there is fairly strong evidence that language areas are present in the brains of primates. Brodmann (1909) concluded that there were overwhelming similarities in the overall organization of the cortex between different species:

> whether we are dealing with a brain with complex sulcal development, like that of man, or one with smooth surfaces, like that of the marmoset or rabbit or ground squirrel, the same fundamental structural subdivisions are always found. (…) The essentials of cerebral cortical areal parcellation are the same in all mammalian orders examined so far; it is influenced by a principle of segmentation. [58]

Despite large differences in size and shape between the brains of different species (Fig. 7.10), regional brain organization is 'surprisingly constant throughout the whole mammalian class, although their arrangement, the number and the shape of their individual areas and, most of all, their size and position, may vary markedly' [59]. It is tempting to relate the increase in the number of cortical areas in mammals

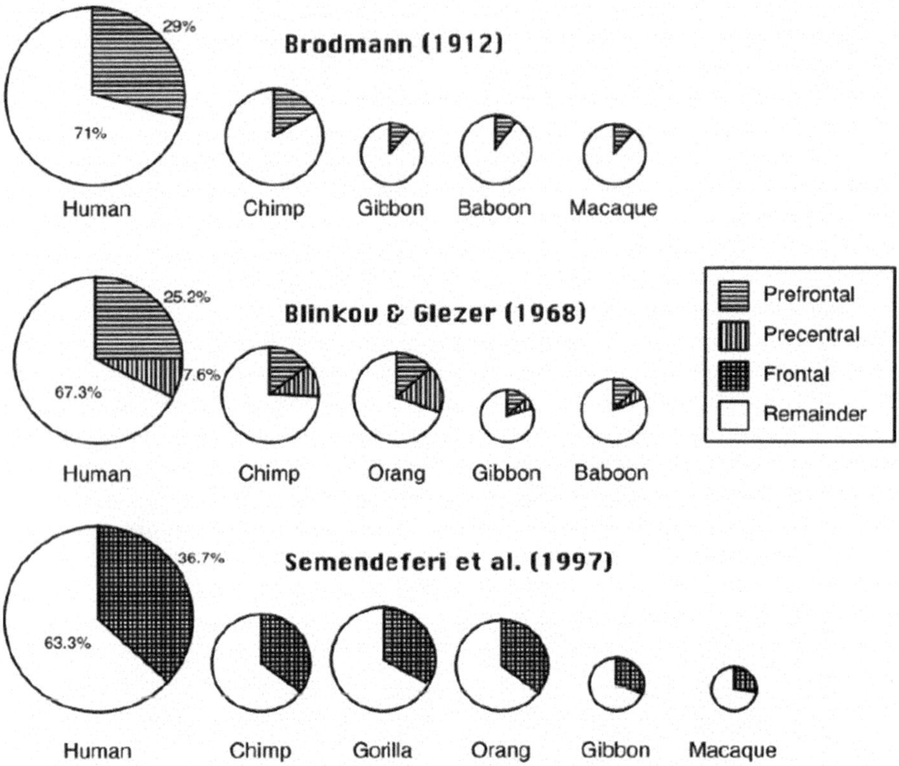

Fig. 7.10 There are reasons to believe that the enlargement of the brain played a significant role in the emergence of cognitive functions, such as language and theory of mind. At the moment, however, we are far from understanding cause and effect in the evolutionary changes that occurred in brain structure and function. For example, there are discussions over fairly basic anatomically related questions, such as whether or not the frontal lobe expanded most as compared with other brain areas. Many scholars seem to accept that the frontal lobes are relatively largest in humans and great apes. However, even this relative enlargement of the human (pre)frontal cortex is not universally accepted [60]. Shown are relative parts of the frontal lobes in apes and monkeys, as indicated by the studies of Brodmann (1912), Blinkow and Glezer (1968) and Semendeferi (1997). Note that Semendeferi did not observe a relative enlargement of the frontal lobe but that may be due to the fact that this study did not distinguish between prefrontal and frontal cortex. When this distinction is made, the evidence is more consistent that there is at least relative enlargement of the prefrontal cortex during evolution (Figure from Preuss (2000) [60])

to the development of new functional abilities [60]. Attractive as such an explanation may be for non-primate animals, however, it may not hold for the development of higher cortical functions because:

> there is at present no good evident that humans possess cortical areas in addition to those found in other primates (…). Several recent frontal-lobe studies bear on this point. Petrides & Pandya (1994) concluded that there is essentially a one-to-one match between the cytoarchitectonic areas of macaque frontal cortex and that of humans. Similarly, recent architectonic studies of human dorsolateral prefrontal cortex [61] and dorsomedial premotor cortex [62, 63] have identified areas known to be present in macaques, but not additional, human-specific areas. [60]

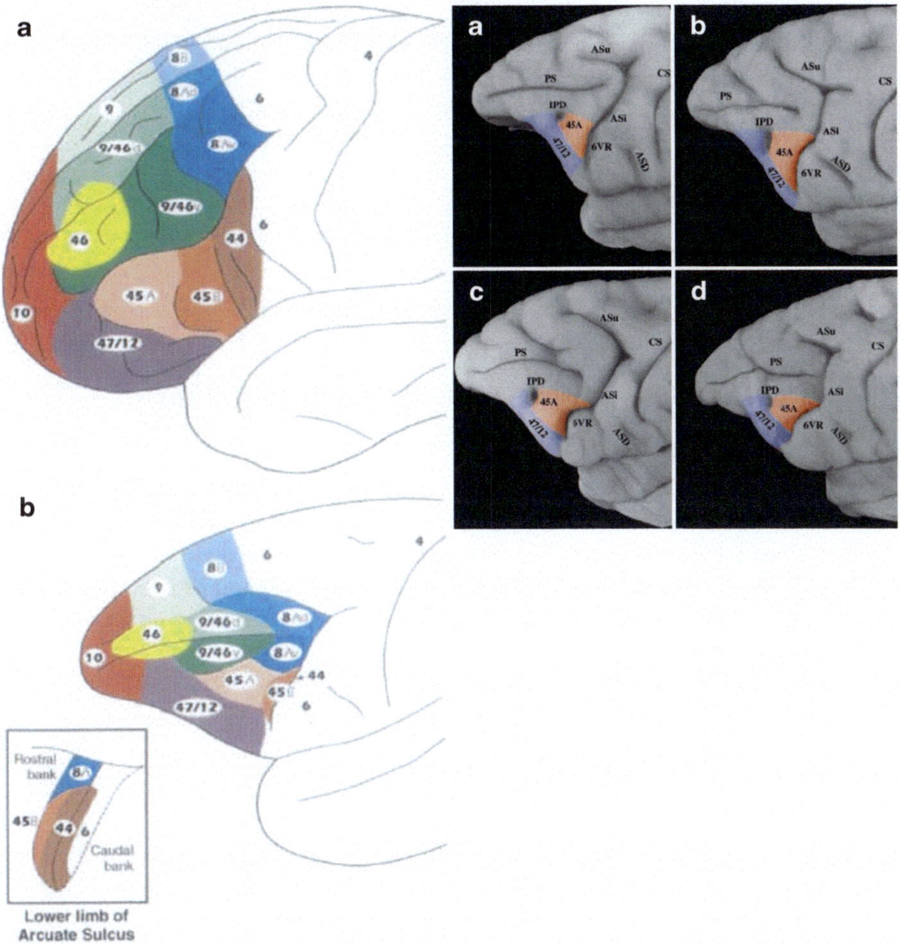

Fig. 7.11 (*Left column*) Architectonic map of the prefrontal cortex of the human and the macaque monkey. Lateral surface of the human prefrontal cortex (**a**) and macaque monkey prefrontal cortex (**b**). The *inset* shows the opened anterior and posterior banks of the lower (inferior) limb of the arcuate sulcus of the monkey to illustrate the position of area 44 that lies in the fundus of this part of the sulcus. Area 45 extends anterior to it as far as the IPD [infraprincipal dimple, GR] on the surface of the inferior frontal convexity. The caudal subdivision of area 45 (i.e. area 45B) lies immediately in front of area 44 on the anterior bank of the sulcus and extends as far as the lip of the sulcus. Area 45A occupies the dorsal part of the inferior frontal convexity and is succeeded ventrally by area 47/12. (*Right column*) Variations (**a–d**) in the morphology of the inferior limb of the arcuate sulcus (ASi) and homologue BA 45. BA 44 cannot be seen on the surface of the brain because it lies hidden in the depth of the sulcus (Figures and text (adapted) from Petrides and Pandya (2009) [137])

Figure 7.11 shows Petrides and Pandya's arrangement of frontal cortical areas in the macaque and human brains, whereby areas have been segregated on cytoarchitectonic grounds. Over time, primate cortical areas seem to have expanded and 'morphed' into other shapes and positions, but kept their relative position (i.e. their cortical neighbours and local connectivity). It has to be said that current understanding of this cortical organization is 'far from complete' [60]. The question whether or not humans have the same number of cortical areas as other primates has not been definitely solved. Preuss (2000) concluded in a review that 'it is evident that neuroscientists have focused too narrowly on the addition of cortical areas as *the* mode of evolutionary change, neglecting possible changes at finer levels of cortical organization' [60]. More research is clearly needed to unravel the many questions regarding the evolutionary emergence of cognition and language. Other forms of anatomical reorganization, such as changes in neuronal density and regional and hemispheric connectivity, likely played an important role in this process and should be taken into account [59].

7.4.1 Homologue Language Areas in Non-human Primates

7.4.1.1 Cortical Areas

Do apes and monkeys have Broca's and Wernicke's areas? Let us look at this in more detail. There has been considerable historical controversy with regard to the identification and localization of homologue language areas in non-human primates. Brodmann, for instance, claimed in 1909 that some of the frontal and temporal areas were unique to humans, in particular those that later became synonyms for the classical language areas:

> In the frontal lobe it is mainly the inferior frontal gyrus that has differentiated into a whole series of new areas, which cannot even be demonstrated in monkeys. These are areas 44, 45, 46 and 47. The parietal lobe is divisible into the four areas 5, 7, 39 and 40 that correspond to only two (5, 7, and) in most other brains, or often to only one (7). Finally the temporal lobe is characterized mainly by the three areas 41, 42 and 52 on the superior aspect of the superior temporal gyrus that are not comparable to any formations in other mammals, even in the closely related higher monkeys. (…) In all these cases there undoubtedly emerges an increase in the specifically differentiated cortical mass that is manifested on the one hand by a greater surface area and on the other hand by a larger number of differentiated areas; in other words, there are new additions to the cortex. [58]

In contrast, other scholars, notably Walker (1940) [64] and Bonin (1949) [65], did identify architectonic homologue areas in the ventral prefrontal region of non-human primates (see Fig. 7.12). For Petrides and Pandya, these discrepancies were a motivation to start their own detailed anatomical–functional studies. They applied the same cytoarchitectonical criteria to both macaques and humans and concluded that primates did have Broca's area [66]. Here is Petrides on his findings (2005):

> The presence of Broca's area in the nonlinguistic monkey brain implies that language is a byproduct of the growth, over millions of years, of the primate brain,' Petrides said. 'As associated cerebral cortical areas expanded, the pre-existing cellular structures found in Broca's area were able to take advantage of the massive upgrade of the brain's computational power.[c]

[c]Quote taken from the website of the Society for Neuroscience, www.sfn.org

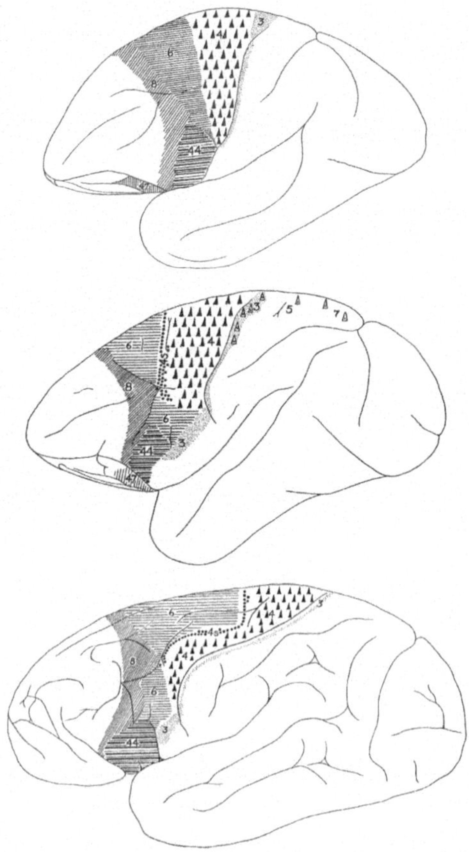

Fig. 7.12 Precentral motor cortex of the cebus (*top*), macaque (*middle*) and chimpanzee (*bottom*) (taken from the chapter by von Bonin in Bucy (ed.) (1949) [65]). Note in particular the presence of BA 44, which Brodmann claimed was not present in non-human primates

As in humans, both BA 44 and 45 in the macaque are located just anteriorly to BA 4 and 6. These latter two areas (i.e. primary motor cortex and ventral premotor cortex) have been consistently identified in primates as agranular cortex (although their exact borders are again a point of discussion [60]). These areas are characterized by an absence of layer IV. Then, moving further in a posterior-to-anterior direction:

> First, a dysgranular area lying just anterior to the ventral premotor cortex (area 6) could be identified in the macaque monkey in the depth of the ventral part of the inferior branch of the arcuate sulcus, and this area had the architectonic characteristics of human area 44 [our Fig. 7.11, GR]. Furthermore, a combined anatomical–physiological study demonstrated that the neurons in the newly identified area 44 of the macaque monkey were involved with the orofacial musculature [67]. Dysgranular area 44 is succeeded, anteriorly, by area 45, which is a clearly granular cortex with the architectonic characteristics of area 45 in the human brain: clusters of unusually large neurons in layer III, a well-developed layer IV, and moderate sized neurons in layer V (see [67–69]). We found that monkey area 45 as defined

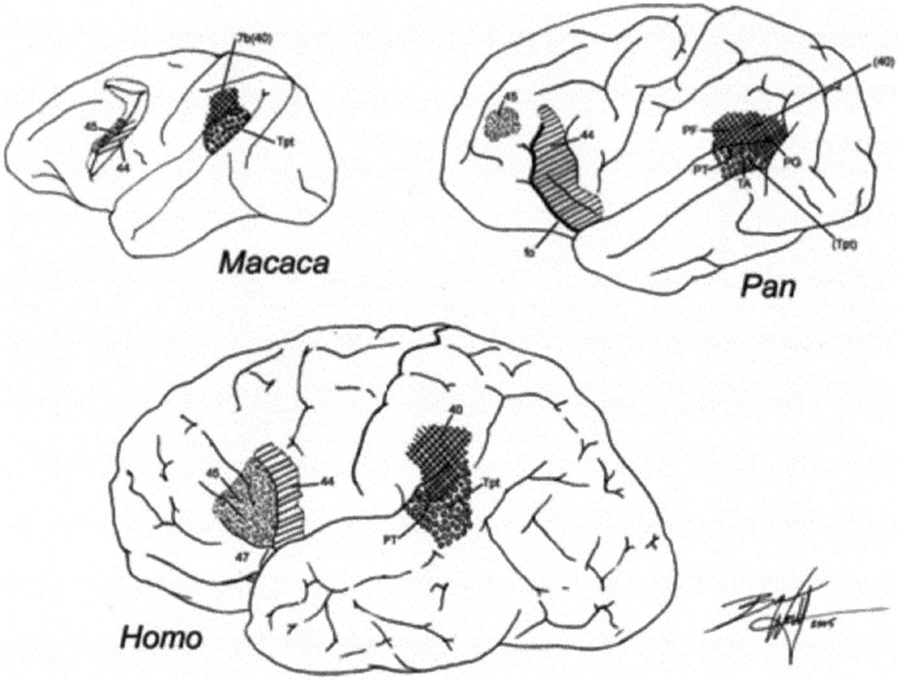

Fig. 7.13 Gross language areas in humans and their proposed homologues in macaques and common chimpanzees, as taken from Falk (2007) [53]. Identifications are based on cytoarchitectonic and functional similarities and should—according to the author—be viewed as tentative

> by criteria comparable to those of the human brain extends anteriorly as far as the infraprincipal dimple (IPD) and that it can be subdivided into a caudal (area 45B) and a rostral (area 45A) part. Furthermore, monkey area 45 when defined by the criteria of human area 45 is not related to oculomotor function as shown by a combined architectonic–microstimulation study that examined this issue. [67]

In a similar manner to Broca's area, anatomical homologue areas have also been described for Wernicke's area (Fig. 7.13). Architectonic similarities have been noted for the inferior parietal cortex (area PG) and the temporal cortex in the posterior part of the Sylvian fissure (area Tpt) [70, 71]. However, as we have seen before, these findings should be put in perspective, as there is little consensus about any anatomical definition of Wernicke's area in humans.

Most scholars nowadays would share the view that precursors of the human language system are present in our ancestors, although the exact functional roles of these homologue language areas remain unclear. Are these areas subserving specific speech functions, or are they only controlling the orofacial muscles that are used during speech? Overall scientific evidence is conflicting here. Some, like Lieberman, plainly state that 'a functional Broca's area is present only in humans' [52]. What we do know is that all attempts to teach chimpanzees to talk have failed. Some recent studies indicate that chimpanzees can use novel sounds to capture the attention of humans, but chimpanzees and other apes are not capable of—or at least have great difficulty—intentionally controlling their speech [52, 72, 73]. In this respect, their homologue Broca's area seems more directly hardwired to the areas for facial motor control than in humans. Electrical stimulation of Broca's area in monkeys and macaques has predominantly resulted in lip and tongue movements, and subsequent resection of this area had no effect on tasks that involve vocalization [74]. On the other hand, there are functional imaging studies that indicate that the left inferior frontal gyrus is involved during the production of communicative signals in chimpanzees [75]. Overall, it is likely that the homologue Broca's area in non-primates is involved in both vocal and gestural communication. How to bridge the gap further to the human situation is unclear, in particular because there remains significant discussion about the functionality of human Broca's area in humans.

7.4.1.2 Subcortical Pathways

What about the subcortical anatomy? Are the homologue language areas in primates connected in a similar manner to those in humans? Although many details about the human language circuitry still need to be filled in, there is fairly convincing evidence that there are two main pathways involved: a ventral and a dorsal one. Both pathways consist of both direct and indirect routes. The dorsal language pathway, for instance, has an indirect route via the inferior parietal region. In animals it is possible to study connecting fibre pathways with much more precision than in humans. Human brain analysis is limited to either the traditional post-mortem dissection methods or to MRI-based fibre tractography. Both methods are unable to determine precisely the origins and terminations of white matter tracts, although MR technology is rapidly advancing towards more accurate identification of smaller fibres [76]. In animals, fibre pathways can be visualized with invasive methods, for instance, by injecting radioactively labelled amino acids in specific cortical areas, or by retrograde tracing of cells that project to areas that were injected with horseradish peroxidase (Fig. 7.14). For obvious reasons these techniques have not been applied in humans. A good introduction to this methodology, as well as the history of fibre tracing in monkey and man, can be found in the seminal book by Schmahmann and Pandya [16]. There is another major advantage of animal studies, namely, that the connectivity of the brain can be studied in detail with invasive neurophysiological (e.g. microrecordings) and lesional techniques. As with the anatomical studies, this has led to many neuroscientific breakthroughs [77–79]. A key question, of course, is to what extent these results can be extrapolated to the human brain, in particular for higher cortical functions.

A good starting point is comparison of the subcortical pathways that are involved in speech perception in both animals and humans, as obviously all vocal communication starts with the analysis of sound [80]. Speech perception refers to the mapping of sounds to internal linguistic representations [81]. It is a precursor and subroutine of language comprehension, whereby the computation of meaning is not required (the distinction is important, as 'speech' and 'language' are

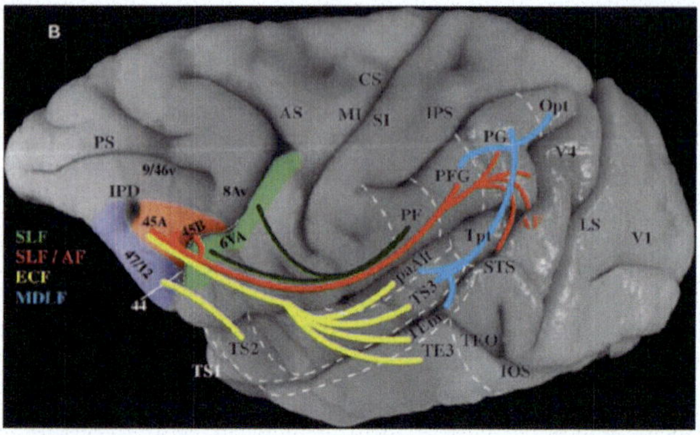

Fig. 7.14 Dorsal and ventral pathways in the macaque monkey as determined via the autoradiographic method, whereby labelled isotopes are injected into a particular cortical area of interest. Both pathways link perisylvian regions with the homologues of Broca's region. The photograph is of the lateral surface of the macaque monkey brain. 'The ventral pathway (in *yellow*) is the extreme capsule fasciculus (ECF), which originates from cortical areas of the superolateral temporal region (i.e. superior temporal gyrus, superior temporal sulcus, and adjacent dorsal inferotemporal cortex), courses through the extreme capsule, and terminates primarily in area 45, with a more moderate projection to area 44. The dorsal pathway (in *red*) is the superior longitudinal fasciculus (SLF), which originates from areas of the inferior parietal lobule and terminates in areas 44, 45B, and 45A. Fibers originating from the caudal part of the superior temporal sulcus arch around the caudal end of the lateral fissure forming the arcuate fasciculus (AF) and blend with the fibers of the SLF in the white matter of the inferior parietal lobule. The ventral premotor cortex (area 6), which controls the orofacial musculature, receives strong input from the most rostral part of the inferior parietal lobule (area PF) via a part of SLF (shown in *green*). (…) The middle longitudinal fasciculus (MDLF, *blue*), links the superolateral temporal region with the inferior parietal lobule. This shows that the suprasylvian inferior parietal lobule and the infrasylvian superolateral temporal regions, which are connected with Broca's region, are themselves massively interconnected. The figure shows the complete homologue of the circuitry that, in the left hemisphere of the human brain, will be used to serve linguistic processing when language develops. Note, however, that the circuitry already exists in the prelinguistic primate brain' (Text and figure from Petrides and Pandya 2009 [137])

sometimes—incorrectly—interchangeably used) [81]. There is much evidence that subcortical anatomy that underlies speech perception is grossly similar in primates. As Poeppel (2012) wrote:

> The connectivity of the human auditory cortex is often presumed to be very similar to that of nonhuman primates. This is probably true for the overall pattern of connectivity, including that of the early auditory stages. (…) Auditory cortex of nonhuman primates is subdivided into core, containing the primary areas (…); the belt, which flanks the core laterally and medially (…); and the parabelt, which is lateral to the (lateral) belt.
>
> Current evidence from nonhuman primates suggests that there is: (1) a cascade of connections from core to belt to parabelt; (2) multisensory input to caudal parabelt; and (3) long-range connections between parabelt and prefrontal cortex. In humans, there is evidence for a similar type of connectivity within early-stage auditory areas; the connectivity of higher-order auditory and related cortex appears to follow the blueprint of nonhuman primates, however with several notable differences. [81]

Both in monkeys and humans, the auditory system is organized into a ventral and dorsal stream, analogous to the visual system (Figs. 7.14 and 7.15). Details differ

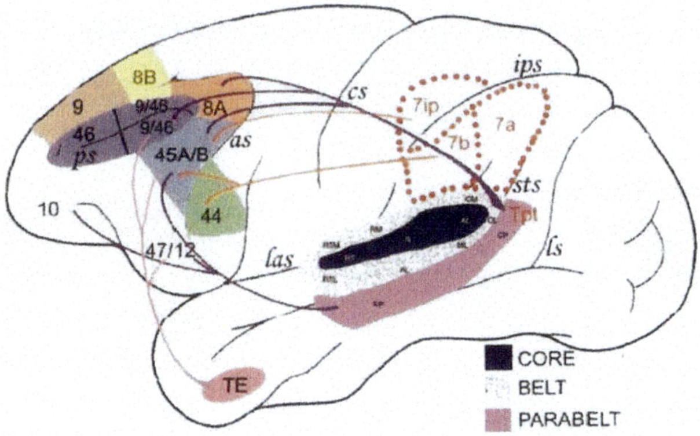

Fig. 7.15 Diagram indicating some of the connections between the temporoparietal and the inferior prefrontal areas in the monkey, taken from Grodzinsky (2006) [66]. The auditory region in the superior temporal lobe is subdivided into core, belt and parabelt regions. There are two main processing streams: (1) the rostral belt and parabelt (the 'what' pathway), which projects to the inferior convexity of the prefrontal lobe; and (2) the caudal belt and parabelt (the 'where' pathway), which project to more dorsolateral areas. The intraparietal and inferior parietal regions (7ip, 7b) project to the inferior convexity. Numbers indicate Brodmann's areas (Data from Hackett (1998) [80], Romanski (1999) [138, 139] and Petrides and Pandya (1999, 2001) [69, 140]). *as* arcuate sulcus, *cs* central sulcus, *ls* lunate sulcus, *ps* principal sulcus, *sts* superior temporal sulcus

between research groups [82–85]. The dorsal or 'where' pathways are involved in spatial auditory information and mainly connect areas in caudal belt and parabelt to dorsolateral areas (BA 8 and 46). The 'what' pathways are more linked to 'intrinsic features of auditory stimuli including speech' and reciprocally connect rostral and orbital regions of the prefrontal cortex to the rostral belt and parabelt areas [81]. There is thus an anatomical–functional differentiation along the anterior–posterior axis of the temporal lobe:

> In general, the caudal parabelt ('where') provides auditory projections to directing eye movements and to the dorsal prefrontal cortex, and the rostral parabelt ('what') projects to ventral/orbital prefrontal cortex and the granular cortex, perhaps more related to human Broca's area. [81]

The graphic representations of the monkey's auditory system resemble modern anatomical schemes of the human language system, where frontal and temporal areas are connected with multiple subcortical pathways [82, 86]. Recently, in an initial study, the anatomical connectivity of frontal lobe tracts was compared in monkeys and humans (Fig. 7.16) [87]. The human results were constructed with non-invasive MRI diffusion techniques, a method with a considerably inferior spatial resolution than axonal tracing. Nonetheless, the overall correspondence was large, with two significant exceptions: the arcuate fasciculus and the inferior fronto-occipital fasciculus (IFOF). Both in monkeys and humans, the arcuate fasciculus terminates in the caudal and posterior part of the superior temporal gyrus (area Tpt). However, this pathway is poorly developed in monkeys, as projections to the middle and inferior temporal gyrus are absent. More or less similarly, the anterior projections of the external capsule/IFOF overlap, while the posterior projections in the

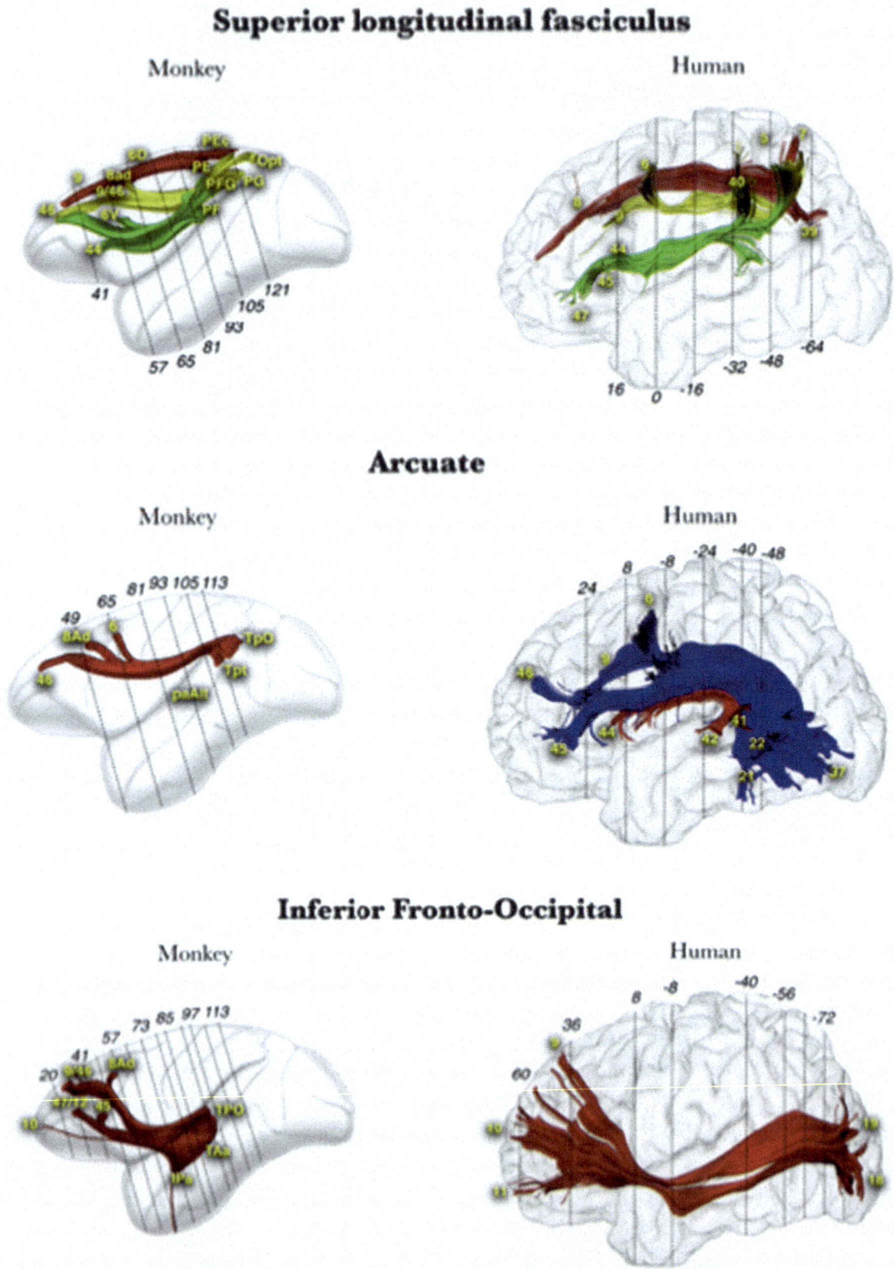
Superior longitudinal fasciculus
Monkey
Human
41
57 65 81 93 105 121
16 0 -16 -32 -48 -64
Arcuate
Monkey
Human
49 65 81 93 105 113
24 8 -8 -24 -40 -48
Inferior Fronto-Occipital
Monkey
Human
20 41 57 73 85 97 113
36 60 8 -8 -40 -56 -72

monkey do not reach the temporal and occipital lobe. In this respect, the IFOF seems uniquely human. These differences, so conclude the authors, may underlie unique human cognitive functions [87].

7.5 Mesulam, Hickok and Poeppel

In 1990, an influential paper was published: 'Large-scale neurocognitive networks and distributed processing for attention, memory and language' [46]. The paper broke with the traditional neurological thinking that functions were either fairly localized (e.g. for motor or language functions) or hardly localizable at all (e.g. for attention or planning). It was written by Marcel Mesulam (1945), an American neurologist and expert in cognitive and behavioural neurology, who had 5 years earlier published his now renowned book *Principles of Behavioral and Cognitive Neurology*. At the time, the notion that behaviour and higher cognitive functions were subserved by complex neural networks was gradually gaining ground. Much of the underlying anatomical evidence came from animal studies, where new techniques (such as axonally transported tracers) were revealing a complex interconnected brain. These studies suggested that brain networks had an internal structure that was 'commensurate with complex computational architectures such as parallel distributed processing', a computer technique that was also developed in that same period [46]. Remember that in the 1990s human brain mapping studies were still in their infancy, and the successful era of functional MRI and MRI-based tractography had yet to begin. A central feature of networks, as Mesulam said, is the absence of a one-to-one correspondence between anatomical site, neural computation and complex behaviour. Behaviour is somehow contained in 'grids of connectivity' that are both localized and distributed (see Fig. 7.17):

> The model (…) helps to explain how anatomical localization is compatible with the fact that lesions in different parts of the brain can yield perturbations of the same overall behavior, why single lesions lead to only partial deficits of a given behavior or to multiple

Fig. 7.16 Thiebaut de Schotten (2012) compared fibre tracts derived from axonal tracing studies in monkeys (*left column*) with those in humans as measured with non-invasive MR tractography (*right column*). The latter technique may be biased due to methodological limitations. The comparison of fibre pathways in the frontal lobe found 'several similarities between human and monkey in the cingulum, uncinate, superior longitudinal fascicles, frontal aslant tract and orbitopolar tract. These similarities suggest preserved functions across anthropoids. In addition, we found major differences in the arcuate fascicles and inferior fronto-occipital fascicles. These differences indicate possible evolutionary changes in the connectional anatomy of the frontal lobes underlying unique human abilities' [87]. (*Top row*) Superior longitudinal fasciculus with three branches (respectively *red, yellow and green*). The comparison suggests similarities between simian and human for the three SLF branches. (*Middle row*) Arcuate fasciculus. This pathway has many fewer temporal connections in monkeys than in humans (common features in *red*, differences in *blue*). (*Bottom row*) Extreme capsule in monkeys versus IFOF in humans. Both tracts project to similar frontal regions. The posterior projections are different, however, as the monkey pathways do not reach posterior-temporal and occipital areas. As such, the IFOF is absent in monkeys

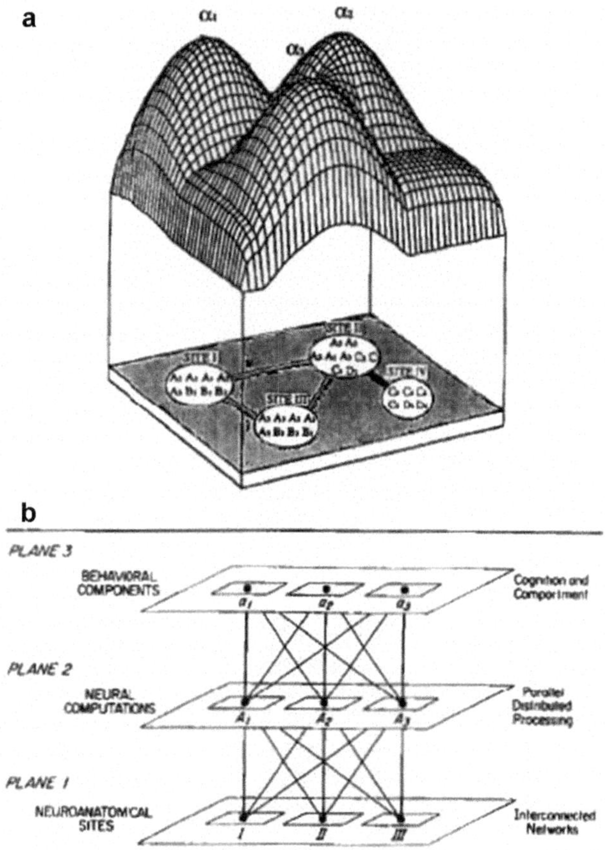

Fig. 7.17 Mesulam (1990) introduced the concept of large-scale neurocognitive networks [46]. The key principle is that there are no one-on-one relationships between anatomical sites, neural computation and behaviour. In Mesulam's words: 'This configuration reconciles reductionism with emergence; reductionism represents a top-down perspective from behavior to computation and to neural structure, whereas emergence results from the same interactions from a bottom-up perspective. The additional features that emerge during the upward ascent from one level to the next represent the relational architecture among the components and cannot be reduced to a simple list of lower-level constituents'. 'Sites I, II, and III collectively constitute a large-scale network underlying behavior alpha. The alpha1, alpha2, and alpha3 components define the behavioral plane. Site I is most closely associated with alpha1, site II with alpha2, and site III with alpha III, but the relationship is not one to one and contains considerable eccentricity' (Text and figures from Mesulam (1990) [46])

> behavioral deficits, and why brain mapping studies (…) are likely to detect multiple areas of activation in association with individual complex behaviors. For the more practical purposes of neuropsychological assessment, this model predicts that no neuropsychological task can ever be entirely specific for a single region of association cortex and that the clinician need not to look for multiple lesions just because the patient shows more than one cognitive deficit. [46]

These large-scale networks consist of widely separated and interconnected local networks, whereby the local networks are 'confined to single cytoarchitectonic fields

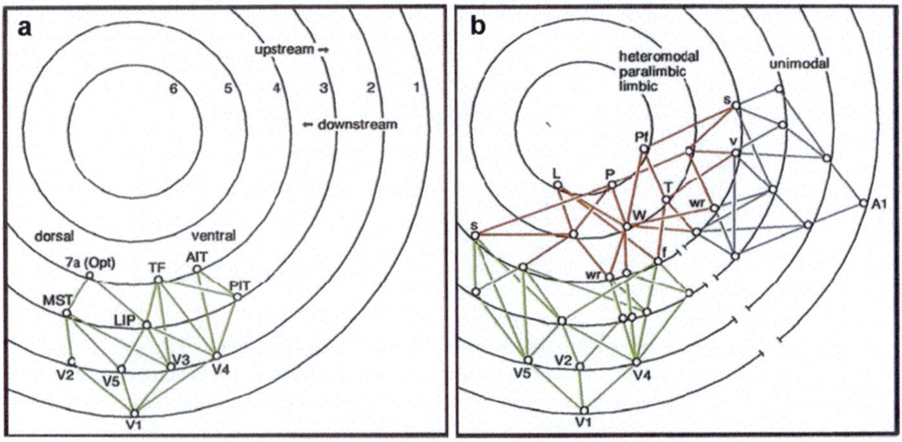

Fig. 7.18 Mesulam's model of the flow of auditory and visual information in the human brain. 'Each concentric ring represents a different synaptic level. Any two consecutive levels are separated by at least one unit of synaptic distance. Level 1 is occupied by primary sensory cortex. *Small empty circles* represent macroscopic cortical areas or "nodes", one to several centimeters in diameter. Nodes at the same synaptic level are reciprocally interconnected by the black arcs of the concentric rings. Colored lines represent reciprocal monosynaptic connections from one synaptic level to another. (**a**) Visual pathways as demonstrated by experimental neuroanatomical methods in the macaque brain. (**b**) Visual (*green*), auditory (*blue*), and transmodal (*red*) pathways in the human brain. Individual pathways are inferred from the experimental work in the monkey. The anatomical identity of many of the nodes is not specified because their exact anatomical location is not critical. The assumption is that these *types* of anatomical interconnections and functionally specialized nodes exist in the human brain even thought their exact location has not yet been determined. The terms 'dorsal' and 'ventral' refer to the separation of visuofugal pathways, especially at the fourth synaptic level, into dorsal and ventral streams of processing. The gaps in the *circles* at the first four levels indicate the absence of monosynaptic connections between modality-specific components of auditory and visual processing'. *A1* primary auditory cortex, *AIT* anterior inferotemporal cortex, *f* area specialized for face encoding, *L* the hippocampal-entorhinal or amygdaloid components of the limbic system, *LIP* lateral intraparietal cortex, *MST* medial superior temporal cortex, *P* heteromodal posterior parietal cortex, *Pf* lateral prefrontal cortex, *s* area specialized for encoding spatial location, *PIT* posterior inferotemporal cortex, *T* heteromodal lateral temporal cortex, *TF* part of medial inferotemporal cortex, *v* area specialized for identifying individual voice patterns, *V1* primary visual cortex, *V2, V3, V5, V5* additional visual cortices, *W* Wernicke's area, *wr* area specialized for encoding word forms, *7a(Opt)* part of dorsal parieto-occipital cortex (Text and figure from Mesulam (2000) [88])

or immediately contiguous areas'. Five different types of large-scale networks are described by Mesulam: for spatial attention, language, memory–emotion, executive functions and face and object identification [88]. He acknowledges that it is difficult to specify the neurobiological features and computational algorithms for these networks but that they 'provide the only opportunity for addressing the neurological basis of complex cognitive domains' [50]. Behaviour is represented at many neural sites and vice versa; individual sites are involved in different cognitive operations. Figure 7.18 schematically shows that the flow of auditory and visual information in the brain runs both upstream and downstream. Higher-order areas, such as Wernicke's area, function as complex nodes in larger networks. The boundaries between low-level auditory processing and high-level cognitive processing in networks are fuzzy, and information that is processed in 'simple' primary auditory areas may also contribute to eventual conceptualization of acoustic information (i.e. language comprehension). Thus, if we want to unravel language processing, we should focus not only on the so-called higher cognitive processes but also consider information processing in primary sensory areas, as this information can interact in important ways with other downstream areas [89]:

> Neurons of A1 of the human brain are sensitive to pure tones and pitch whereas those of the mid-to-anterior parts of the superior temporal gyrus are relatively unresponsive to pure tones and nonlinguistic noises but respond to specific phonetic parameters of spoken language [90–92]. The superior temporal gyrus neurons are broadly tuned to the segmentation and sequencing of phonemes as well to their coherence within polysyllabic and compound words [93]. They encode speech at a presemantic level since they respond to real spoken words as readily as to distorted backward speech. [93]
>
> Approximately half of the neurons in parts of the middle temporal gyrus (BA21) give highly selected responses, mostly in the form of suppression, to understandable speech but not to distorted speech [93]. (…) it appears that upstream auditory areas in the human brain tend to encode more elementary features of sound such as frequency and pitch, whereas downstream areas may encode neuronal groups that encode more composite features related to the identification of words ['wr' in Fig. 7.18], the localization of sound sources for attentional targeting ['s' in Fig. 7.18], the categorization of object-specific sounds, and perhaps also the characterization of individual voice patterns. [area 'v' in Fig. 7.18]

Mesulam's model has several different auditory areas, comparable to the many visual areas that are already well described in primate brains. Information is analysed in various bits and pieces at multiple places in multiple times. Consequently, lesions in auditory areas can lead to various complex functional impairments:

> Lesions of unimodal auditory association cortex or of its connections give rise to complex 'auditory perceptual impairments' (such as the inability to identify variations in timber or sound sequences), 'cortical deafness' (inability to recognize meaningful verbal and nonverbal auditory patterns despite normal brain stem auditory potentials), 'pure word deafness' (inability to understand or repeat spoken language despite good recognition of environmental sounds and no other language deficit), 'auditory agnosia for environmental sounds' (inability to identify sounds characteristic of objects despite good speech comprehension), and 'phonagnosia' (inability to recognize the identity of familiar voices despite preserved recognition of spoken words and environmental sounds) [94, 95]. The first two deficits are caused by damage to upstream unimodal auditory association cortex (usually in the presence of A1 involvement) whereas the latter three may reflect damage to downstream auditory association cortex or functional disconnections of auditory association areas from transmodal cortices related to language comprehension and object recognition. [88]

7.5.1 Epicentres

The best studied network at the time was that of directed spatial attention (Fig. 7.19). The ability to direct attention to parts of the extra-personal space is an important prerequisite for adaptive behaviour [46]. This functionality is linked to three major

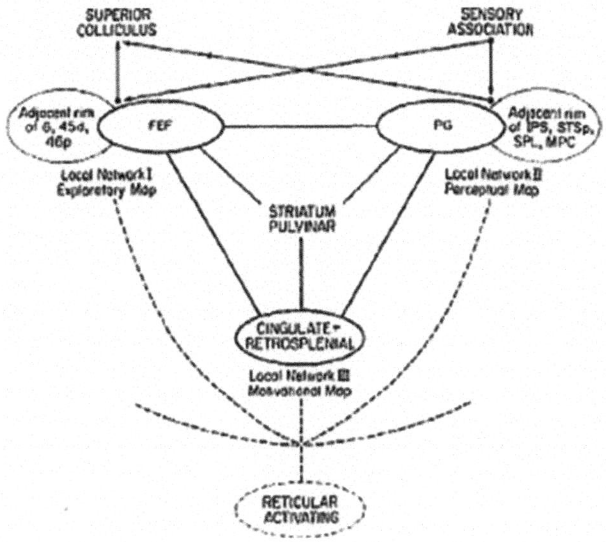

Fig. 7.19 A large-scale network for directed attention. Lesions in any one major site or pathway can cause neglect. For example, frontal lesions can cause neglect just as readily as parietal lesions. *FEF* frontal eye fields, *PIS* intraparietal sulcus, *MPC* medial parietal cortex, *SPL* superior parietal lobe, *STSp* posterior part of cortex within the superior temporal sulcus, *45d* dorsal part of area 45, *46p* posterior part of area 46, *PG* heteromodal association cortex of the inferior parietal lobule (Text and figures adapted from Mesulam (1990) [46])

regions in the right hemisphere: dorsolateral posterior parietal cortex, dorsolateral premotor–prefrontal cortex and cingulate gyrus. Spatial attention is well studied in animals, and experiments have shown that lesions in each of the three regions or of important interconnections cause neglect, albeit with different behavioural disturbances that relate to the specific lesion site

> Each of the three cortical components (…) serves a dual purpose; that is, it provides a local network for regional neural computations and it also provides a nodal point for the convergence and reentrant accessing of distributed information. All three core components are probably engaged simultaneously and interactively by attentional tasks, and it is unlikely that there is a temporal or processing-level hierarchy among them. The resultant phenomenon of directed attention is not the sequential sum of perception plus motivation plus exploration but an emergent (i.e. relational) quality of the network as a whole. It is also important to realize that this large-scale network is implicated primarily in the distribution of spatially-addressed attention. The distribution of object-addressed attention requires the additional contribution of visual association areas in the temporal lobe. [88]

Mesulam's ideas are strongly based on the results of animal studies. For the language network, good animal models are obviously lacking. There is, however, abundant lesion data from human studies that points to the left hemisphere and in particular to left perisylvian areas. Clinical observations suggest that Broca's area is involved most in articulation and syntax and that Wernicke's area is involved in semantics and lexical retrieval. Still, lesions in either area can produce both production and comprehension deficits, in contrast to classic language theories. Mesulam sees the areas of Broca and Wernicke as the two major nodes ('cortical epicentres') in the language network. These areas are not strictly defined, as there are no definite cytoarchitectonic, topographic or physiological criteria for their exact delineation. Mesulam acknowledges that, in general, anatomical landmarks in the brain do not allow for a reliable classification of the function of particular areas [88]. Cytoarchitectonic maps or topographical patterns do not consistently match with similar functions across individuals.

When analysing the history of brain cartography, Mesulam distinguishes two different 'schools'. Proponents of the first school, he wrote, primarily constructed maps on the basis of structural features (e.g. Exner, Brodmann, the Vogts, von Economo and Flechsig). Proponents of the second school combined results from anatomical, physiological and behavioural experiments (e.g. Campbell, Broca, Filimonoff, Yakovlev and Sanides). This led to a subdivision of the cerebral cortex into five major functional subtypes: limbic, paralimbic, heteromodal association, unimodal association and primary sensorimotor [88]. Mesulam combined the cytoarchitectonical information of Brodmann with this concept of primary, unimodal and heteromodal cortical areas to characterize the function of brain regions (see Fig. 7.20). Unimodal Brodmann area 44 is positioned as the core of Broca's area, with adjacent areas 45, 47, 12 and 6 (these are, with the exception of BA 6, heteromodal prefrontal areas) [88]. For Wernicke's area Mesulam sees no universally accepted boundaries: 'It is usually defined as "the region which causes Wernicke's aphasia when damaged"' [88]. He stresses that Wernicke's area is not a 'word bank', but:

> a nodal bottleneck for assessing a distributed grid of connectivity that contains information about sound-word-meaning relationships. At the output stage, Wernicke's area constitutes a final common pathway for the chunking of thoughts into words that are commensurate with the underlying meaning. [50]

Mesulam links this functionality to a potential larger area in the left posterior-temporal and parietal region:

> Some investigators would confine Wernicke's area to auditory association cortex in the posterior third of the superior temporal gyrus (BA 22). (…) the multimodal nature of Wernicke's aphasia argues against this possibility. There are numerous reasons for concluding that Wernicke's area includes not only the posterior third of BA 22 but also the immediately adjacent parts of heteromodal cortex in BA 39-40 and perhaps also parts of the middle temporal gyrus. [88]

According to one of Mesulam's more recent studies, axonal pathways are likely also included, as well as the anterior temporal lobe [96]. Others also have failed to find any common ground for Wernicke's area. Wise wrote in 2001: 'Over time, both the functional and anatomical boundaries of "Wernicke's area" have become so broad as to be meaningless' [97].

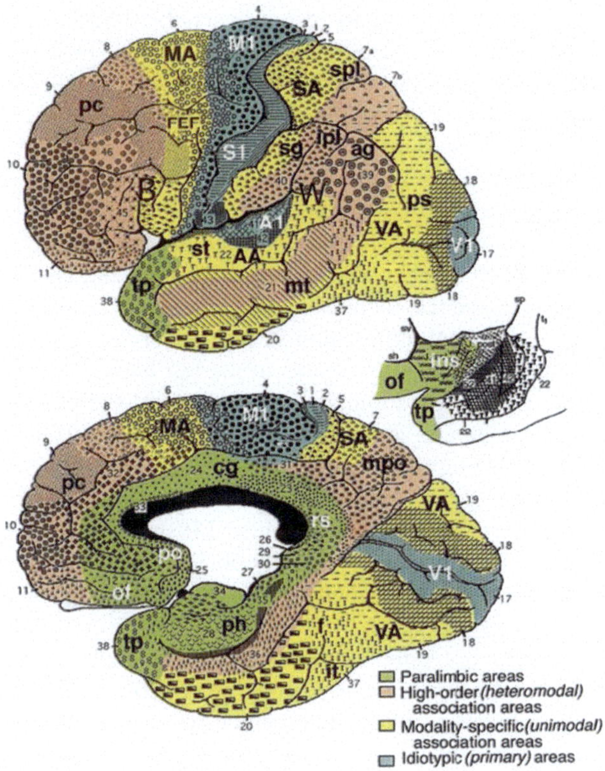

Fig. 7.20 Mesulam's distribution of functional regions within the brain. Results from Brodmann's cytoarchitectonic studies are combined with the concept of unimodal and heteromodal cortical areas. Lesions in unimodal areas yield deficits only in tasks guided by that single modality, whereas lesions in heteromodal areas can lead to more complex functional deficits. Mesulam added that 'The boundaries are not intended to be precise. Much of this information is based on experimental evidence obtained from laboratory animals and needs to be confirmed in the human brain'. *AA* auditory association cortex, *ag* angular gyrus, *A1* primary auditory cortex, *B* Broca's area, *cg* cingulate cortex, *f* fusiform gyrus, *FEF* frontal eye field, *ins* insula, *ipl* inferior parietal lobule, *it* inferior temporal gyrus, *MA* motor association cortex, *mpo* medial parieto-occipital area, *mt* middle temporal gyrus, *M1* primary motor area, *of* orbitofrontal region, *pc* prefrontal cortex, *ph* parahippocampal region, *po* parolfactory area, *ps* peristriate cortex, *rs* retrosplenial area, *SA* somatosensory association cortex, *sg* supramarginal gyrus, *spl* superior parietal lobule, *st* superior temporal gyrus, *S1* primary somatosensory area, *tp* temporopolar cortex, *VA* visual association cortex, *V1* primary visual cortex, *W* Wernicke's area (Figure taken from Mesulam, 2000 [88])

7.5.2 Dual-Stream Models

Since around the year 2000, evidence has grown in favour of a dual-route model for language. Although the various proposed models differ, there is general agreement that there is a dorsal route that is implicated in phonological processing (this is the classic Broca–Wernicke connection) and an additional ventral route that subserves semantic processing [89, 98, 99]. In recent years, MRI-based tractography has greatly contributed to the anatomical specifications of these fibre tracts [100]. The ventral route consists of multiple pathways that together connect regions in four different cerebral lobes [101]. All of these tracts were initially described well over a century ago. To date, their functional roles within the various networks are to a large extent still unclear and in need of clarification. Ventral route connections run via the inferior fronto-occipital fasciculus (IFOF), uncinate fasciculus (UF), middle longitudinal fasciculus (MLF) and inferior longitudinal fasciculus (ILF). More specifically, the UF has been associated with proper naming and also with emotional and memory processing, the ILF with reading and face and visual recognition [102]. Results from direct electrical stimulation in awake surgical patients support the existence of two major language pathways in the left hemisphere (see for details Chap. 6). Subcortical electrical stimulation of the ventral stream, and in particular the IFOF, consequently impairs semantic language processing [103]. Stimulation of the dorsal stream usually elicits speech arrest or phonological paraphasias. There is only limited data available from right hemisphere procedures, prohibiting generalization of stimulation results to both hemispheres. It should be noted, though, that language disturbances have been convincingly obtained after stimulation of homologue pathways in the right hemisphere, albeit in a small sample of left-handed patients [104].

Hickok and Poeppel not only have an impressive track record in language research, they also moderate an inspiring blog on which they post new ideas and comments on papers from the scientific community: www.talkingbrains.org. And they have a keen eye for history. They realized that their dual-stream model was not a new idea, but was 'a central feature of Wernicke's model of the functional anatomy of language':

> We have written that our dual stream model of speech processing builds on research on cortical models of the visual system, particularly the distinction between dorsal and ventral processing streams. And it does. But there is a much older precedent both to our own proposal, and to current dual-stream vision theories: Wernicke's classic 1874 language model. As we all know, Wernicke proposed that sensory representations of speech ('auditory word images') interfaced with two distinct systems, the conceptual system, which he believed was broadly distributed throughout cortex, and the motor system located in the frontal lobe. The interface with the conceptual system supported comprehension of speech, whereas the interface with the motor system helped support the production of speech. Thus, one stream processes the meaning of sensory information (the 'what' stream), while the other allows for interaction with the action system (the 'how' stream). This is basically identical to what David and I have been claiming in terms of broad organization of our dual stream model, and identical to what folks like Milner and Goodale have proposed in the vision domain. When will those vision folks get an idea of their own. ;-)[d]

In Hickok and Poeppel's model, the dorsal stream provides an interface for speech with the motor system (Fig. 7.21). The critical region for these

[d]Quote from Greg Hickok, taken from www.talkingbrains.org, 25 May 2007

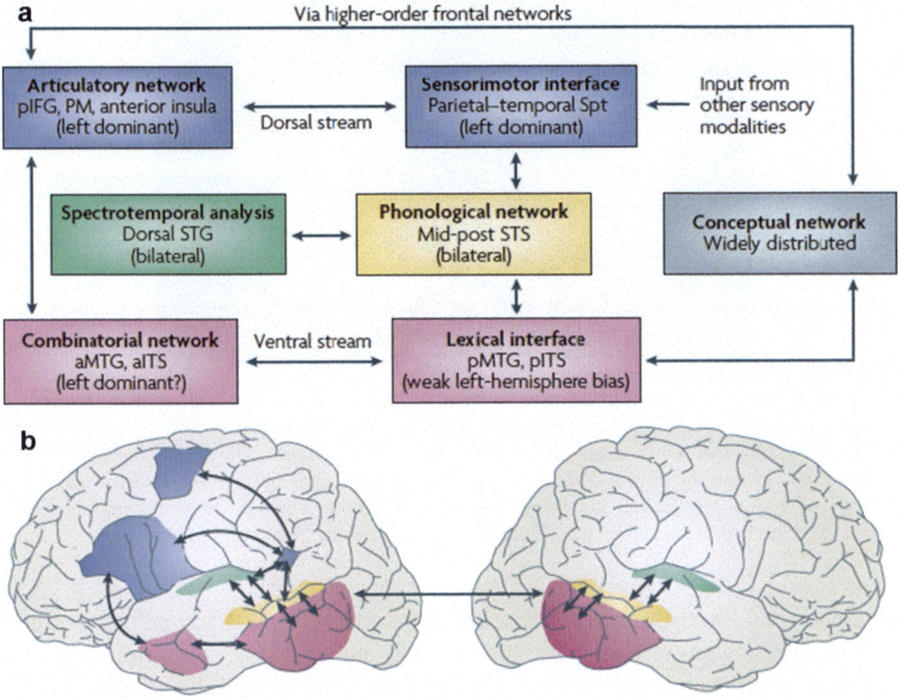

Fig. 7.21 The dual-stream model of the functional anatomy of language according to Hickok and Poeppel (2007) [86]. The interface with the motor system is left lateralized. The classic sensory part of language processing (the ventral stream in this model) has to a large extent a bilateral representation. Spectrotemporal and phonological analyses are performed in the dorsal STG and the mid-to-posterior part of the STS. *pIFG* posterior part of inferior frontal gyrus, *PM* premotor cortex, *aITS* anterior inferior temporal sulcus, *aMTG* anterior middle temporal gyrus, *STG* superior temporal gyrus, *STS* superior temporal sulcus

auditory–motor transformations lies deep within the posterior part of the Sylvian fissure, in so-called area Spt (Sylvian parietal–temporal) [89]. The ventral stream provides an interface with conceptual representations, which are necessary for language comprehension [89, 99, 105]. In contrast to many other anatomical language models, Hickok and Poeppel hypothesize that retrieval of lexical information and conceptualization are bilaterally represented processes, although they suggest 'a weak left-hemisphere bias at this level of processing' [86]. The dorsal stream is strongly left lateralized, in accordance with clinical findings that usually only damage to the left hemisphere results in speech production deficits. (For a different opinion, see, for instance, Cogan (2014) [106].) It is involved in articulation and phonology and seems embedded in the sensorimotor circuits that run between the inferior frontal region and ventral premotor cortex and the inferior parietal lobe (Fig. 7.22) [107]. This pathway corresponds to the anterior segment of the arcuate fasciculus, as described by Catani (2005) [107]. Other authors refer to it as a

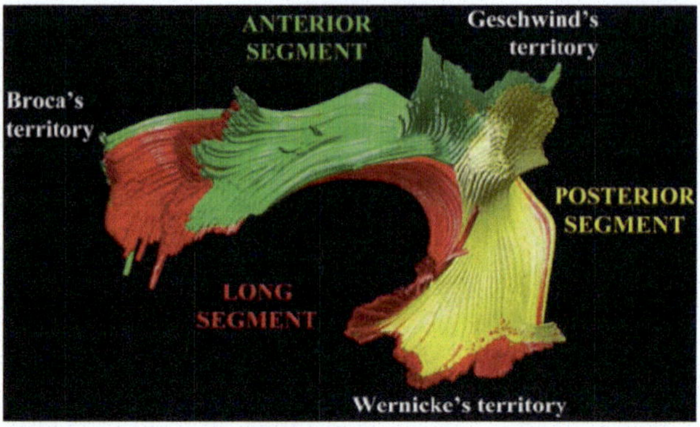

Fig. 7.22 MRI-based tractographic reconstruction of the arcuate fasciculus. Frontal and temporo-parietal areas are connected through direct and indirect pathways. The direct pathway (long segment shown in *red*) runs medially and corresponds to classical descriptions of the arcuate fasciculus. The indirect pathway runs laterally and is composed of an anterior segment (*green*) connecting the inferior parietal cortex (Geschwind's territory) and Broca's territory and a posterior segment (*yellow*) connecting Geschwind's and Wernicke's territories (Figure from Catani (2005) [107])

subdivision of the superior lateral fasciculus (part III) [108–110]. Note that nomenclature of these tracts remains confusing as some authors use the terms interchangeably, although the arcuate fasciculus is generally seen as part of the superior lateral fasciculus.

7.5.3 Phonological Loop

Left fronto-parietal connections have not only been associated with language (speech perception, mapping of sounds onto motor representations) but also with verbal working memory [111]. This has led researchers to question the extent to which these functional systems overlap. The term *working memory* was first used by Baddeley and Hitch (1974) to describe a system that is involved in the temporary processing and storage of information [112, 113]. Part of their original model was a phonological loop that consists of a 'phonological store' (to hold verbal information for approximately 2 s per item) and an 'articulatory rehearsal process' (to refresh the decaying contents of the phonological store by subvocal rehearsal) [114].

Some researchers have proposed that the phonological loop is part of the language system and that assistance of verbal working memory is demanded, for instance, during processing of complex sentences. It is indeed appealing to think that this loop can be used to hold words and sentences online until a proper (syntactic) judgement has been made [111]. Several studies have provided evidence for this hypothesis, by showing that both language and verbal working memory tap into the same neural resources [113, 115–117]. Lesion studies have indicated that Broca's aphasics, for example, not only have problems with the comprehension of more complex sentences but also have a reduced working memory span for digits and sentences. Functional imaging studies indicate that the inferior frontal region (including the ventral premotor cortex) and the inferior parietal lobe (i.e. classic language regions) are engaged during phonological working memory tasks [109, 118–120, 111, 121].

On the other hand, there is also considerable evidence that there are different working memory systems and that working memory consists of multiple separate subsystems [122, 123]. For the sake of argument, it should be noted that working memory is a slow and somewhat inaccurate system. It is under attentional control and is, by definition, not involved in automatic processing. Language, in contrast, is a much faster and more automatized process. Studies in healthy subjects and patients have found that verbal working memory span and language performance are dissociable [122]. Individuals with a severe reduction in the capacity of the phonological loop typically have normal spontaneous speech and just a few difficulties with language comprehension [124–126]. Lesion studies suggest that there are separate systems within the verbal memory system for different types of linguistic stimuli [122]. Aboitz (2010) reached a similar conclusion when reviewing the literature on the phonological loop:

> It is likely that Baddeley's restricted concept of phonological working memory may not be sufficient to account for the memory requirements of complex language processing because it is likely that other forms of short-term memory also participate in this process. For example, syntactical processing in the adult is probably automatic to a large extent (Endress and Hauser 2009) and partly depends on corticostriatal networks that have become stabilized as procedural memories that need to interact with episodic memory networks conveying meaning (Ullman 2004). [109]

It is of (historical) interest to note that Baddeley, later in his career (1998), proposed that the phonological loop primarily evolved as a system to store unfamiliar sound patterns until more permanent representations have been formed. He attributed an important role to the phonological loop for the *learning* of language, representing 'the processes and mechanisms by which the sound patterns of the words of the native language are learned by the child' [126, 127]. Interestingly, this idea strongly resembles that of Wernicke (1874), who argued that the arcuate fasciculus is only of critical importance during childhood, when language is learned and associative connections are formed between auditory and motor images (see Chap. 2).

7.5.3.1 The Problem of Definitions

The question thus remains whether or not working memory is involved in aspects of language processing and to what extent there are common neural pathways [113, 117]. The main reason for bringing up this (ongoing) controversy here is to illustrate that a discussion in terms of *brain functions* can easily lead to confusion and misunderstanding. This is caused by the inability to construct proper operational definitions. Definitions of brain functions are often so broad that a comparison of experimental results across studies or against gold standard techniques is next to useless. They can differ significantly between different textbooks and between different research groups. Instead of 'functions', it is better and more practical to study and compare the results of specific tasks.[e] Tasks, as opposed to functions, can be better defined and controlled and—when adequately constructed—will target a specific function of interest.

This process—designing tasks to study brain function—is the basis of functional neuroimaging techniques. Task-based functional MRI, in particular, is increasingly used to formulate new theories on brain function and is moving the neuroscientific field forward at a rapid pace. The next chapter provides a detailed look into the related methodological and fundamental issues in language mapping with functional imaging techniques and specifically focuses on development of clinical applications.

References

1. Tesak J, Code C. Milestones in the history of aphasia: theories and protagonists. Hove: Psychology Press; 2008.
2. Wepman JM. Recovery from aphasia. New York: Ronald Press; 1951.
3. Geschwind N. Selected papers on language and the brain. Berlin: Springer; 1974.
4. Myers RE, Sperry RW. Interocular transfer of a visual form discrimination habit in cats after section of the optic chiasm and corpus callosum. Anat Rec. 1953;115:351–2.
5. Geschwind N, Devinsky O, Schachter SC. Norman Geschwind: selected publications on language, epilepsy, and behavior. Boston: Butterworth-Heinemann; 1997.
6. Galaburda AM. Norman Geschwind 1926-1984. Neuropsychologia. 1985;23:297–304.
7. Geschwind N, Kaplan E. Random reports: human split-brain syndromes. N Engl J Med. 1962;266:1013.
8. Geschwind N, Kaplan E. A human cerebral deconnection syndrome. Neurology. 1962;12:675–85.
9. Geschwind N. Carl Wernicke, the Breslau school and the history of aphasia. In: Carterette EC, editor. Brain function, vol 3: speech, language, and communication. Berkely: University of California Press; 1963.
10. Lichtheim L. On aphasia. In: Grodzinsky Y, Amunts K, editors. Broca's region. Oxford: Oxford University Press; 2006. p. 318–33.
11. Freud S. On aphasia; a critical study. New York: International Universities Press; 1953.
12. Catani M, ffytche DH. The rises and falls of disconnection syndromes. Brain. 2005;128:2224–39.
13. Geschwind N. Disconnexion syndromes in animals and man. Part I. Brain. 1965;88:237–94.
14. Geschwind N. Disconnexion syndromes in animals and man. Part II. Brain. 1965;88:585–644.
15. Flechsig P. Developmental (myelogenetic) localisation of the cerebral cortex in the human subject. Lancet. 1901;2:1027–29.
16. Schmahmann JD, Pandya DN. Fiber pathways of the brain. New York: OUP; 2006.
17. Finger S. Origins of neuroscience: a history of explorations into brain function. Oxford: Oxford University Press; 2001.
18. Anderson JM, Gilmore R, Roper S, et al. Conduction aphasia and the arcuate fasciculus: a reexamination of the Wernicke-Geschwind model. Brain Lang. 1999;70:1–12.

[e] Remember that Luria defined functions more practically as a form of goal-directed behaviour. Such a definition emphasizes the fact that task performance underlies behaviour and functionality.

19. Hagoort P. MUC (memory, unification, control) and beyond. Front Psychol. 2013;4:416.
20. Damasio AR, Geschwind N. Anatomical localization in clinical neuropsychology. In: Vinken PJ, Bruyn GW, Klawans HL, editors. Handbook of clinical neurology. Amsterdam: Elsevier; 1985. p. 7–22.
21. Goodglass H, Kaplan E. The assessment of aphasia and related disorders. Philadelphia: Lea & Febiger; 1972.
22. Geschwind N. Review of traumatic aphasia by A.R. Luria. Language. 1972;48:755–63.
23. Cole M, Cole S. The making of mind: a personal account of Soviet psychology. Cambridge: Harvard University Press; 1979.
24. Luria AR. Neuropsychology in the local diagnosis of brain damage. Cortex. 1964;1:3–18.
25. Vocate DR. The theory of A.R. Luria: functions of spoken language in the development of higher mental processes. Hillsdale: Lawrence Erlbaum Associates; 1987.
26. Cole M, Levitin K, Luria AR. The autobiography of Alexander Luria. A dialogue with the making of mind. New York: Psychology Press; 2006.
27. Luria AR. Traumatic aphasia: its syndromes, psychopathology and treatment [in Russian]. Izd Akad Ped Nauk RSFSR. 1947.
28. Luria AR. Traumatic aphasia: its syndromes, psychology and treatment. The Hague: Walter De Gruyter; 1970.
29. Luria AR. Neuropsychological studies in the USSR. A review (part I). Proc Natl Acad Sci U S A. 1973;70:959–64.
30. Anokhin PR. Problems of the centrum and the periphery in the physiology of the nervous activity [in Russian]. State Publish House Gorkij. 1935.
31. Bernstein NA. Problems of interrelation of coordination and localization [in Russian]. Arch Biol Sci. 1935;38:1–43.
32. Luria AR. Higher cortical functions in man. 2nd ed. New York: Basic Books Inc.; 1980.
33. Critchley M. In memoriam. A. R. Luria. Brain Lang. 1978;5:v–vi.
34. Luria AR. The man with a shattered world: the history of a brain wound. Cambridge: Harvard University Press; 1987.
35. Luria AR. Language and cognition. New York: Wiley; 1982.
36. Nissl V, Meyendorf E. Vom Lokalisations Problem der Artikulierten Sprache. Leipzig: Deutike; 1930.
37. Alajouanine T, Ombredane A, Durant M. Le syndrome desintegration phonetique. Paris: Masson; 1939.
38. Robinson G, Blair J, Cipolotti L. Dynamic aphasia: an inability to select between competing verbal responses? Brain. 1998;121:77–89.
39. Satoer D, Kloet A, Vincent A, et al. Dynamic aphasia following low-grade glioma surgery near the supplementary motor area: a selective spontaneous speech deficit. Neurocase. 2014;20(6):704–16.
40. Christensen A-L. Luria's legacy in the 21st century. New York: Oxford University Press; 2009.
41. Bailey CH, Giustetto M, Huang YY, et al. Is heterosynaptic modulation essential for stabilizing Hebbian plasticity and memory? Nat Rev Neurosci. 2000;1:11–20.
42. Goddard GV, McIntyre DC, Leech CK. A permanent change in brain function resulting from daily electrical stimulation. Exp Neurol. 1969;25:295–330.
43. Bliss TV, Lomo T. Long-lasting potentiation of synaptic transmission in the dentate area of the anaesthetized rabbit following stimulation of the perforant path. J Physiol. 1973;232:331–56.
44. Brown RE, Milner PM. The legacy of Donald O. Hebb: more than the Hebb synapse. Nat Rev Neurosci. 2003;4:1013–9.
45. Rummelhart DE, McClelland JL. Parallel distributed processing. Cambridge: MIT Press; 1986.
46. Mesulam MM. Large-scale neurocognitive networks and distributed processing for attention, language, and memory. Ann Neurol. 1990;28:597–613.

47. McClelland JL, Rumelhart DE, Hinton GE. The appeal of parallel distributed processing. Cambridge: MIT Press; 1986.
48. Hopfield JJ. Neural networks and physical systems with emergent collective computational abilities. Proc Natl Acad Sci U S A. 1982;79:2554–8.
49. Montgomery SH, Geisler JH, McGowen MR, et al. The evolutionary history of cetacean brain and body size. Evolution. 2013;67:3339–53.
50. Mesulam M. The evolving landscape of human cortical connectivity: facts and inferences. Neuroimage. 2012;62:2182–9.
51. Darwin C. On the origins of species by means of natural selection. London: John Murray; 1859.
52. Lieberman P. Uniquely human: the evolution of speech, thought, and selfless behavior. Cambridge: Harvard University Press; 1993.
53. Falk D. The evolution of Broca's area. IBRO History of Neuroscience; 2007.
54. Pollick AS, de Waal FB. Ape gestures and language evolution. Proc Natl Acad Sci U S A. 2007;104:8184–9.
55. Warden CJ, Warner LH. The sensory capacities and intelligence of dogs, with a report on the ability of the noted dog "fellow" to respond to verbal stimuli. Q Rev Biol. 1928;3:1–28.
56. Chomsky N. Powers and prospects: reflections on human nature and the social order. Boston: South End Press; 1996.
57. Tecumseh Fitch W, de Boer B, Mathur N, Ghazanfar AA. Monkey vocal tracts are speech-ready. Sci Adv. 2016;2:e1600723.
58. Brodmann K. Vergleichende lokalisationslehre der Grosshirnrinde in ihren prinzipien dargestellt auf grund des Zellenbaues. Leipzig: J.A. Barth; 1909.
59. Falk D, Gibson KR. Evolutionary anatomy of the primate cerebral cortex. Cambridge: Cambridge University Press; 2008.
60. Preuss TM. What's human about the human brain. New Cognit Neurosci. 2000;2:1219–34.
61. Rajkowska G, Goldman-Rakic PS. Cytoarchitectonic definition of prefrontal areas in the normal human cortex: I. Remapping of areas 9 and 46 using quantitative criteria. Cereb Cortex. 1995;5:307–22.
62. Baleydier C, Achache P, Froment JC. Neurofilament architecture of superior and mesial premotor cortex in the human brain. Neuroreport. 1997;8:1691–6.
63. Zilles K, Schlaug G, Matelli M, et al. Mapping of human and macaque sensorimotor areas by integrating architectonic, transmitter receptor, MRI and PET data. J Anat. 1995;187:515.
64. Walker AE. A cytoarchitectural study of the prefrontal area of the macaque monkey. J Comp Neurol. 1940;73:59–86.
65. Von Bonin G. Architecture of the precentral motor cortex and some adjacent areas. In: Bucy PC, editor. The precentral motor cortex. Urbana: University of Illinois Press; 1949. p. 7–82.
66. Grodzinsky Y, Amunts K. Broca's region. New York: Oxford University Press; 2006.
67. Petrides M, Cadoret G, Mackey S. Orofacial somatomotor responses in the macaque monkey homologue of Broca's area. Nature. 2005;435:1235–8.
68. Petrides M, Pandya DN. Comparative cytoarchitectonic analysis of the human and the macaque frontal cortex. In: Boller F, Grafman J, editors. Handbook of neuropsychology. Amsterdam: Elsevier; 1994. p. 17–58.
69. Petrides M, Pandya DN. Comparative cytoarchitectonic analysis of the human and the macaque ventrolateral prefrontal cortex and corticocortical connection patterns in the monkey. Eur J Neurosci. 2002;16:291–310.
70. Eidelberg D, Galaburda AM. Inferior parietal lobule: divergent architectonic asymmetries in the human brain. Arch Neurol. 1984;41:843.
71. Galaburda AM, Pandya DN. Role of architectonics and connections in the study of primate brain evolution. In: Falk D, Armstrong E, editors. Primate brain evolution. Berlin: Springer; 1982. p. 203–16.
72. Taglialatela JP, Savage-Rumbaugh S, Baker LA. Vocal production by a language-competent *Pan paniscus*. Int J Primatol. 2003;24:1–17.
73. Hopkins WD, Taglialatela J, Leavens DA. Chimpanzees differentially produce novel vocalizations to capture the attention of a human. Anim Behav. 2007;73:281–6.

74. Sutton D, Jurgens U. Neural control of vocalization. In: Steklis HD, Erwin J, editors. Comparative primate biology, neurosciences. New York: Alan R. Liss, Inc; 1988. p. 625–47.
75. Taglialatela JP, Russell JL, Schaeffer JA, Hopkins WD. Communicative signaling activates 'Broca's' homolog in chimpanzees. Curr Biol. 2008;18:343–8.
76. Caverzasi E, Papinutto N, Amirbekian B, et al. Q-ball of inferior fronto-occipital fasciculus and beyond. PLoS One. 2014;9:e100274.
77. Milner B, Squire LR, Kandel ER. Cognitive neuroscience and the study of memory. Neuron. 1998;20:445–68.
78. Oleksiak A, Postma A, van der Ham IJ, et al. A review of lateralization of spatial functioning in nonhuman primates. Brain Res Rev. 2011;67:56–72.
79. Wurtz RH. Recounting the impact of Hubel and Wiesel. J Physiol. 2009;587:2817–23.
80. Hackett TA, Stepniewska I, Kaas JH. Subdivisions of auditory cortex and ipsilateral cortical connections of the parabelt auditory cortex in macaque monkeys. J Comp Neurol. 1998;394:475–95.
81. Poeppel D, Overath T, Popper AN. The human auditory cortex. New York: Springer; 2012.
82. Rauschecker JP. Ventral and dorsal streams in the evolution of speech and language. Front Evol Neurosci. 2012;4:7.
83. Kaas JH, Hackett TA. 'What' and 'where' processing in auditory cortex. Nat Neurosci. 1999;2:1045–7.
84. Tian B, Reser D, Durham A, et al. Functional specialization in rhesus monkey auditory cortex. Science. 2001;292:290–3.
85. Rauschecker JP, Tian B. Mechanisms and streams for processing of "what" and "where" in auditory cortex. Proc Natl Acad Sci U S A. 2000;97:11800–6.
86. Hickok G, Poeppel D. The cortical organization of speech processing. Nat Rev Neurosci. 2007;8:393–402.
87. Thiebaut de Schotten M, Dell'Acqua F, Valabregue R, Catani M. Monkey to human comparative anatomy of the frontal lobe association tracts. Cortex. 2012;48:82–96.
88. Mesulam MM. Principles of behavioral and cognitive neurology. Oxford: Oxford University Press; 2000.
89. Hickok G, Poeppel D. Dorsal and ventral streams: a framework for understanding aspects of the functional anatomy of language. Cognition. 2004;92:67–99.
90. Demonet JF, Chollet F, Ramsay S, et al. The anatomy of phonological and semantic processing in normal subjects. Brain. 1992;115:1753–68.
91. Petersen SE, Fox PT, Posner MI, et al. Positron emission tomographic studies of the cortical anatomy of single-word processing. Nature. 1988;331:585–9.
92. Zatorre RJ, Evans AC, Meyer E. Neural mechanisms underlying melodic perception and memory for pitch. J Neurosci. 1994;14:1908–19.
93. Creutzfeldt O, Ojemann G, Lettich E. Neuronal activity in the human lateral temporal lobe. I Responses to speech. Exp Brain Res. 1989;77:451–75.
94. Mazzucchi A, Marchini C, Budai R, Parma M. A case of receptive amusia with prominent timbre perception defect. J Neurol Neurosurg Psychiatry. 1982;45:644–7.
95. Polster MR, Rose SB. Disorders of auditory processing: evidence for modularity in audition. Cortex. 1998;34:47–65.
96. Mesulam MM, Thompson CK, Weintraub S, Rogalski EJ. The Wernicke conundrum and the anatomy of language comprehension in primary progressive aphasia. Brain. 2015;138:2423–37.
97. Wise RJ, Scott SK, Blank SC, et al. Separate neural subsystems within 'Wernicke's area'. Brain. 2001;124:83–95.
98. Saur D, Kreher BW, Schnell S, et al. Ventral and dorsal pathways for language. Proc Natl Acad Sci U S A. 2008;105:18035–40.
99. Hickok G, Poeppel D. Towards a functional neuroanatomy of speech perception. Trends Cogn Sci. 2000;4:131–8.
100. Catani M, Thiebaut de Schotten M. Atlas of human brain connections. Oxford: Oxford University Press; 2012.

101. Bajada CJ, Lambon Ralph MA, Cloutman LL. Transport for language south of the Sylvian fissure: the routes and history of the main tracts and stations in the ventral language network. Cortex. 2015;69:141–51.
102. Papagno C, Casarotti A, Comi A, et al. Long-term proper name anomia after removal of the uncinate fasciculus. Brain Struct Funct. 2016;221(1):687–94.
103. Duffau H, Gatignol P, Mandonnet E, et al. Intraoperative subcortical stimulation mapping of language pathways in a consecutive series of 115 patients with Grade II glioma in the left dominant hemisphere. J Neurosurg. 2008;109:461–71.
104. Duffau H, Leroy M, Gatignol P. Cortico-subcortical organization of language networks in the right hemisphere: an electrostimulation study in left-handers. Neuropsychologia. 2008;46:3197–209.
105. Hickok G, Buchsbaum B, Humphries C, Muftuler T. Auditory-motor interaction revealed by fMRI: speech, music, and working memory in area Spt. J Cogn Neurosci. 2003;15:673–82.
106. Cogan GB, Thesen T, Carlson C, et al. Sensory-motor transformations for speech occur bilaterally. Nature. 2014;507:94–8.
107. Catani M, Jones DK, ffytche DH. Perisylvian language networks of the human brain. Ann Neurol. 2005;57:8–16.
108. Makris N, Kennedy DN, McInerney S, et al. Segmentation of subcomponents within the superior longitudinal fascicle in humans: a quantitative, in vivo, DT-MRI study. Cereb Cortex. 2005;15:854–69.
109. Aboitiz F, Aboitiz S, García RR. The phonological loop. Curr Anthropol. 2010;51:S55–65.
110. Bernal B, Altman N. The connectivity of the superior longitudinal fasciculus: a tractography DTI study. Magn Reson Imaging. 2010;28:217–25.
111. Duffau H, Moritz-Gasser S, Mandonnet E. A re-examination of neural basis of language processing: proposal of a dynamic hodotopical model from data provided by brain stimulation mapping during picture naming. Brain Lang. 2014;131:1–10.
112. Baddeley AD, Hitch GJ. Working memory. Psychol Learn Motiv. 1974;8:47–89.
113. Gathercole SE, Baddeley AD. Working memory and language. Hove: Psychology Press; 1993.
114. Baddeley AD. Working memory. Oxford: Oxford University Press; 1986.
115. De Renzi E, Nichelli P. Verbal and non-verbal short-term memory impairment following hemispheric damage. Cortex. 1975;11:341–54.
116. Acheson DJ, Hamidi M, Binder JR, Postle BR. A common neural substrate for language production and verbal working memory. J Cogn Neurosci. 2011;23:1358–67.
117. Rogalsky C, Matchin W, Hickok G. Broca's area, sentence comprehension, and working memory: an fMRI study. Front Hum Neurosci. 2008;2:14.
118. Vigneau M, Beaucousin V, Herve PY, et al. Meta-analyzing left hemisphere language areas: phonology, semantics, and sentence processing. Neuroimage. 2006;30:1414–32.
119. Paulesu E, Frith CD, Frackowiak RS. The neural correlates of the verbal component of working memory. Nature. 1993;362:342–5.
120. Awh E, Schumacher E, Smith E et al. Investigation of verbal working memory using PET. 1994. Cognitive Neuroscience Society Meeting (conference proceedings).
121. McGettigan C, Warren JE, Eisner F, et al. Neural correlates of sublexical processing in phonological working memory. J Cogn Neurosci. 2011;23:961–77.
122. Caplan D, Waters GS. Verbal working memory and sentence comprehension. Behav Brain Sci. 1999;22:77–94.
123. Miyake A, Shah P. Models of working memory: mechanisms of active maintenance and executive control. New York: Cambridge University Press; 1999.
124. Shallice T, Butterworth B. Short-term memory impairment and spontaneous speech. Neuropsychologia. 1977;15:729–35.
125. Vallar G, Shallice T. Neuropsychological impairments of short-term memory. Cambridge: Cambridge University Press; 1990.
126. Baddeley A, Gathercole S, Papagno C. The phonological loop as a language learning device. Psychol Rev. 1998;105:158–73.

127. Szmalec A, Brysbaert M, Duyck W. Working memory and (second) language processing. Memory, language, and bilingualism: theoretical and applied approaches; 2012. p. 74–94.
128. Geschwind N. Specializations of the human brain. Sci Am. 1979;241:180.
129. Poliakov GI. Neuron structure of the brain. Cambridge: Harvard University Press; 1972.
130. Blinkov SM. Structural peculiarities of the human cerebrum. Moscow: Medgiz; 1955.
131. Kagan A, Saling MM. An introduction to Luria's aphasiology: theory and application. Baltimore: Paul H Brookes Publishing Company; 1988.
132. Hebb DO. The organization of behavior: a neuropsychological theory. New York: Wiley; 1949.
133. Berlucchi G, Buchtel HA. Neuronal plasticity: historical roots and evolution of meaning. Exp Brain Res. 2009;192:307–19.
134. Gage N, Hickok G. Multiregional cell assemblies, temporal binding and the representation of conceptual knowledge in cortex: a modern theory by a "classical" neurologist. Carl Wernicke Cortex. 2005;41:823–32.
135. Coolen ACC, Jonker HJJ. Introduction to neural networks [Dutch]. Utrecht: University Utrecht; 1991.
136. Rutten GJ, Ramsey NF. Functional neuroimaging in neurosurgical practice. In: Duffau H, editor. Brain mapping: from neural basis of cognition to surgical applications. Berlin: Springer; 2011. p. 207–227.
137. Petrides M, Pandya DN. Distinct parietal and temporal pathways to the homologues of Broca's area in the monkey. PLoS Biol. 2009;7:e1000170.
138. Romanski LM, Tian B, Fritz J, et al. Dual streams of auditory afferents target multiple domains in the primate prefrontal cortex. Nat Neurosci. 1999;2:1131–6.
139. Romanski LM, Bates JF, Goldman-Rakic PS. Auditory belt and parabelt projections to the prefrontal cortex in the rhesus monkey. J Comp Neurol. 1999;403:141–57.
140. Petrides M, Pandya DN. Dorsolateral prefrontal cortex: comparative cytoarchitectonic analysis in the human and the macaque brain and corticocortical connection patterns. Eur J Neurosci. 1999;11:1011–36.

Functional MRI

8

Functional neuroimaging techniques, first positron emission tomography (PET) and later functional MRI (fMRI), have revolutionized cognitive neuroscience. These tools have also greatly improved our understanding of how language is implemented in the brain. Almost from the beginning, fMRI was also applied for language mapping in surgical practice because of its obvious benefits: high-resolution whole-brain mapping without the need for invasive procedures. Other clinical applications that have been investigated, although less frequently, are the use of fMRI as a tool to help diagnose or understand diseases that lack clear neuroanatomical characteristics or as a predictor for language outcome after stroke (see Chap. 9).

But in contrast to its success in the cognitive neurosciences, it has proven to be more problematic than expected to turn fMRI into an instrument that clinicians want to use for their patients. This chapter explores fMRI from both a neuroscientific and a clinical perspective. First, key principles of functional neuroimaging are listed, together with some methodological considerations. Next, the history of functional neuroimaging is described and the scientific progress that has been made in understanding language with the help of neuroimaging. Finally, we will explore the potential of fMRI in a clinical setting and discuss strategies to further develop it as a clinical instrument.

8.1 Brief Introduction to the Method

Before the 1980s, the neural basis of brain functions was largely inferred from the results of post-mortem lesion studies and the small number of neurosurgical patients that had been investigated with electrocortical stimulation mapping. The problem was that this data had been gathered from diseased and damaged brains, and the

This chapter was co-authored by Martijn Jansma (Department of Neurosurgery, St Elisabeth-Tweesteden Hospital, Tilburg, the Netherlands)

G.-J. Rutten, *The Broca-Wernicke Doctrine*, DOI 10.1007/978-3-319-54633-9_8

presence of the lesion had probably already altered the functional organization of the brains of these patients or at least induced some form of compensation. PET provided, for the first time, an instrument that could study the normal functional neuroanatomy of brains unaffected by pathology or potential functional reorganization with good spatial precision [1]. PET is a slightly invasive technique because it requires the intravenous injection of radioactively labelled water (with a half-life of 2 min). At the beginning of the 1990s, it was discovered that changes in brain activity can be visualized without any contrast agent, using an MRI machine. This technique became known as 'functional' MRI or fMRI. The first fMRI studies were published in 1992 by Bandettini [2], Ogawa [3] and Kwong [4], and these groups are usually credited with the discovery.

The principle upon which PET and fMRI rest is that neurons that become more active increase their energy consumption. The brain is a particularly greedy consumer of energy, as it takes up to 20% of the bodily energy consumption with only 2% of its body weight [5]. To supply this energy, local blood vessels dilate to increase the inflow of nutrients. This leads to regional changes in the brain's physiology and metabolism, of which some can be mapped with surprisingly precise spatial resolution. Current functional imaging techniques mostly use differences in the concentration of glucose, oxygen or haemoglobin as an indirect means to construct three-dimensional images of brain activity.

fMRI is the only neuroimaging technique that is completely non-invasive. It exploits the fact that the actual inflow of oxygen to regions with increased neuronal activity is much more than locally used by neurons [6, 7]. Malonek and Grinvald very eloquently described this as the brain 'watering the entire garden for the sake of one thirsty flower' [6]. This phenomenon is also called the 'blood oxygen-level dependent', or 'BOLD' effect (Fig. 8.1), and is the basis of the contrast that is used in almost all fMRI measurements to visualize changes in brain activity.[a]

8.1.1 Task Conditions

fMRI provides the investigator with a three-dimensional map of signal changes in the brain. These changes can be spontaneous or associated with different task conditions (see Fig. 8.2 for an example of a block-designed experiment). Language studies have often focused on the distinction between productive and

[a]The BOLD contrast is sensitive to the level of deoxygenated haemoglobin in the blood. Haemoglobin is a large protein that contains iron and transports oxygen. When the molecule releases its oxygen, and gets deoxygenated, it becomes much more magnetic; it now acts as a little magnet that distorts the local magnetic field of the MR scanner. With the inflow of new oxygenated blood, in response to increased neural activity, deoxygenated blood is washed out, and the ability to measure signals from the tissue with the MR scanner improves. Although BOLD imaging is the most frequently used fMRI technique, there are several other methods that rely on other physiological changes (e.g. arterial, capillary or venous flow) [13].

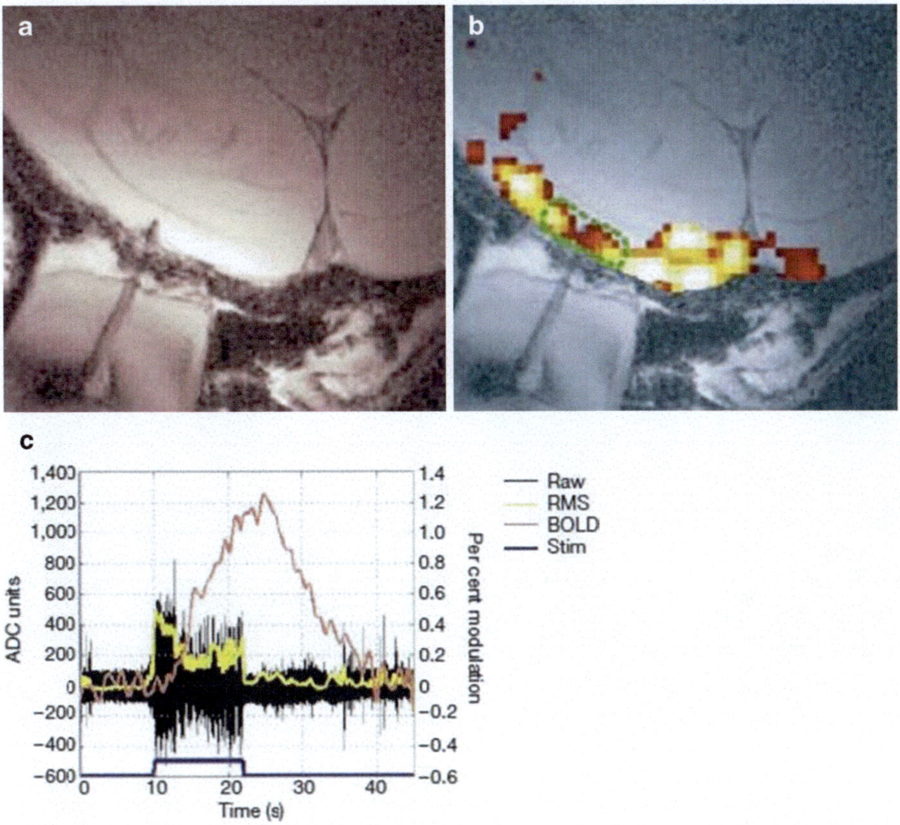

Fig. 8.1 The exact relationship between neural and cerebrovascular changes remains unknown, but microelectrode recordings in both animals and humans strongly suggest that the BOLD signal corresponds to local field potentials (LFPs) [8, 9]. LFPs reflect the input and intracortical processing of a population of neurons, rather than their output. Several studies have also reported a correlation between fMRI signals and an increase in the power spectrum as measured with electrocorticography for a variety of tasks [10, 11]. Note that a typical voxel (3–5 mm in each dimension) contains over 5 million neurons, 50 billion synapses and 200 km of axon [12]. (**a**) MRI scan shows the position of a microelectrode within the visual cortex of a monkey (Logothetis 2001) [9]. (**b**) Results of an fMRI experiment in the same monkey with visual stimuli (rotating chessboard for 10 s). BOLD responses (colour) are measured from the visual cortex. (**c**) Graph shows that the BOLD response starts several seconds after neural activity within the same region. Logothetis (2001) concluded from the experiments that 'results show unequivocally that a spatially localized increase in the BOLD contrast directly and monotonically reflects an increase in neural activity. (…) In a first approximation BOLD and neural responses are shown to have a linear relationship for short stimulus presentation durations' [9] (Figures from Logothetis (2001) [9])

receptive functions, but this dichotomy has not always been helpful in designing fMRI experiments to unravel the anatomical basis of language functions, nor in understanding aphasic problems. Binder suggested the use of a classification

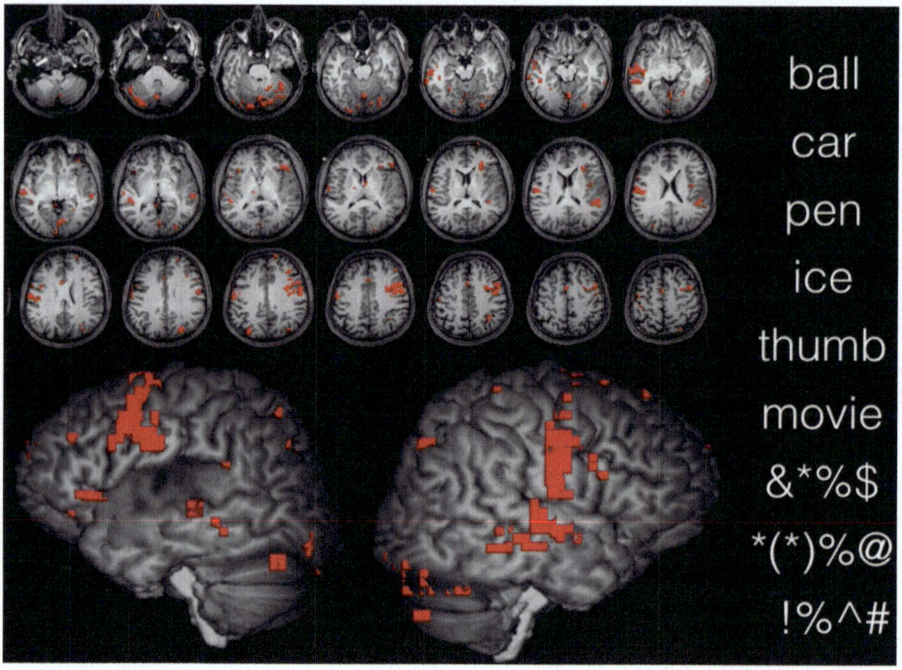

Fig. 8.2 BOLD-fMRI language experiment in a patient with a left temporal lobe tumour (low-grade glioma located predominantly in the left superior temporal gyrus). Maximal surgical removal of such a glioma requires information on critical language-related structures located nearby. BOLD-fMRI can only measure relative signal changes and does not provide the investigator with absolute measurements of brain activity. As such it relies upon a contrast of two or more experimental conditions (i.e. it requires a reference condition against which signal changes can be measured). In its simplest and most commonly used form, as shown here, the experiment has a 'block-design' with two conditions. In the 'active' condition, single nouns were presented, and the patient was instructed to think of a related verb (e.g., ball → throw, car → drive). In the control condition, symbols were presented; it was hypothesized that these stimuli do not engage the language system. There were ten epochs of 30 s (five for the active and five for the control condition). Spatial resolution (voxel dimension) is 3 × 3 × 3 mm. Red voxels denote areas where the signal is significantly higher during the active than during the control condition. Note that the results from all different stimuli condense in a single functional brain map. To provide an anatomical frame of reference, fMRI results are projected onto detailed anatomical MRI images (here, a cortical rendering of a T1-weighted MRI scan). In this patient, considerable activation of frontal and temporal language areas in both hemispheres was observed, suggesting bilateral representation of language functions. (Sub)cortical mapping during awake surgery confirmed language areas in the left hemisphere (for obvious reasons, no information was acquired from right fronto-temporal areas during surgery). These single-task fMRI experiments are increasingly used for the planning of brain surgery. Point of discussion is to what extent fMRI areas are truly critical for normal language functions and have clinical relevance. In other words, it is not known whether removal of these areas will lead to lasting language problems for patients

from linguistics for fMRI experiments, which splits up language into five subcomponents instead of two:

> (1) phonetics, the processes governing production and perception of speech sounds; (2) phonology, the processes by which speech sounds are represented and manipulated in abstract form; (3) orthography, the processes by which written characters are represented

and manipulated in abstract form; (4) semantics, the processing of word meanings, names, and other declarative knowledge about the world; and (5) syntax, the processes by which words are combined to make sentences and sentences analysed to reveal underlying relationships between words [14].

Although this classification seems a better approach for the study of language, it should be noted that it cannot always be expected that different brain functions add up linearly. Cognitive processes can interact in a complex and non-linear fashion, creating unpredictable effects [15]. Block-designed experiments are rooted in a form of modular thinking and rely on the principle of *pure insertion*, implicitly assuming that perception, cognition and action are largely independent brain processes [16]. Modern theories, on the other hand, propose that the same neurons can be active during both perception and action and that sensorimotor systems form an integral part of cognitive processes [17].

8.1.2 Detection Power

The signals that can be picked up with fMRI only increase a few percent after brain regions become neurally more active. These changes are so small that they easily get lost in huge amounts of background activity [13, 18]. There is an ongoing discussion about the exact physiological processes that take place when the brain is engaged in certain tasks [19]. The term *active* or *activated* is frequently used in functional neuroimaging studies to denote an area that is responsive during a particular sensorimotor or cognitive task. The term implies that brain areas change from a dormant state into a functional one. However, brain areas are probably never truly 'inactive'. In fact, most of the brain's energy is needed to sustain a certain baseline neural activity. It is therefore better said that areas become more (or less) active and that the level of activation—whatever that means—varies over time.

Investigators may use different strategies to increase the detection power for the task-related signals that they are interested in. These come at a cost, though, and impose various constraints on the design of experiments. Common strategies include repetition of stimuli and spatial smoothing of the data.[b] What raises the complexity of fMRI analyses further is that signals are measured from several thousands of voxels and that for each voxel hundreds of measurements are performed in a time series. As is the case with noisy data, statistics are subsequently needed to decide whether or not the measured signals contain information. Functional brain maps are therefore, by definition, statistical representations of the outcome of the

[b] When fMRI images are spatially smoothed, the signal from any given voxel is averaged with that of its neighbours. This results in an image that is blurred and has less anatomical detail. The spatial extent of smoothing is determined by the experimenter. In mathematical terms, a convolution is done with a Gaussian kernel of which the full width at half of its maximum (FWHM) determines the spatial extent of smoothing. Smoothing accounts for local differences in anatomy across individuals so that images can more easily be aligned with those of other brains or with a standard template. This process generally also increases the signal-to-noise ratio. However, there are important downsides: small areas of activation are lost in the process (resulting in false-negative activation), and the spatial detail may not be good enough for surgical purposes [20, 21].

experiment.[c] Although there are a number of popular software packages for fMRI analysis (e.g. SPM, AFNI, FSL), standardization is lacking at this point [22].

fMRI can be performed with a regular MRI scanner, which is nowadays widely available in hospitals. At most, it requires relatively inexpensive adaptations to software and hardware, making it much more accessible than PET. Current state-of-the-art fMRI techniques are sensitive enough to pick up signals related to a single stimulus in individual subjects at a millimetre resolution. However, a typical cognitive fMRI experiment provides brain maps reflecting average activity of groups of subjects with a resolution of about 1–2 cm.

In a relative short time, fMRI has become an immensely popular tool in neuroscience. fMRI protocols and techniques have continuously evolved, and by 2010, scientists published about it in more than 1500 articles [24]. Other brain mapping techniques, such as electroencephalography (EEG) or magnetoencephalography (MEG) are used much less frequently because of their lack of spatial precision (EEG, MEG) or very high costs (MEG). In stark contrast, clinical fMRI investigations have not changed their basic methodology much since the pioneering language studies of Petersen or Binder or the first neurosurgical patient studies that came out in the 1990s [25, 26]. This is remarkable because there have always been significant discrepancies between the results of fMRI and those of the invasive clinical techniques. One would have expected this to be a strong impulse to re-evaluate both new and established clinical methods and to understand why these results are so different (as they both, of course, aim for a similar goal: a map of critical language areas). In reality, we think, relatively little effort has been made to close this barrier. Clinicians still rely on the Wada test and direct electrical stimulation (DES) for language mapping, although they are starting to acknowledge that these techniques have significant flaws and that they are not the ideal gold standards against which to judge fMRI maps (see also Chap. 6) [1, 14]. Neuroscientists, on the other hand, never really turned their attention to the mapping of functions in individuals—an important prerequisite for clinical use. For some reason, the same fMRI methods that were developed for group studies were also used for precise localization of functions in individual patients. However, single-subject fMRI requires a different approach. Before this discussion is entered in more detail, we will first summarize what neuroscientific studies, and in particular those that used functional imaging techniques, have taught us about the neural basis of language.

8.2 Historical Perspective

The underlying assumption of BOLD imaging is that regional blood flow changes are correlated to local neural activity. Such a relationship was already suspected at the end of the nineteenth century [27, 28]. Angelo Mosso (1846–1910), an Italian physiologist, was the first to actually gain experimental evidence of this phenomenon [29]. He measured pulsations of the brain in a patient with a skull defect after a neurosurgical intervention. A farmer named Bertino had bone pieces removed after a traumatic skull

[c] The validity of many published fMRI studies has been seriously questioned over the past years (e.g. Vul 2009, Eklund 2016), suggesting that these had too high rates of false positives or have reported correlations between brain and behaviour that are 'impossibly high' [22, 23]. This reflects the ongoing discussion in search for valid methods to analyse and interpret fMRI data.

fracture, and that had left him with a bony opening in the skull. Mosso designed a device that could simultaneously measure pulsations from the forearm and from the skin overlying the bone defect. He hypothesized that these latter measurements reflected changes in brain volume and intracranial pressure. Then, and this was truly a scientific inquiry, Mosso asked Bertino to perform certain cognitive tasks (i.e. multiplication) and noted that—after some delay—the pulsations from the surface of the brain increased, whereas those of the forearm did not (Fig. 8.3). This indicated, or at

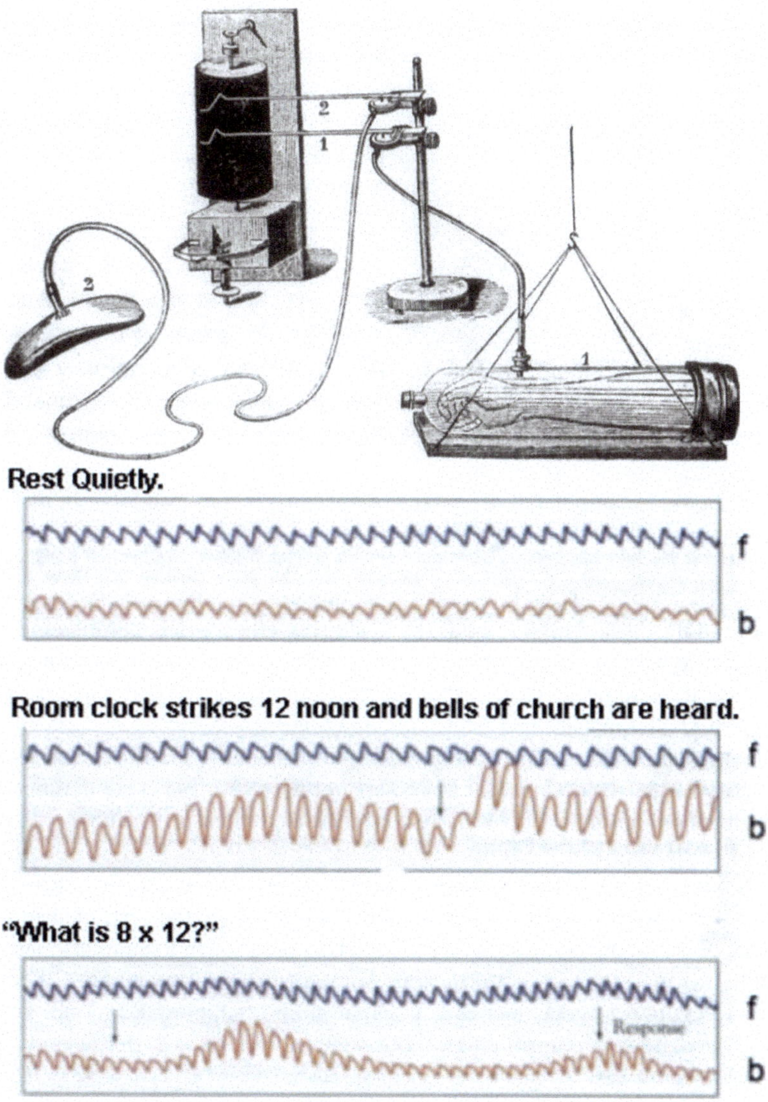

Fig. 8.3 Mosso's device for simultaneously recording pulsation in the arm and brain. The brain showed stronger pulsations after an event that stimulated brain activity (shown with *arrow*) [29]

least suggested to him, that blood rushes to the brain when mental performance increases. As such, he was one of the first researchers to develop a technique for *in vivo brain imaging*. Mosso also performed other experiments that—as we read it now—require somewhat more wishful thinking:

> The subject to be observed lay on a delicately balanced table which could tip downwards either at the head or the foot if the weight of either end were increased. The moment emotional or intellectual activity began in the subject, down went the balance at the head-end, in consequence of the redistribution of blood in his system.[d]

At the time, many other researchers were interested in the psychophysiological relationships between the mind and brain. Some, including Mosso, studied temperature changes of the brain and performed various experiments in animals and man to study the causal effects of these changes. Mosso described them in his book *Die Temperatur des Gehirns* (1894). He concluded that 'the fluctuation in the temperature of the brain was independent of blood temperature and was likely related to the metabolic activity of the brain itself' [29]. There are several historical notes of interest here. For instance, Hans Berger (1873–1941), the inventor of electroencephalography, was inspired by Mosso's work. In his clinic in Jena, he adapted Mosso's technique of plethysmography and this eventually led to development of the EEG. As Schiller wrote: 'One day in 1924 this none too rewarding pursuit [of Berger] gave him the idea of using electrodes, to replace "thermoencephalography" with the E.E.G.' [30]. Broca also tried to localize brain lesions via recording of the temperature of the skin. He had learned from his general surgical work that blood flow changes affected the temperature of a limb. Broca used this information to determine at what exact place a diseased extremity should be amputated. In an analogous way, he proposed that brain lesions led to changes in the cerebral vasculature and subsequently to changes in local temperature. To measure these changes, Broca had devised a 'thermometric crown' that he considered sensitive enough to pick up temperature changes despite the physical boundaries of the dura, skull and skin. Broca also hypothesized that brain temperature should increase with the execution of cognitive tasks, in particular in the frontal areas, and he described some of these experiments in his work [31]. Again, these were among the first attempts at functional brain imaging.

Following up on these pioneering studies at the end of the nineteenth century, Charles Roy (1854–1897) and Charles Sherrington (1857–1952) studied brain volume changes in animals via a more sophisticated measuring device that was implanted in the skull of an animal (Fig. 8.4). This allowed them to record these changes under various controlled experimental circumstances, for instance, during the stimulation of peripheral nerves or medulla oblongata, the restricted inflow or outflow of blood to the cranium or asphyxia. This resulted in a landmark paper that was published in 1890 and that described some of the basic principles of cerebral blood flow regulatory mechanisms [33]. Their studies supported a model whereby cerebral blood flow is controlled by both 'extrinsic' factors (arterial and venous blood pressure) as well as 'intrinsic' (local) factors:

[d] Quote taken from *The principles of psychology* by William James, 1890 [107]

> These facts seem to us to indicate the existence of an automatic mechanism by which the blood supply of any part of the cerebral tissue is varied in accordance with the activity of the chemical changes which underlie the functional action of that part. Bearing in mind that strong evidence exists of localization of function in the brain, we are of opinion that an automatic mechanism, of the kind just referred to, is well fitted to provide for a local variation of the blood-supply in accordance with local variations of the functional activity. (Roy & Sherrington, 1890 [33])

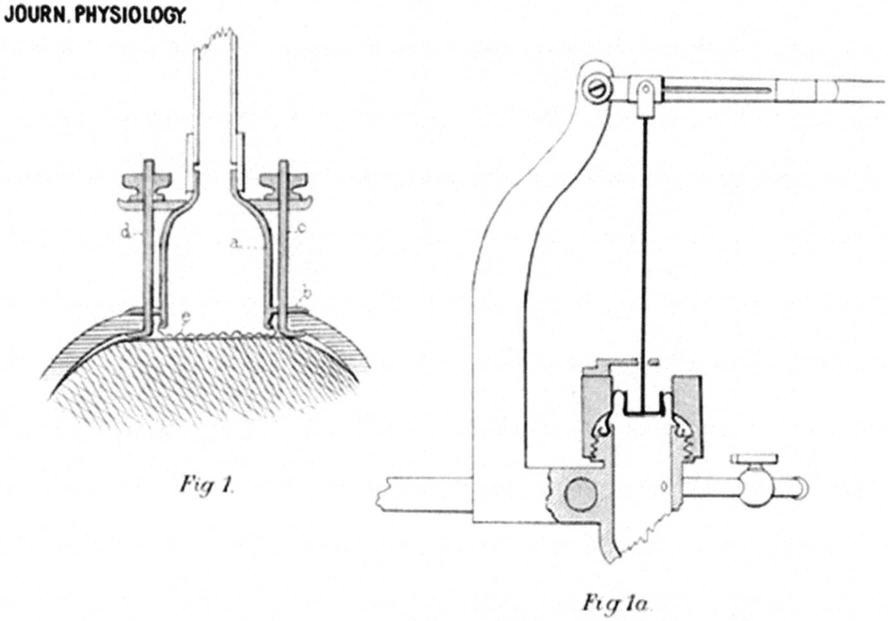

Fig. 8.4 'A trepan hole, about 22 mm in diameter in the case of dogs, but smaller when a cat or a rabbit was used, was made as near the middle line of the vertex of the cranium as is compatible with avoidance of the longitudinal sinus, after which the subjacent dura was removed by a circular incision. After any oozing of blood from the diploë had ceased, a small metal capsule, of a size corresponding with that of the trepan hole, was fixed over the aperture by means of screws. The shape of the capsule and the mode of fixing it firmly to the skull can be seen on reference to Fig. 1. The lower opening of the bell-shaped capsule (*a*) is closed by a very flexible, delicate, animal membrane (*e*), of the kind already used by one of us (R.) in other apparatus. It is tied on in such a way that it readily follows all changes in the level of the part of the cortex on which it rests, while it prevents any escape of the air with which the capsule is filled. Outside the capsule, about two mm from its lower edge, is a projecting rim (*b*), which rests on the external surface of the cranial bone. This rim has in it two notches, in which fit two metal pins (*c* and *d*), bent at right angles at their lower ends, so that they can hook under the bone on opposite sides of the hole. By means of small thumbscrews on the upper parts of these pins, the capsule is held firmly in position. The upper opening of the capsule is connected by means of rigid-walled tubing with the recording apparatus. This latter consists of an arrangement similar to that which one of us has described as useful for studying the form of the pulse wave and which is shown in Fig. 1a. A light piston, escape of fluid by the side of which is prevented by a flexible membrane of the kind already referred to, conveys to a recording lever any changes in the volume of the brain (Text and figures taken from Roy and Sherrington (1890)' [33])

The principles that were laid out by Roy and Sherrington are still the basis of modern functional neuroimaging techniques, although controversy remains about the exact underlying mechanisms. See for an overview Fox (2012) [7].

8.3 What Neuroscientific Studies Taught Us About the Neural Basis of Language

An important goal of neuroscience is to unravel and, if possible, to understand the neural architecture that underlies brain functions. The traditional approach, that started with the lesion-deficit studies at the end of the nineteenth century, is to search for consistent and meaningful structure–function relations among subjects. This is also the approach that is still often used in functional imaging studies. Data from individual subjects are transposed to some standard brain template, after which they are averaged to form groups. Averaging is beneficial in the sense that it reduces the influence of noise and individual variations that are considered to be of no interest. Although meaningful differences between subjects may get lost in this process, this trade-off is generally accepted by the neuroscientific community in search for overarching theories and models [34].[e]

Functional neuroimaging studies have greatly improved our understanding of how language is implemented in the brain, and have provided alternatives for the classic convictions on language localization. It is worth taking a closer look at some of the older studies first, as these already yielded several observations that conflicted with the classic Broca-Wernicke model, both from a conceptual and anatomical point of view.

8.3.1 Some Landmark Studies

One of the first functional imaging studies on language processing was published in 1988 by Petersen and colleagues: *Positron emission tomographic studies of the cortical anatomy of single-word processing* [37]. In the introduction of their paper in

[e] There are, however, critics who state that group studies of patients, and even of normal subjects, have no relevance to the understanding of brain function. Read Caramazza (1986) for an elegant overview of arguments [35]. Caramazza strongly proposed that single-case studies are the only valid manner to study brain–behaviour relations. Others have reasoned that this could potentially lead 'to the logical absurdity of there being as many theories as there are patients' (Halai 2016) or warn that individual patient measurements may be too specific to allow meaningful generalization to a reference population [35, 36]. It is indeed difficult to draw general conclusions when experimental conditions and performance among patients are not homogeneous. However, deviant responses from brain-damaged patients are, of course, not entirely unconstrained and as such can be used to test new hypotheses and models. As Caramazza wrote 'the performance of all individual patients (as well as the performance of normal subjects) must be considered in the evaluation of a proposed model of a cognitive system' [35]. In recent years, new and complementary methods are being developed that not only generate a model for the group as a whole but also capture individual differences (e.g. [36]).

Nature they refer to Geschwind's model as the clinical model that was most widely accepted at that time, and that 'argues for serial processing, with an early recoding of visual input into an auditory-based code which is used in turn for semantic and articulatory access' [37]. The results from Petersen's functional imaging studies did not support this model, but were more consistent with models of parallel processing that had already been suggested in contemporary studies [38, 39]. The authors concluded that there were three main findings that were inconsistent with any model of serial language processing:

> First, there is no activation in any of our visual tasks near Wernicke's area or the angular gyrus in posterior temporal cortex. Visual information from occipital cortex appears to have access to output coding without undergoing phonological recoding in posterior temporal cortex. Second, tasks calling for semantic processing of single words activate frontal, rather than posterior, temporal regions. Third, sensory-specific information appears to have independent access to semantic codes and output codes; simple repetition (output tasks) of a presented word failed to activate the left-frontal semantic areas (association tasks) [37].

Petersen studied both auditory and visual processing of single words. They used four behavioural conditions in a three-level hierarchical block design. 'Each task state was intended to add a small number of operations to those of its subordinate (control) state' [34]. A description of the tasks and results is given in Fig. 8.5. The authors proposed a model whereby there are multiple routes between areas that code articularly, phonological or semantic information. The importance of the early PET studies was far-reaching, as, for instance, stated by Price in an extensive review of 20 years of PET and fMRI language studies (1992–2011) [40].

> They [the first PET studies] illustrated that functional imaging could provide anatomical localization with a precision that far exceeds that attainable with human brain lesion studies. Moreover, the study of healthy subjects avoids possible confounding effects of brain lesions, such as compensatory reorganization of brain function [41–43]. Methodological challenges were also well appreciated, particularly when the results appeared to contradict classic axioms of language organization. For example, Steinmetz and Seitz (1991) [44] argued that data should not be averaged over subjects because intraoperative stimulation showed diversity in location of language functions and morphometrical imaging studies showed diversity of brain shape and gyral patterns that would be difficult to correct with anatomical normalization techniques. Many other concerns were succinctly addressed in a review by Petersen and Fiez (1993) [45], who pointed out that functional neuroimaging results should be viewed as evolutionary, rather than revolutionary and that they were most interpretable when they were backed up by supporting data from other studies.(…) Petersen and Fiez (1993) [45] also emphasized that complex language functions were not localized in specific brain regions; they were distributed across networks of regions with each area making a specific contribution to the performance of the task, which depends on its connections to other areas in a parallel distributed hierarchy. In this context, understanding the functional anatomy of language cannot be deduced from a single experiment; rather, it requires the integration of results from multiple experiments using multiple techniques [40].

When fMRI was further developed, it gradually became the most frequently used tool to study the neural basis of language and other cognitive functions. One of the first fMRI studies that targeted language areas was from Binder and colleagues (1997)

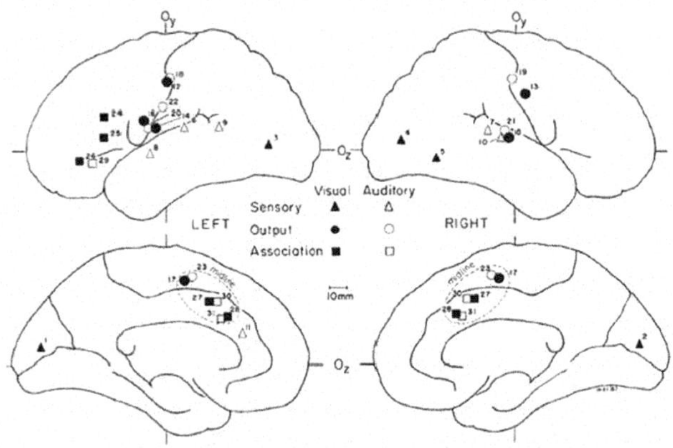

Table 1 Paradigm design

Subtraction	Control state	Stimulated state	Task
Sensory task	Fixation point only	Passive words	Passive sensory processing Modality-specific word code
Output task	Passive words	Repeat words	Articulatory code Motor programming Motor output
Association Task	Repeat words	Generate uses	Semantic association Selection for action

Fig. 8.5 Schematic results of a PET language experiment in 17 healthy subjects from the paper of Petersen (1988) [37]. There are four behavioural conditions, with both auditory and visual presentation of stimuli (looking at fixation point—listening or reading passive words—repeating visually or auditory presented words—generating a verb from a given noun). The researchers calculated three different contrasts from these task conditions (the right column in the table shows the cognitive processes that are hypothesized to be different in the stimulated state versus the control state). The figures denote the 'activated' areas that were found

[46]. His group studied 30 right-handed volunteers with a semantic decision task, with auditive presentation of words (via MRI-compatible headphones). There were two control conditions: one in which subjects had to perform a tone decision task and one where they were asked to remain relaxed and motionless (i.e. without explicit instructions) (see Fig. 8.6 for details and results). As in Petersen's and other studies, some of the results were clearly incongruent with the classic language view [37, 47, 48]. For instance, despite abundant activation of left temporoparietal areas on the group maps, Wernicke's area was not clearly activated. Most of the temporal activation was found in the middle temporal gyrus. Another remarkable finding, also found in Petersen's study, was that the semantic decision task not only activated temporoparietal areas, as expected from classic teaching, but also left frontal language areas. In general, frontal areas seem more easily activated during fMRI language tasks than temporoparietal areas. These frontal areas often extend well beyond the classic Broca's area and include large parts of the medial and lateral prefrontal cortex.

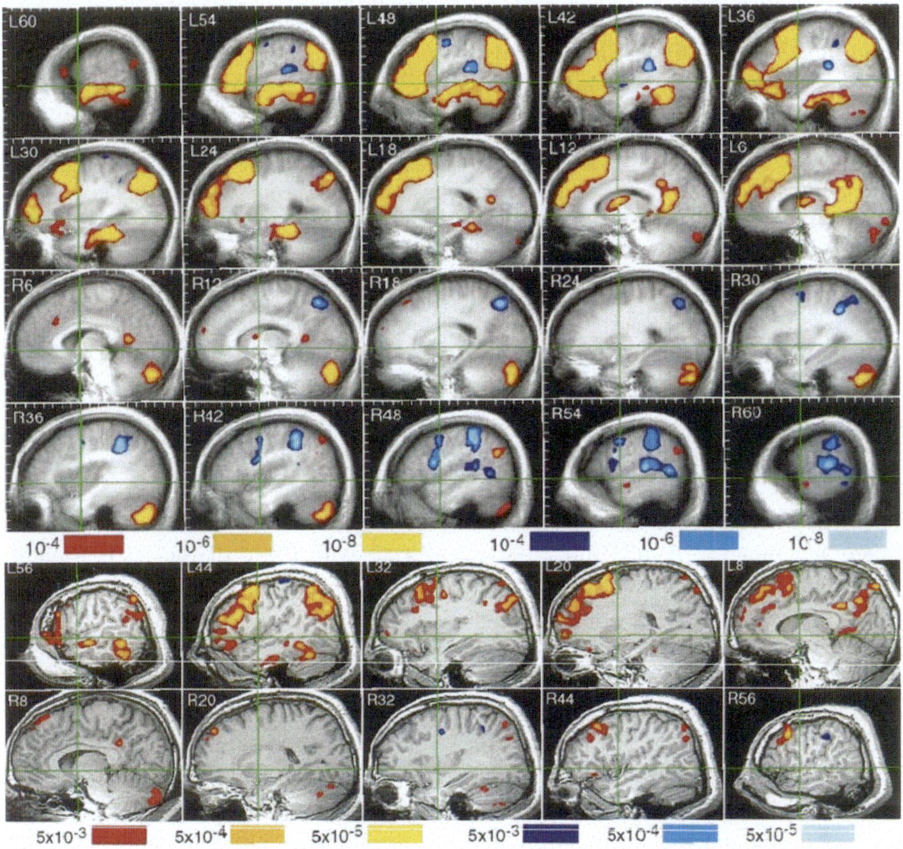

Fig. 8.6 Block-designed fMRI language experiment taken from the paper of Binder (1997) [46]. Experimental conditions included a 'rest' state and two behavioural tasks. Stimuli were given via headphones and were either tones or sampled male speech sounds. In the semantic decision condition, subjects had to decide (via a button press) whether a spoken English noun was an animal that was both 'native to the United States' and 'used by humans'. In the control condition, two different tones were presented (500 and 750 Hz). Subjects had to respond when they heard two consecutive 750 Hz tones. (*Top*) Group results for 30 right-handed healthy volunteers. fMRI activation maps are shown for the semantic decision versus tone decision comparison, whereby the results were scaled to an averaged standard brain. Note strict left-sided lateralization and extensive involvement of large parts of frontal, temporal and parietal areas outside classic language areas. *Yellow-red colour* scale denotes the probability that voxels are activated in the semantic decision task relative to the tone decision task. Cyan-blue voxel scale with the reverse contrast (note that these areas are strongly lateralized to the non-dominant hemisphere). (*Bottom*) fMRI language areas in an individual subject (26-year-old male)

But judgements on how well functional neuroimaging results correspond to those of classic language models depend on the investigator's perspective. In 2000, Price reviewed the functional neuroimaging studies that had been performed thus far—focusing on single-word processing tasks—and specifically compared them against classic nineteenth- and twentieth-century lesion-deficit models [49].

Although Price suggested modifications to the classic models, based on the findings from modern imaging studies and cognitive psychology, she concluded that there were more commonalities than differences:

> The correspondence to the 19th Century neurological model illustrated in Figure 1 [our Fig. 8.7] is clear although a few refinements have been made. First, the site that corresponds to the function of Wernicke's area is the upper bank of the posterior superior temporal sulcus. Second, the critical site for articulatory planning is the anterior insula, not the third frontal convolution (Broca's area). Third, the angular gyrus is not specific to visual word forms but is engaged when semantic associations are made. Fourth, the meaning of words is also distributed along the left inferior and middle temporal cortices. Fifth, reading and name retrieval tasks activate the left posterior inferior temporal lobe. This region is thought to have monosynaptic connections to Broca's area (DiVirgilio & Clarke, 1997) thereby providing the semantic reading route that was missing from the 19th Century model. In brief, the only anatomical regions that were missing from the 19th Century neurological model were in the inferior temporal cortices, areas that are relatively resistant to the ischemic damage that the lesion deficit model is dependent upon [49].

Price, as a neuroscientist, speaks of 'only a few refinements' [49]. Overall, functional imaging results indeed show overlap with those of the lesion-deficit models at a generic level (Fig. 8.7). However, when judged from a more clinical perspective, there are important differences. It is easily seen that some of the areas of Price's model do not accord with clinical experience. For example, surgery within the left inferior temporal lobe generally does not result in language deficits, and electrical stimulation finds language functions in a far wider temporal region than the superior temporal sulcus alone. Neurologists are interested in language representation in the individual patient and differ in their questions from neuroscientists, who want to generalize results across populations. Neurosurgeons are even more exacting and require individualized information on 'eloquent' and 'non-eloquent' areas with sub-centimetre accuracy for the planning of their operations. They are interested in the precise anatomical organization of language and want to know what areas are truly essential for normal language functioning. In this respect, the generalized models that are drawn from functional imaging results are not very helpful.

An important point—and warning—here is that most people will probably look at these fMRI images, or any brain map for that matter, with the implicit assumption that all of the highlighted areas play a critical role in normal language functioning. This is a very understandable but wrong assumption. These experiments were not designed, and thus not meant, to provide us with answers to such questions (nor may we assume that the uncoloured areas are *not* involved in language processing). Even for me, with a fair amount of background in functional neuroimaging, it is sometimes difficult to suppress this most intuitive reaction. We may not compare fMRI maps directly to those of lesion-deficit studies. Graphical images, as we have repeatedly seen before, easily speak for themselves (think of the schemes of Wernicke or Penfield). But it is in fact impossible to 'see' what they actually represent without sufficient background information.

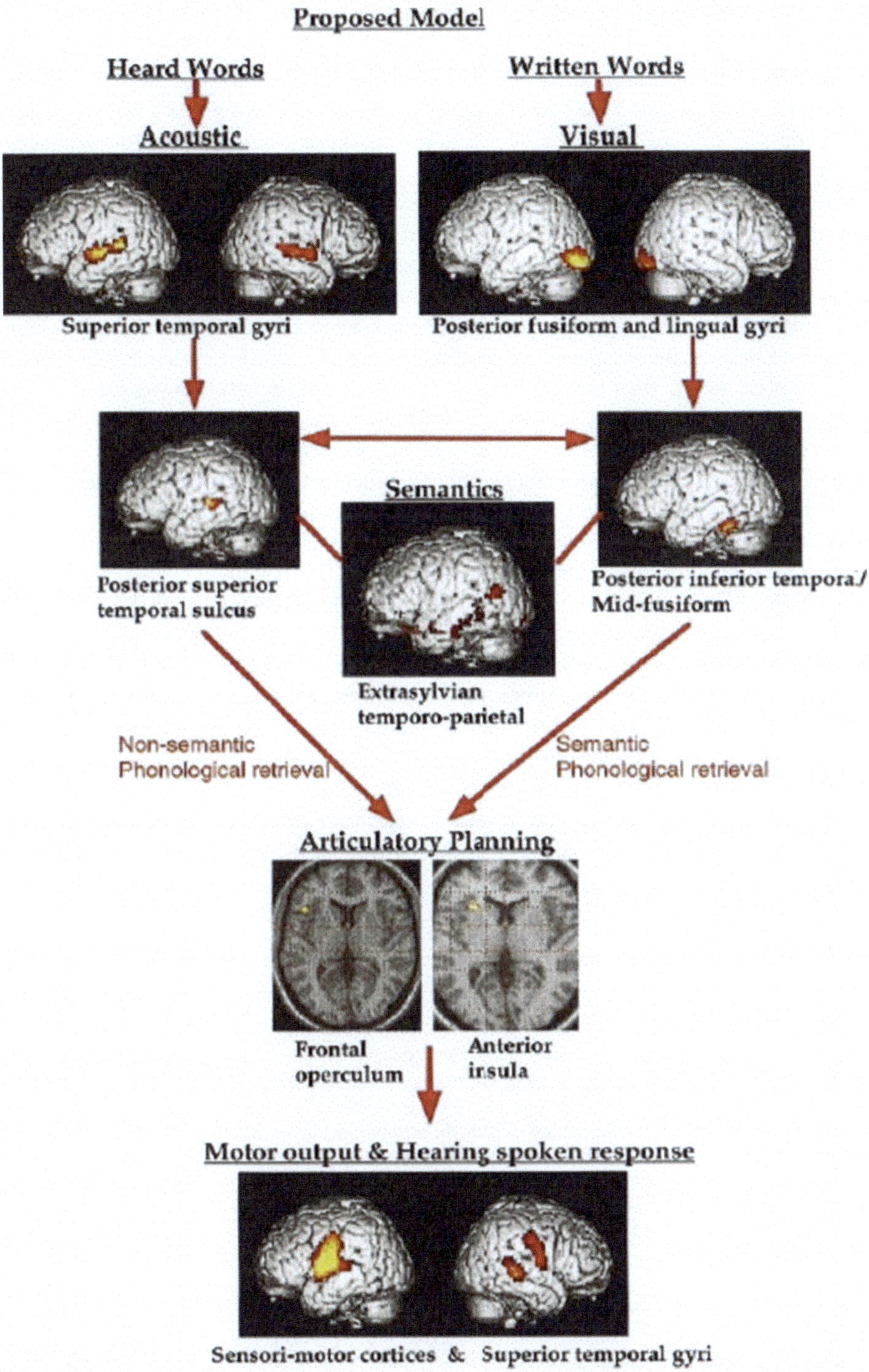

Fig. 8.7 Proposed language model as derived from functional imaging studies by Price (2000) [49]. The model was considered to be largely consistent with the classic models of language processing. Semantics is described at word level

8.3.2 From Single Words to Sentences

In the meantime, a wealth of fMRI studies has been published. A large part of these studies have investigated the processing of single stimuli, using tasks such as word generation or picture naming (see for some of the earlier reviews, for instance, Bookheimer (2002) [50] or Demonet (2005) [51]). Clinical studies predominantly stuck to these 'simpler' tasks, whereas neuroscientists and linguists moved further and also began studying the neuroanatomical basis of more complex tasks, such as sentence processing. For obvious reasons, the neural processes that are responsible for the production and understanding of sentences are even more complex than those of single words. In order to comprehend a sentence or an utterance, it is not enough to deduce the meaning of individual words in the linear manner to which they are presented to the listener or reader. The relationship between words, in terms of the overall meaning of a sentence, is often non-linear, as parts of sentences can be embedded or otherwise have a different hierarchical structure (e.g. the cat that chases the dog is black). Information therefore needs to be kept 'online' before its meaning can be grasped, and additional resources (such as selective attention and working memory) may be required to understand and apply grammatical and syntactical relationships at sentence level.

In 2006, Vigneau and colleagues published a meta-analysis of fMRI and PET studies that went beyond single-word processing [52]. They specifically investigated phonology, semantics and sentence processing. Despite the methodological limitations that are inherent to meta-analyses (the authors had no access to the raw data and had to deal with differences in spatial normalization and data analyses methods across studies), the authors claimed that spatial resolution was 'under the gyral level'. Several functionally specialized networks were identified that covered extensive areas in the frontal and temporal lobes (see Fig. 8.8). A number of interesting propositions were extracted from these results. For instance, it was found that phonological and semantic processing have separate networks within the left inferior frontal gyrus, confirming earlier work by Poldrack (1999), who described an

Fig. 8.8 Meta-analysis of 129 fMRI studies of healthy volunteers that specifically investigated phonology (*blue*), semantics (*red*) or sentence processing (*green*), as studied by Vigneau (2006) [52]. (*Top*) 730 activation peaks are shown on a cortical rendering of the left hemisphere (in stereotactic space). (*Middle*) For further analyses, peaks were clustered. The semantic network is shown, which includes a dorsal and a ventral component in the temporal lobe. The ventral component is dedicated to visual material and includes T3p at the interface between phonological and semantic processes for audio-visual processing (*yellow*). The dorsal component is dedicated to auditory material and includes the voice area (*yellow*) at the interface between phonological and semantic processing. In the frontal lobe, the semantic areas are located in the anterior part of the inferior frontal gyrus. (*Bottom*) Three different working memory loops. The working memory loop for phonological material is shown in *blue* and connects inferior frontal areas to those in the parietal lobe. The working memory loop for semantics (*red*) includes a frontal area at the junction of the precentral gyrus and opercular part of the inferior frontal gyrus (PrF3op) and the angular gyri. The working memory network for sentence and text comprehension includes the posterior part of the middle frontal gyrus (F2p) and the posterior part of the superior temporal sulcus (STSp, *green*)

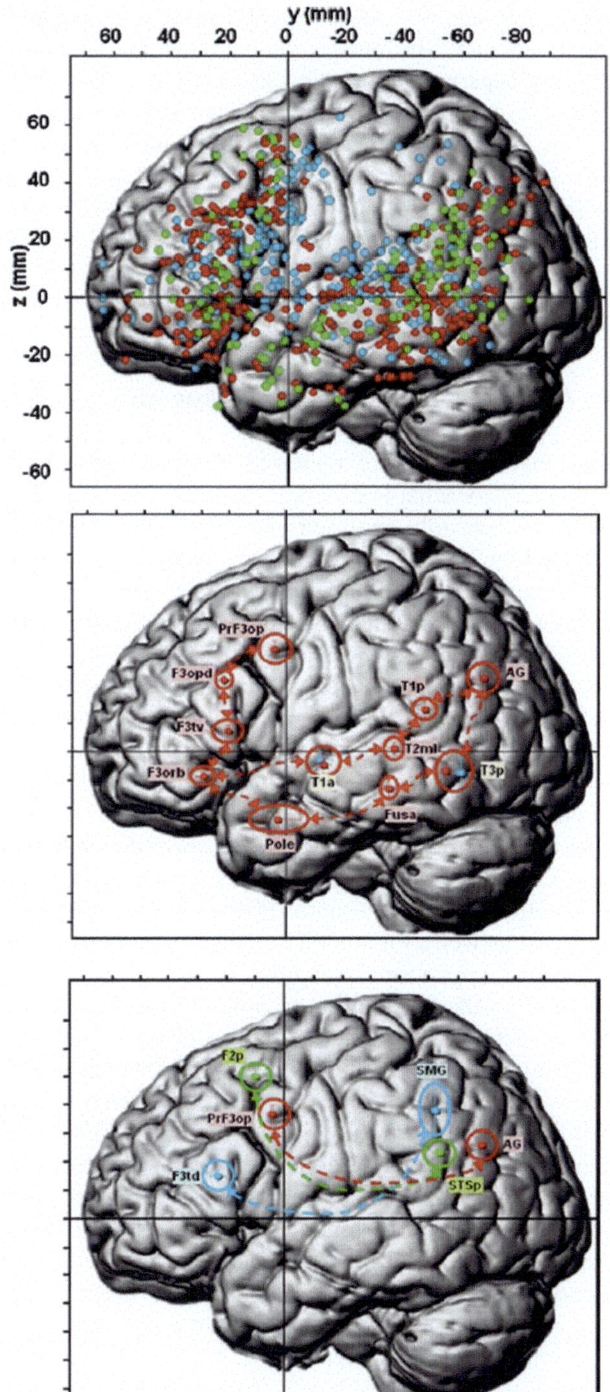
y (mm)
60 40 20 0 -20 -40 -60 -80
z (mm)
60
40
20
0
-20
-40
-60
PrF3op
F3opd
F3tv
F3orb
T1a
Pole
T1p
AG
T2ml
T3p
Fusa
F2p
PrF3op
F3td
SMG
AG
STSp

anterior–posterior dissociation of phonological and semantic areas [53]. Also, the semantic network was found to be much larger than traditionally envisioned, including the angular gyrus, superior and middle temporal gyrus, fusiform gyrus, temporal pole and clusters in the left inferior frontal gyrus. 'This semantic network can be considered to construct an overall meaning on the basis of the association of integrated knowledge issued from the main domain of external (audition, vision) and internal (long-term memory, emotion) messages; this construction of sense forms the foundation of language communication' [52]. Finally, the authors describe three different working memory loops: for phonology, semantics and sentence processing.

8.3.3 A New Anatomical–Functional Perspective

The four studies that were briefly reviewed before (Petersen, Binder, Price, Vigneau) are exemplary for what is generally found in functional imaging experiments when these are carefully analysed and interpreted. These results are not sufficiently explained by the older lesion-deficit models and contrast with what is generally taught in medical school. In fact, they make a strong case that the neural basis of language needs to be redefined. One of the more consistent findings in both fMRI and modern lesion studies is that language production and comprehension are not restricted to, respectively, left inferior frontal and temporoparietal regions. Broca's area is clearly involved in both production and comprehension, as was already observed in the early functional imaging studies of Petersen and Binder. A similar conclusion holds for Wernicke's area; when defined as the posterior part of the superior temporal gyrus, this area seems predominantly involved in phonological processes that facilitate both language production and comprehension. Before words can be spoken and before relevant muscles are innervated for this purpose, the neural representations of speech sounds (phonemes) need to be made available. Phonological retrieval as well as temporary storage of phonetic sequences is thus an important prerequisite for normal execution of speech [54]. Phonological processing is also central to the 'acquisition of long-term lexical memories of novel words' [54, 55]. As such it plays a vital but indirect role in language comprehension. The actual meaning of words and sentences is represented in a much wider (and bilateral) network that is located outside of classic Wernicke's area (Fig. 8.9). This view is perfectly in line with the original ideas of Wernicke and Lichtheim, but not very commonly held in today's clinical practice, where lesions in Wernicke's area are still largely synonymous with comprehension disorders [54, 56–58].

Another common observation in functional imaging studies is that classic language regions are involved in multiple functions and, in the case of Broca's area, also in different cognitive domains [59].[f] In this respect, Broca's area differs from Wernicke's area, as it shows up as a component of many different nonlinguistic functions: motor imagination and preparation, music, visuospatial recognition,

[f]This has also been repeatedly shown with electrical stimulation mapping. Individual sites can be involved in more than one function, for instance, auditory and visual naming [60], reading and naming [61], writing and naming [62] or different languages in bilingual patients [63, 64].

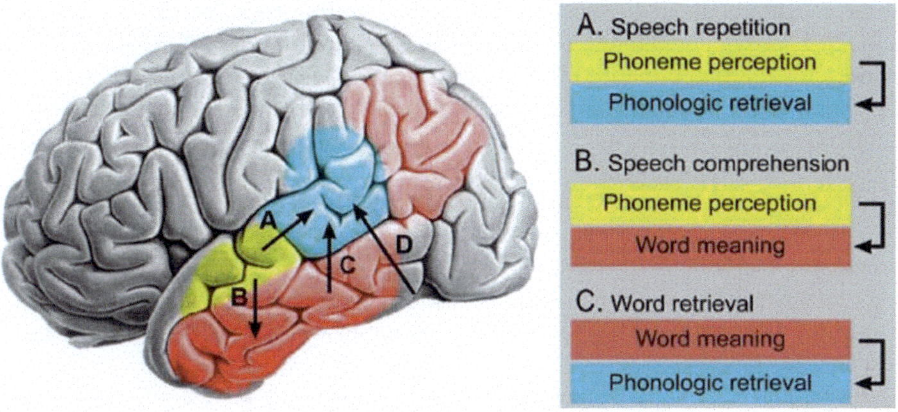

Fig. 8.9 Binder's model (2015) of the major posterior language systems, based on modern lesion and functional imaging studies [54]. Classic Wernicke's area (posterior temporal gyrus) is not directly associated with verbal comprehension, but instead with phonological processing. '*Yellow* indicates a bilateral speech phoneme perception system. *Blue* indicates the Wernicke area, which supports prearticulatory phonologic retrieval. *Red* indicates the temporal and parietal components of a distributed system for word meaning (semantic) representations. Speech repetition requires the pathway designated *A* in the figure, as well as more anterior parietal and frontal regions [not shown in colour] that support articulatory preparation and execution. Spoken word comprehension involves the pathway marked *B* in the figure, which maps perceived phoneme sequences to word concepts. Communicative speech production, in which the speaker retrieves words to express concepts, requires the pathway marked *C*, which maps concept representations onto phonologic representations. Pathway *D* indicates a direct mapping from visual word forms to phonologic representations, required for reading aloud' (Figure and text taken from Binder (2015) [54])

working memory and executive control (i.e. the organization of action and thought) [65–68]. Wernicke's various subfunctions are basically all linguistic in nature [54, 55]. The fact that brain areas participate in multiple different functions implies that sets of brain areas can be temporarily bound together to perform a specific function [69]. Such a view would be consistent with the fact that different language tasks generally result in different brain activity maps (see Fig. 8.10 for an example). Conjunction analyses make use of this principle, hypothesizing that areas that are activated by different tasks play a more crucial role for the particular function that they (broadly) target [72]. Such an approach has been advocated for presurgical planning, in order to differentiate between areas that are supportive for a particular function and those that are critically needed or essential for normal performance [73].

Finally, fMRI studies consistently indicate that language is organized in networks that exceed the borders of the classic language territories [74, 75]. The spatiotemporal profiles of different language functions are laid out across a large part of the brain. These observations go back to the time of Wernicke, Lichtheim and many others, who first described 'connectionist' models. Similar concepts were more recently—in the 1990s—introduced into clinical practice by Mesulam, in the aftermath of pioneers such as Luria. Mesulam provided a framework to better understand the neural basis of sensorimotor and cognitive impairments (see also Chap. 7). In his view, there are numerous sets of interconnected brain areas that are dedicated to a specific function. Within these networks there are hubs that 'provide nodal points for receiving and distributing information that is critical for the functionality of the relevant domain' [76]. Importantly, many of the more recent anatomy-based models also include language areas in the right hemisphere. Friederici's model (2002) of auditory sentence comprehension is a bilateral fronto-temporal network, with right-sided areas involved in the processing of sentence melody and prosody [77]. Indefrey and Levelt (2004) describe four areas in the right hemisphere that are related to the core process of word processing (see Fig. 8.10) [70]. Hickok and Poeppel's model of speech processing (2007), as we have seen in Chap. 7, consists of a dorsal and a ventral stream. The ventral stream, which maps sound to meaning, is essentially bilaterally represented [57].

It can thus be concluded that the results from the many different fMRI studies cannot be condensed into a single and invariant representation of the language system. Rather, language is the product of different subsystems that are activated depending on task requirements [14]. Although there remains unexplained

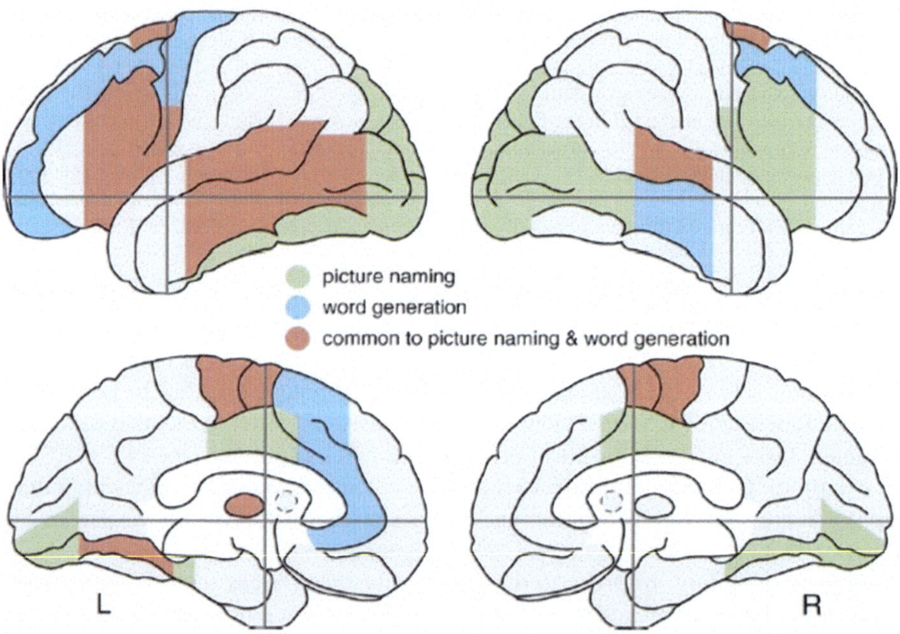

Fig. 8.10 Results of a meta-analysis of 82 (mostly) functional neuroimaging studies of word production (Indefrey and Levelt, 2004) [70]. The figure shows regions activated by picture naming (*green*) and word generation (*blue*). Regions shared by both tasks (*red*) are assumed to be involved in core process of language production. Indefrey and Levelt refer to these differences and commonalities as, respectively, *lead-in* and *core* processes [70, 71]. 'Lead-in processes are task-specific cognitive processes, such as visual object recognition in picture naming, taking place before the core word production pathway is entered. These processes are not well understood for all tasks, but they always contribute essentially to the neuroimaging results' [70] (Figure and text (modified) taken from Indefrey and Levelt (2004) [70])

variability across studies and subjects, the core findings and principles regarding brain–language relationships seem reliable and coherent. These are supported by converging evidence from other modalities (lesional and electrophysiological methods) and disagree with the basic principles of classic and clinical models of language organization [78, 79]. Findings may be summarized as follows:

1. Language-related activity is found in many more brain regions than the classic left inferior frontal and posterior temporoparietal areas.
2. The classic dichotomy of frontal/production versus temporoparietal/comprehension is not true.
3. A significant amount of language-related activity is found in the right hemisphere; its functional role is not clear.
4. Different language tasks are in part executed by spatially different neural systems.
5. Language areas participate in multiple functional networks, and these areas are not necessarily involved in language functions per se (e.g. Broca's area).
6. Language-related brain maps differ between subjects. It is unclear to what extent these differences reflect true variability in functional–anatomical organization and to what extent there is an underlying consistency across subjects.
7. There is significant evidence that structure and function are not necessarily uniquely coupled.

8.4 What fMRI Can Contribute to Clinical Care

Clinicians have other interests to those of neuroscientists and have specific requirements when it comes to functional brain mapping that logically follow from their daily work: predicting outcome in the individual patient. Think of the neurosurgeon who tries to estimate the chance that surgery will affect the patient's survival and functional outcome or the neurologist who is considering whether it is worthwhile to refer an older and impaired stroke patient for rehabilitation therapy. Fortunately, there is an increasing awareness in the medical community that the classical canonical knowledge that has been obtained from textbook models and group-based neuroimaging studies is not always relevant in clinical care for individuals [80]. Although the location of primary sensorimotor functions can be predicted with reasonable accuracy with the use of averaged data, this is not the case for the precise location of cognitive functions, including language. Likely, this is because these functions are more complex and distributed over larger regions.[g] Classical language theories cannot explain why patients can have large tumours in classic language

[g] This is best explained by the fact that primary cortices have a direct relationship with large subcortical fibre bundles, such as the corticospinal tract or optical radiation, which probably restricts variability and plasticity.

regions without language impairments or that patients can recover from a stroke after initial severe neurological deficits.

There are two areas in which fMRI should improve to become a relevant technique for neurosurgeons and neurologists. As a first requirement, fMRI should provide information on an individual level that is reliable enough so that it can be trusted by clinicians. This means that the precision and reproducibility of maps should be sufficient to detect meaningful functional information and to identify the 'odd-one-out', even if such a case is infrequently encountered in clinical practice (think of patients with a language-dominant right hemisphere). The second, even more challenging requirement is that language maps for clinical purposes should indicate what the *risk* is that damage to a region leads to permanent and significant deficits. Clinical maps should not so much represent the level of brain activity or the functionality of a brain region, but its ability to withstand or recover from damage (i.e. the redundancy).

8.4.1 From Significance to Relevance

One important area that may have limited the clinical value of fMRI so far is the way statistics are applied. Statistical analyses play a crucial role in fMRI. In contrast to a regular anatomical MRI, an fMRI 'scan' represents a large number of scans taken over a longer period of time (~5–60 min). Each of these scans provides information on 20–30,000 voxels (cube-shaped small brain regions of about 1–5 mm). To produce a three-dimensional brain map of activity, each voxel is tested separately for a significant change in activity.

In science, it is especially important to prevent incorrect support for a theory. From a scientific perspective, it is important to limit the number of false-positive errors and thus to reduce the chance that a hypothesis is falsely supported.[h] As a single fMRI experiment contains many thousands of statistical tests (i.e. for each voxel), a correction is applied to limit the overall number of false-positive results.[i] The consequence of this stringent statistical approach, however, is that there will also be an increase in the number of false negatives (i.e. 'missed' significant results).[j] While this is a sensible approach for scientific purposes, it may pose a problem in a clinical setting, where there is no uniform rule about the 'importance' of false negatives and false positives. For instance, when clinical images are used for presurgical planning, it is vital not to accidentally miss brain activity (as this could potentially

[h] In science it is generally considered less of a problem when the hypothesis is not rejected on false grounds (i.e. a false-negative result).

[i] Even if an experiment has no effect on the signals that are measured from the brain, due to pure noise, on average one in a hundred voxels will show a significant result if all voxels are tested with a *p*-value of 1%.

[j] In scientific papers on fMRI, unfortunately, non-significant regions are usually reported as 'not active'. Instead, it would be better to label these areas as 'unknown', as there is a fair chance that these areas in fact are false negatives.

result in postoperative neurological deficits when these areas are surgically removed). In these cases, first priority is to reduce the number of false negatives, even at the expense of a higher number of false positives [81]. Depending on the clinical question that is asked, minimization of the number of false negatives can thus be at least as important as minimization of the number of false positives.

The standard statistical fMRI approach to create brain maps is therefore often not optimal for clinical applications. For clinical use of fMRI, we need to consider alternative ways to produce and visualize fMRI brain maps. These maps should be more tailored to provide useful clinical information instead of useful scientific information. From this point of view, it is good to realize that the main goal of statistics in science is to analyse and compare groups that are considered to be a sample of a larger population. In contrast, a clinical test result from one subject only provides information about that one subject, and thus the use of similar statistical methods may be less relevant here. Radiologists, for example, largely decide on qualitative grounds whether or not X-rays or anatomical MRI scans contain pathological abnormalities in a particular subject (i.e. these clinical decisions are taken without the use of formal statistical methods). An alternative approach to use fMRI in a clinical setting may be to clearly visualize all the actual measurements, from highly active to nonactive regions and everything in between. As the human brain is an expert in recognizing and understanding patterns, clinicians may be able to learn what the fMRI brain maps represent and how they correlate with normal and impaired behaviour and cognition of patients. This, in fact, is similar to the way in which clinicians have learned, over time, to interpret clinical tests such as MRI scans or X-rays.

8.4.2 Some Remarks About the Reliability and Spatial Precision of Brain Maps

Obviously, good reliability is a sine qua non for use of any data in clinical and scientific practice.[k] Ideally, statistical tests should give researchers a good impression about the likelihood that similar results are obtained when their experiment is repeated. Unfortunately, test–retest studies appear to indicate that results obtained with fMRI are not very reliable. However, the topic has not been addressed extensively, as for instance remarked by Bennett and Millner in their review paper (2010) [82]:

> Surprisingly, most functional magnetic resonance imaging (fMRI) researchers have only a vague idea of how reliable their results are. Reliability is not a typical topic of conversation between most investigators and only a small fraction of papers investigating fMRI reliability have been published.

We will shortly discuss these problems here. The first test–retest studies focused on the amount of brain activity across different sessions, by counting the numbers of suprathreshold voxels or calculating the overlap between sessions [83]. Surprisingly,

[k] Note that there is virtually no data available on the reliability of the Wada test and electrocortical stimulation mapping, despite being the current clinical gold standards for language mapping.

the amount of voxels varies enormously between sessions for the same subject, and the percentage of overlapping voxels hardly exceeds 30–40%, regardless of the height of the statistical threshold [18, 83].[l] McGonigle et al. (2000) scanned an individual on 33 different occasions with three different fMRI tasks [85]. These authors found such a large intersession variability that they explicitly warned that single fMRI experiments in individuals may lead to erroneous conclusions. 'This result demonstrates that session context effects have a significant effect on fMRI data and illustrates that a single session should be considered merely as a single sample of a subject's responses to the experimental intervention employed' [85].

What may—at least in part—underlie the disappointing results of these test–retest studies is an incorrect interpretation of the statistical results in fMRI studies. As noted before, fMRI results are often published as a spatial pattern of supra-threshold ('significant') voxels. Although represented and visualized as a pattern, each voxel is in fact tested independently using a univariate test. After correction for multiple comparisons—to reduce false-positive errors—a binary map of significant and non-significant voxels is produced. In test–retest studies of fMRI, the reliability of this entire binary pattern of significant voxels is often tested and interpreted as an indication of the reliability of fMRI as a technique. Unfortunately, in such an approach, the level of reproducibility will almost always be low as it is based on an incorrect interpretation of results. First of all, significance tests do not provide information about the reproducibility of non-significant results, only about significant results. Second, the goal of any correction method for multiple comparisons (such as Bonferroni or the less stringent Family Wise Error method) is to ensure that the *originally* chosen uncorrected significance threshold is valid for all significant results. Thus, when the experiment is repeated, only the significant voxels should be retested, and they should be retested at the originally chosen *uncorrected* threshold in order to calculate test–retest reliability of fMRI. This explains why we cannot expect the entire map of both significant and non-significant voxels to replicate across different fMRI sessions: it is a direct result of the chosen statistical approach and not some kind of inherent unreliability of fMRI.

To underline this statement, remarkable results come from a recent multisite study that applied a different statistical approach.[m] For this study, data from seven European fMRI research sites were gathered from five different countries. Each site scanned six healthy right-handed volunteers (three males, three females) with a standardized verb-generation task. In this study, a single statistical test was applied that simply tested the correlation over all voxels between an image of single subject and the average of all other images. All single-subject language maps showed a high correlation to the average image of all other subjects, despite different MR scanners,

[l]Variability can be significantly lowered by calculating relative (and not absolute) measures. A lateralization index is a reliable measure to assess language representation when a verb-generation task is used, for instance [18, 84].

[m]This study was conducted among members of the European Low-Grade Glioma Network (www.braintumours.eu). Data were included from Frankfurt (Elke Hattingen), Graz (Gord von Campen, Margit Jehna), Madrid (Mar Jiménez de la Peña), Milan (Alberto Bizzi), Regensburg (Katharina Rosengarth, Frank Dodoo-Schittko), Tilburg (Martijn Jansma, Geert-Jan Rutten) and Utrecht (Nick Ramsey).

different pulse sequences and different native languages (Dutch, German, Italian, Spanish). So, at least on a spatial scale of 2–3 cm, single-subject language maps acquired with fMRI in healthy subjects appear to have a strong similarity (see Fig. 8.11 for an impression of results). Our group (Jansma 2015) even obtained

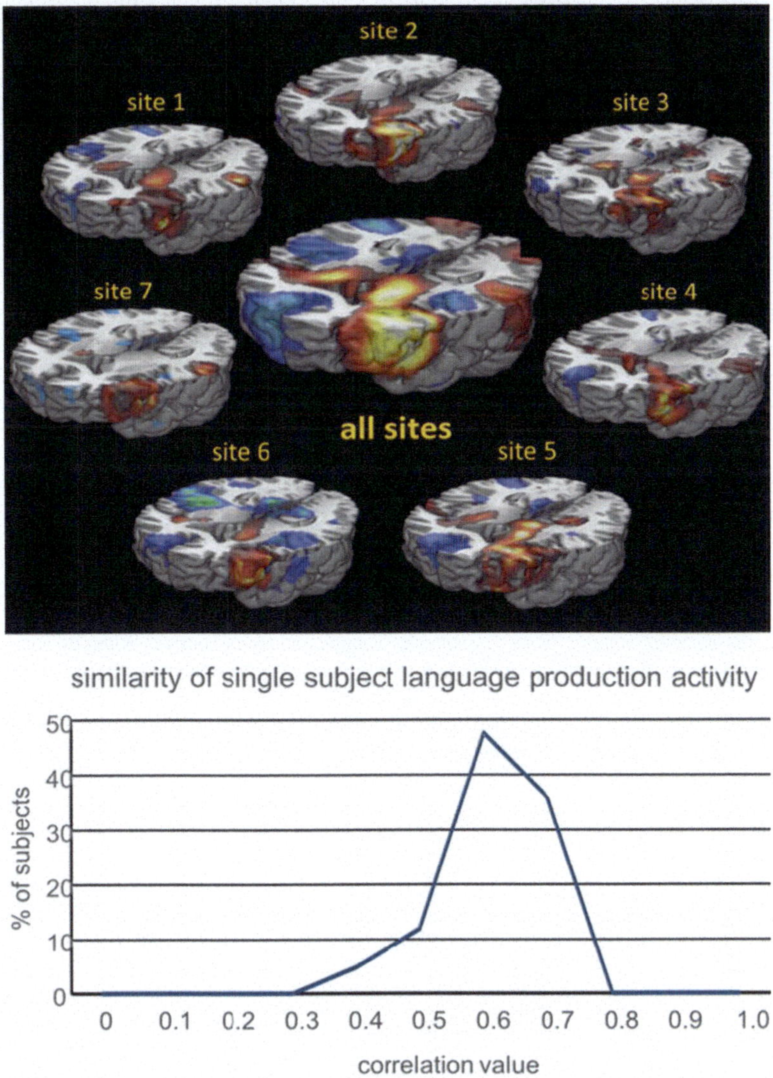

Fig. 8.11 Clinical use of fMRI requires standardized protocols for acquisition and analysis of data. As a first step, this study from the European Low-Grade Glioma Network compared fMRI results from seven different research sites. Each site scanned six healthy subjects with a standardized verb-generation task that was provided to them on a DVD. Individual results were smoothed and registered to a standard brain. Fairly comparable left-lateralized activity is seen across sites (*yellow-orange*). Correlation analyses yielded no significant differences between subjects or sites. The *blue* areas are more active during the control condition than during verb generation and are clearly lateralized to the right hemisphere. Correlation values are plotted for the brain map of each subject as compared to the average brain maps of all other subjects. Courtesy of Martijn Jansma, St Elisabeth-TweeSteden Hospital Tilburg, and Nick Ramsey, UMC Utrecht (submitted for publication)

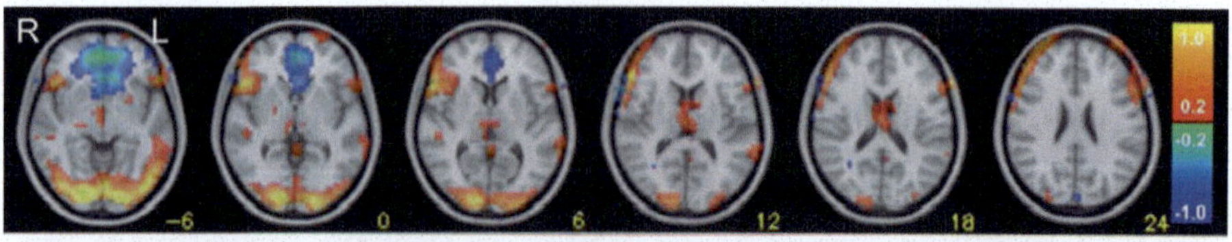

a. RL

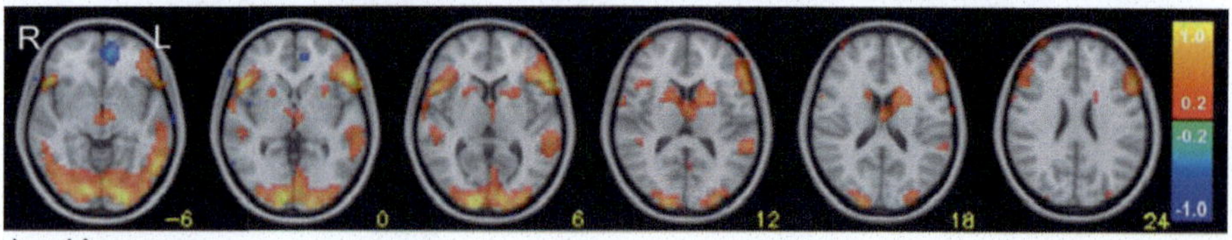

b. mLL

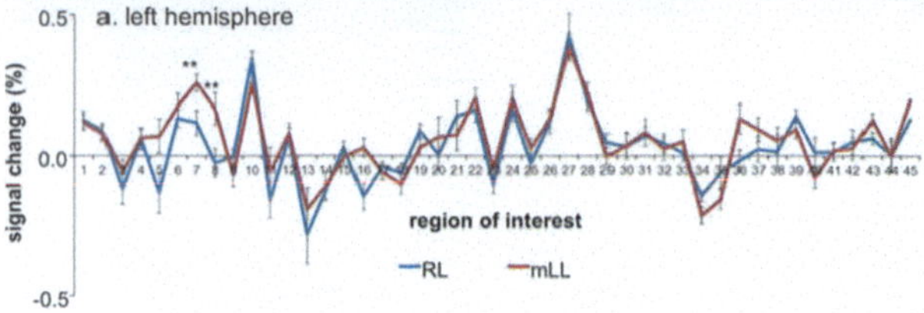

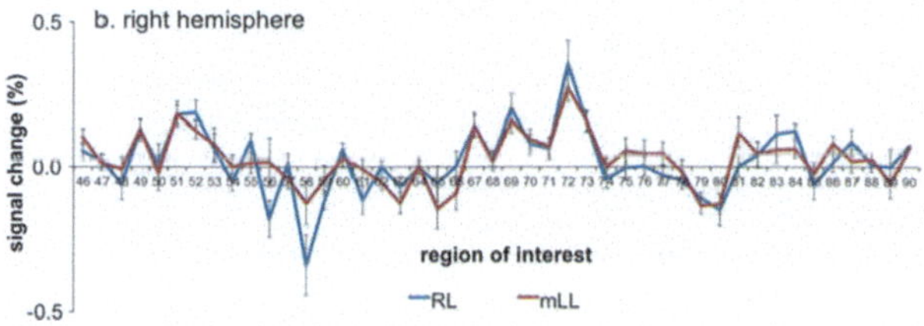

N.	Abb	Full name	No	Abb	Full name
1	PCG	Precentral gyrus	24	LIMG	Lingual gyrus
2	SFGdl	Superior frontal gyrus, dorsolateral part	25	OCC	Superior occipital gyrus
3	SFGor	Superior frontal gyrus, orbital part	26	MOG	Middle occipital gyrus
4	MFG	Middle frontal gyrus	27	IOG	Inferior occipital gyrus
5	MFGor	Middle frontal gyrus, orbital part	28	FFG	Fusiform gyrus
6	IFGoper	Inferior frontal gyrus, opercular part	29	PoCG	Postcentral gyrus
7	IFGtri	Inferior frontal gyrus, triangular part	30	SPG	Superior Parietal gyrus
8	IFGor	Inferior frontal gyrus, orbital part	31	IPG	Inferior Parietal gyrus
9	RolOp	Rolandic operculum	32	SMG	Supramarginal gyrus
10	SMA	Supplementary motor area	33	AG	Angular gyrus
11	Olf	Olfactory cortex	34	pCUN	Precuneus
12	SFGmd	Superior frontal gyrus, medial part	35	ParaCL	Paracentral lobule
13	SFGmdor	Superior frontal gyris, medial orbital part	36	CauNuc	Caudate nucleus
14	Grect	Gyrus rectus	37	PUT	Lenticular nucleus, Putamen
15	INS	Insula	38	Pall	Lenticular nucleus, Pallidum
16	ACC	Anterior cingulate	39	Thal	Thalamus
17	MCC	Median cingulate	40	Heschl	Heschl Gyrus
18	PCC	Posterior cingulate	41	STG	Superior temporal gyrus
19	Hipp	Hippocampus	42	STPtp	Temporal pole, superior temporal gyrus
20	PHipp	Parahippocampal gyrus	43	MTG	Middle temporal gyrus
21	AMY	Amygdala	44	MTGtp	Temporal pole, middle temporal gyrus
22	CALC	Calcarine fissure	45	ITG	Inferior temporal gyrus
23	CUN	Cuneus			

Abb: abbreviation.

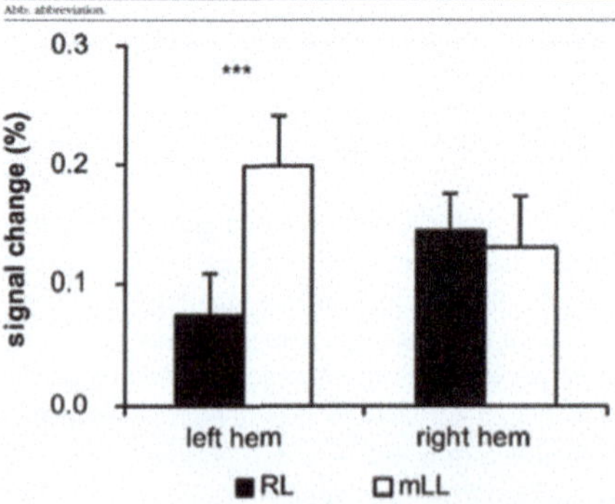

similar findings in a heterogeneous group of brain tumour patients (Fig. 8.12) [86]. These observations are promising and can be seen as a first step towards the development of normative maps and standards for quality control.

8.4.3 Language Laterality and Language Dominance

Most language fMRI studies yield left-lateralized activation patterns, confirming clinical experience that most subjects have a language-dominant left hemisphere [14, 87]. But population-based findings are not very helpful in clinical practice, where a number of patients have atypical language dominance (i.e. a right hemisphere that contains essential language functions).[n] And there is another difficulty, namely, that even with stringent statistical criteria, fMRI shows bilateral language activity in almost all subjects, either healthy or diseased (see Fig. 8.13). This has been replicated across many studies and is probably not an artefact of the methodology [86, 88, 89]. Bilateral language functions are also consistently demonstrated in modern lesion-deficit studies or with other mapping techniques such as transcranial magnetic stimulation (TMS) [90, 91]. In the words of Binder (2010): 'language lateralization has come to be seen as continuously graded rather than an all-or-nothing phenomenon, with relative degrees of dominance rather than distinct categories' [92]. Binder's view, himself a neurologist, clearly differs from the opinion still held by most medical doctors. To better understand these different points of view, the terms 'dominant' and 'lateralized' should first be clarified. These terms are often used interchangeably in the literature, but are in fact not similar:

> A brain function is considered dominant if a unilateral lesion produces a behavioral deficit that subtends to both sides of space, a criterion easily met by the various aphasic and aprosodic syndromes. For a function to be strongly lateralized, however, it must also be shown that the behavioral deficit does not occur following lesions of the opposite hemisphere. In this regard, soon after his discovery that damage to the left third frontal convolution caused loss of articulate speech, Broca reported that similar lesions in the right hemisphere did not

[n] A few percent of the normal population has a right-dominant hemisphere, but this becomes known only after sudden damage to the right hemisphere. There are no methods (yet) that accurately establish hemispheric dominance in healthy individuals.

Fig. 8.12 Average fMRI language maps in 42 brain tumour patients. From a series of 163 patients with a glioma or meningioma who performed a verb-generation task, a subset of 21 patients was selected with right-lateralized language activity (average LI was −0.32). The LI was calculated from activity in Broca's region and its homologue. From the remaining patients, an equal number of patients were selected with equal but opposite LI (0.32). Comparing these groups yielded two interesting findings: (1) the pattern of brain activity was similar for both the right-lateralized (RL) and moderate left-lateralized (mLL) patients (except, of course, for the left and right inferior frontal gyrus that had been used to determine the LI); (2) right-sided language laterality was associated with a significant decrease of signal from Broca's region and not so much by an increase in the right-sided homologue. In fact, signal changes in this latter area were not different for RL and mLL patients (see graph) (Figures from Jansma (2015) [86])

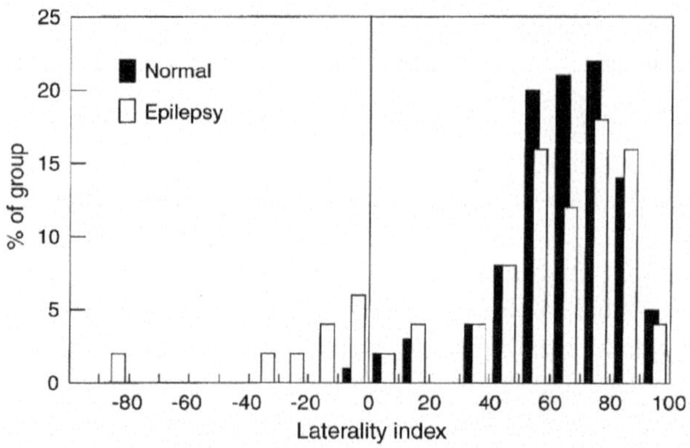

Fig. 8.13 Springer (1999) found a continuum of asymmetric language activation maps in a sample of 100 healthy subjects and 50 epilepsy patients. Almost all subjects showed some degree of right-hemisphere activation. Subjects performed a semantic auditory decision task, compared to a nonlinguistic discrimination task as a control. The figure shows the frequency distribution of the language lateralization index (LI). The index was calculated using the amount of voxels (N) that exceeded a statistical threshold in the left and right hemisphere, according to the formula $LI = 100*(N_{left} - N_{right})/(N_{left} + N_{right})$ (Figure taken from Springer et al. 1999 [88])

impair articulate speech, thus establishing that articulation was a dominant and lateralized function of the left hemisphere (Ross 2000) [91].

It is important also to clarify what is meant with the term ‘bilateral language functions’. Two scenarios can be envisioned here that are fundamentally different. In the first scenario, which is in line with long-standing clinical concepts, language functions are normally confined to one hemisphere (the left one) and only occasionally to the other. The ‘exceptions’ are caused by pathological lesions and are traditionally seen as the result of a reorganization of function [93, 94]. In the second scenario, language functions are commonly sustained by both hemispheres, although for each function the degree of lateralization may vary. The last model better fits with modern language theories [57, 90].

Over the years, clinical fMRI studies have assumed as a ground truth that hemispherical language functions are dichotomized (i.e. the first scenario). Clinical opinion and Wada test results have not been questioned much as gold standards, despite their known limitations (see Chap. 6 for details) [14]. From this rather dogmatic starting point, algorithms were constructed to match fMRI results as closely as possible to those of the Wada test. Most studies calculated a lateralization index (LI) for this purpose. Such an index reduces the rich information that is contained in fMRI maps into a single number that is subsequently compared to a cut-off value to decide whether or not a hemisphere is sufficiently involved in language functions to consider it ‘dominant’ [86, 95]. Researchers quickly noted that the LI shows significant variability both among subjects, tasks and studies and that the use of fixed cut-off values is not a good method to reliably separate patients with typical language (i.e. left-lateralized) from those with atypical language [95]. As a result of this variability, agreed-upon criteria for standardization were never developed. Each study or centre basically formulated their own criteria to match fMRI as closely as possible to outcomes of the Wada test.

Still, fMRI does offer relevant clinical information in a subset of patients [96]. Many experts agree that there is no need for an additional Wada test to confirm fMRI results if activation is robust and strongly lateralized to the left hemisphere (some activity in the right hemisphere is accepted, although there is no consensus on this amount in relative or absolute terms) [89, 97].[o] However, in atypical cases, there is clear disagreement between both methods (again presuming that the Wada test is a valid gold standard). fMRI and Wada test results are concordant in roughly only 50% of cases, effectively reducing the predictive value of fMRI for these patients to chance level [96]. Some methodological improvements have been shown to increase the predictive value of fMRI (e.g. conjunctional analyses), but it remains too low to replace the Wada test as a clinical tool [98].

8.4.4 Involved and Critical Language Areas

fMRI generally finds many more language areas than DES.[p] Despite this surplus of language-related fMRI activity, the results of single-task experiments (e.g. verb generation or picture naming) only partly match those of DES. This means that there is a relatively high rate of false negatives and false positives if one accepts DES as the gold standard. When results of several different fMRI tasks are combined, reasonable detection power can be achieved [81, 99]. However, such a multi-task approach will likely further raise false-positive fMRI activity. This is not very helpful to neurosurgeons, as the more widespread and abundant the activation patterns are, the more these will restrict possible surgical options. There is, however, one promising observation: when no activity is found with a battery of different fMRI tasks, the chance of finding language areas with DES is very small. This means that these areas can be safely resected without the need for DES [81, 100].

The main argument that is usually given for the poor match between fMRI and the clinically accepted invasive techniques is that of activation versus disruption of brain areas. The Wada test and DES measure behavioural effects after temporarily induced brain lesions, whereby the resulting impairments are usually severe and easily detected upon clinical examination or with simple tests. fMRI, on the other hand, measures task-related changes in brain activation. It potentially shows all of the brain areas that are involved in a language task and not necessarily those that are

[o] In patients with typical language representation according to the Wada test, there is agreement with fMRI results in approximately 90–95% of cases [96].

[p] DES is the reference technique for functional mapping in neurosurgery [12]. In general a low morbidity is observed after DES-guided surgical procedures, and this argument is frequently used to confirm its status as a gold standard technique. However, there is little evidence that resection of DES-positive sites leads to permanent language impairments. The method suffers from important conceptual and practical drawbacks, making its gold status questionable. See for a discussion Chap. 6.

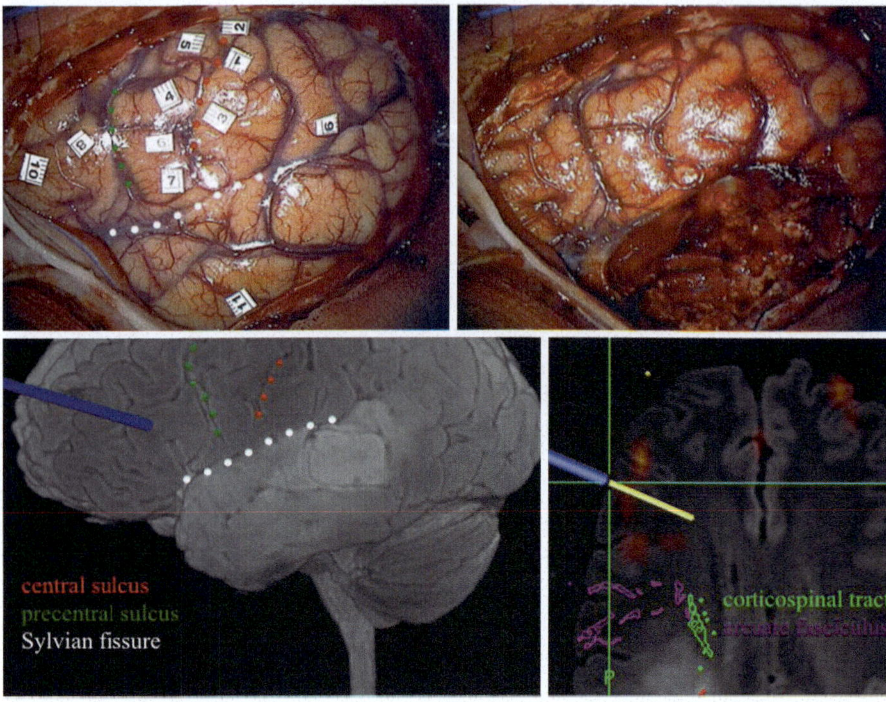

Fig. 8.14 Forty-eight-year-old patient with a clinical debut of epilepsy and a left temporo-insular diffuse low-grade glioma. Neurological and neuropsychological examination revealed subtle word-finding difficulties and impairments in verbal memory. Awake surgery was performed to map functional areas and subcortical language tracts. A counting and picture-naming task was used for this purpose, as is common practice for language mapping during surgery [101]. Clear responses were obtained from primary sensorimotor cortex (nos. *1–3*, sensations in, respectively, tongue, lip and cheek; nos. *4–5*, motor movements of mouth) and ventral premotor cortex (nos. *6–7*, speech arrest). Responses from parietal (no. *9*) and temporal (no. *11*) sites were less convincing (i.e. there were more negative than positive responses). What was remarkable in this case was that one of the inferior frontal language areas (no. *10*) was only detected when we reverted to a rather unusual intraoperative task: verb generation. During stimulation at marker *10*, the patient was able to read the word, but unable to think of a related verb (a similar behavioural response was obtained three times at this location). We selected this specific task because a verb-generation fMRI task indicated language activity in this area. The *bottom images* show screenshots of the surgical guidance system, with fMRI results in orange and the pointer (*blue*) indicating site no. *10*

critically needed for normal performance. Other arguments that help explain the discrepancy between fMRI and the invasive mapping techniques are differences in the tasks that are used (see for an example Fig. 8.14), the fact that intrasulcal fMRI areas can be inaccessible to electrical stimulation, and pathological change in haemodynamics that may affect BOLD responses [20, 100].

The current lack of concordance with clinical methods is still generally seen as an impediment to the clinical use of fMRI. However, it would be presumptuous to conclude that fMRI results are necessarily untrue or that they have no clinical value [81, 100, 102]. No straightforward comparison between fMRI and another (invasive) technique can decide which one is more correct in terms of clinical relevance, as that requires a third independent measure. For obvious reasons, results should be compared to patient outcome, as all of these methods strive to minimize postoperative impairments after surgical treatment. Few studies, however, have truly determined the predictive value of any of these techniques after surgical or endogenous damage of presumed language-critical tissue. The ones that did, have indicated that fMRI outperformed the conventional methods. Sabsevitz (2003) showed that a LI that was based on fMRI activation in the temporal lobe strongly correlated to naming outcome in patients who underwent anterior temporal lobectomy for the relief of epilepsy [103]. In this study, a LI towards the left hemisphere was predictive of worsening of naming outcome, despite the use of DES during surgery. Janecek (2013), also from the Binder group, found that in discordant cases, fMRI better predicted naming outcome after surgical intervention than the Wada test [104]. Clearly, many more of these studies are needed to establish the true clinical potential of fMRI and to develop protocols that have a generic value for clinical practice.

8.4.5 How to Move Further

Despite the valuable results that fMRI can produce in expert hands, the method has remained experimental and has never really entered clinical practice. Some of the reasons and objections that hindered routine clinical use have already been discussed in the previous sections. We will summarize and review them here again, emphasizing that the future evaluation of experimental and clinical mapping techniques needs to be done more open-mindedly than before. In our opinion, fMRI will eventually serve as a better predictor for any neurological or cognitive impairments than the current clinical methods do.

1. *fMRI results do not match very well those of the Wada test or DES for language mapping*
   In fact, they probably never will. This, however, should not be a reason to withhold fMRI from clinical practice. To escape from this impasse, fMRI language maps should be directly compared to functional outcome of patients. The few studies that have done so provided arguments in favour of fMRI over the existing clinical techniques [104]. It is therefore important to abandon the classic views regarding language localization and lateralization or at least to take them not too literally. Clinical opinion still heavily favours a few cortical centres in a language-dominant left hemisphere. It has become clear, though, that the neural basis of language is much more complex and consists of many cortical and subcortical components that together make up a distributed and complex interconnected system in both hemispheres.
2. *fMRI is not a lesional technique*
   fMRI is unable to assess the effect of a lesion in the direct manner of the Wada test and DES. As such, it is often said that it cannot distinguish between critical and noncritical language areas. However, it should be remembered that these latter invasive techniques can only assess the *immediate* functional consequences of a lesion and that they are unable to foresee postlesional recovery and long-term functional reshaping. fMRI offers more potential in this respect. Measurements of whole-brain

function can be used to gain understanding of the effects of loss or removal of particular brain regions on performance and activity in other regions of the brain [64, 105]. One of the largest challenges of fMRI is to develop theories and methods that can model lesions and plasticity in individual patients, and to use the level and location of brain activity to predict the functional consequences of planned surgical interventions or treatment [105, 106]. Although that is a far-reaching goal, it offers a perspective that is lacking in the existing clinical techniques.

3. *The reliability of single-subject fMRI maps is too low*
This is likely not true. At a resolution of 0.5–1 cm, fMRI maps seem to give a reliable impression of brain activity. Previous results in the literature have probably underestimated test–retest reproducibility by focusing on voxels as independent units of measurement, instead of patterns (i.e. calculating the overlap between numbers of voxels or clusters of activation). For clinical applications, pattern reproducibility may be more relevant than single-voxel value reproducibility. Single-subject fMRI methods should clearly be further developed and should be less grounded in the traditional group-based analyses and statistical methods [80]. In addition, clinicians should learn to 'read' fMRI brain maps, based on feedback from clinical practice
4. *fMRI is too complex for use in a clinical setting*
At the moment, fMRI is difficult and cumbersome to use in daily clinical practice. Data acquisition and analysis is complex and requires considerable expertise. In contrast to regular CT and MRI investigations, there are no standardized or turnkey protocols. But even if you are an expert in functional imaging, it is difficult to get things going in a hospital environment.[q] Clinical research (i.e. development and validation of new fMRI protocols) is really only possible for clinicians that work within a dedicated research group. Consequently, the amount of patient data that is published or shared among researchers remains rather limited. This has led to a sort of catch-22 situation: data is required to develop validated and turnkey protocols, but these are only acquired once easy-to-use protocols have been implemented in clinical practice. Large-scale and multicentre clinical databases are needed to tackle this problem.

At the end of this chapter, it is fair to say that in the absence of standardized and validated fMRI language protocols, the interpretation of fMRI brain maps should be left to experts. In any case, fMRI maps should not be automatically seen as surgical roadmaps or as *the* explanation for language disturbances in patients. And clinicians should never be provided with these images without proper background information. Still, careful research with new structural and functional MRI techniques currently seems the best method to further increase our understanding of the complex and dynamic representation of brain functions. As a spin-off, we may expect that

[q]A few companies nowadays facilitate this process and supply equipment and software to run fMRI experiments or perform analyses for clinical customers. Even with their help, though, it is cumbersome to get the data at the doctor's desk or in the operating room in a routine fashion.

more accurate clinical applications will emerge, but these will only gain popularity in clinical practice when they are implemented in turnkey systems.

References

1. Rutten GJ, Ramsey NF. The role of functional magnetic resonance imaging in brain surgery. Neurosurg Focus. 2010;28:E4.
2. Bandettini PA, Wong EC, Hinks RS, et al. Time course EPI of human brain function during task activation. Magn Reson Med. 1992;25:390–7.
3. Ogawa S, Tank DW, Menon R, et al. Intrinsic signal changes accompanying sensory stimulation: functional brain mapping with magnetic resonance imaging. Proc Natl Acad Sci U S A. 1992;89:5951–5.
4. Kwong KK, Belliveau JW, Chesler DA, et al. Dynamic magnetic resonance imaging of human brain activity during primary sensory stimulation. Proc Natl Acad Sci U S A. 1992;89:5675–9.
5. Raichle ME, Gusnard DA. Appraising the brain's energy budget. Proc Natl Acad Sci U S A. 2002;99:10237–9.
6. Malonek D, Grinvald A. Interactions between electrical activity and cortical microcirculation revealed by imaging spectroscopy: implications for functional brain mapping. Science. 1996;272:551–4.
7. Fox PT. The coupling controversy. Neuroimage. 2012;62:594–601.
8. Ojemann GA. Effect of cortical and subcortical stimulation on human language and verbal memory. Res Publ Assoc Res Nerv Ment Dis. 1988;66:101–15.
9. Logothetis NK, Pauls J, Augath M, et al. Neurophysiological investigations of the basis of the fMRI signal. Nature. 2001;412:150–7.
10. Crone NE, Miglioretti DL, Gordon B, Lesser RP. Functional mapping of human sensorimotor cortex with electrocorticographic spectral analysis. II. Event-related synchronization in the gamma band. Brain. 1998;121:2301–15.
11. Sinai A, Bowers CW, Crainiceanu CM, et al. Electrocorticographic high gamma activity versus electrical cortical stimulation mapping of naming. Brain. 2005;128(Pt 7):1556–70.
12. Duffau H. Brain mapping: from neural basis of cognition to surgical applications. Wien: Springer; 2011.
13. Moonen CT, Bandettini PA. Functional MRI. Berlin: Springer; 2000.
14. Binder JR. fMRI of language systems: methods and applications. In: Faro SH, Mohamed FB, editors. Functional MRI: basic principles and clinical applications. New York: Springer; 2006. p. 245–77.
15. Poeppel D. A critical review of PET studies of phonological processing. Brain Lang. 1996;55:317–51. Discussion 352
16. Friston KJ, Price CJ, Fletcher P, et al. The trouble with cognitive subtraction. Neuroimage. 1996;4:97–104.
17. Gallese V, Lakoff G. The brain's concepts: the role of the sensory-motor system in conceptual knowledge. Cogn Neuropsychol. 2005;22:455–79.
18. Rutten GJ, Ramsey NF, van Rijen PC, van Veelen CW. Reproducibility of fMRI-determined language lateralization in individual subjects. Brain Lang. 2002;80:421–37.
19. Gusnard DA, Raichle ME, Raichle ME. Searching for a baseline: functional imaging and the resting human brain. Nat Rev Neurosci. 2001;2:685–94.
20. Rutten GJ, van Rijen PC, van Veelen CW, Ramsey NF. Language area localization with three-dimensional functional magnetic resonance imaging matches intrasulcal electrostimulation in Broca's area. Ann Neurol. 1999;46:405–8.
21. Mikl M, Marecek R, Hlustik P, et al. Effects of spatial smoothing on fMRI group inferences. Magn Reson Imaging. 2008;26:490–503.

22. Eklund A, Nichols TE, Knutsson H. Cluster failure: why fMRI inferences for spatial extent have inflated false-positive rates. Proc Natl Acad Sci U S A. 2016;113(28):7900–5.
23. Vul E, Harris C, Winkielman P, Pashler H. Puzzlingly high correlations in fMRI studies of emotion, personality, and social cognition. Perspect Psychol Sci. 2009;4:274–90.
24. Smith K. Brain imaging: fMRI 2.0. Nature. 2012;484:24–6.
25. Fried I, Nenov VI, Ojemann SG, Woods RP. Functional MR and PET imaging of rolandic and visual cortices for neurosurgical planning. J Neurosurg. 1995;83:854–61.
26. Mueller WM, Yetkin FZ, Hammeke TA, et al. Functional magnetic resonance imaging mapping of the motor cortex in patients with cerebral tumors. Neurosurgery. 1996;39:515–20.
27. Raichle ME. Human brain, functional organisation, altered states of consciousness and the assessment of brain death. Pontifical Academy of Sciences, Scripta Varia 110. 2007.
28. Lin A-L, Gao J-H, Fox PT. Neurovascular and neurometabolic uncoupling in the visual cortex. In: Molotchnikoff S, Rouat J, editors. Visual cortex—current status and perspectives. Rijeka: InTech; 2012.
29. Mosso A. Ueber den kreislauf des blutes im menschlichen gehirn: untersuchungen. Verlag von Veit & Comp.; 1881
30. Millett D. Hans Berger: from psychic energy to the EEG. Perspect Biol Med. 2001;44:522–42.
31. Schiller F. Paul Broca: founder of French anthropology, explorer of the brain. New York: Oxford University Press; 1992.
32. Cohen L, Smith MJ, Leroux-Hugon V. Paul Broca's thermometric crown. J Neurol Neurosurg Psychiatry. 2004;75:32.
33. Roy CS, Sherrington CS. On the regulation of the blood supply of the brain. J Physiol. 1890;11:85–108.
34. Shallice T. From neuropsychology to mental structure. booksgooglecom. 1988.
35. Caramazza A. On drawing inferences about the structure of normal cognitive systems from the analysis of patterns of impaired performance: the case for single-patient studies. Brain Cogn. 1986;5:41–66.
36. Halai AD, Woollams AM, Lambon Ralph MA. Using principal component analysis to capture individual differences within a unified neuropsychological model of chronic post-stroke aphasia: revealing the unique neural correlates of speech fluency, phonology and semantics. Cortex. 2017;86:275–89.
37. Petersen SE, Fox PT, Posner MI, et al. Positron emission tomographic studies of the cortical anatomy of single-word processing. Nature. 1988;331:585–9.
38. McClelland JL, Rumelhart DE, Hinton GE. The appeal of parallel distributed processing. Cambridge: MIT Press; 1986.
39. LaBerge D, Samuels SJ. Toward a theory of automatic information processing in reading. Cogn Psychol. 1974;6:293–323.
40. Price CJ. A review and synthesis of the first 20 years of PET and fMRI studies of heard speech, spoken language and reading. Neuroimage. 2012;62:816–47.
41. Haxby JV, Grady CL, Ungerleider LG, Horwitz B. Mapping the functional neuroanatomy of the intact human brain with brain work imaging. Neuropsychologia. 1991;29:539–55.
42. Raichle ME. Memory mechanisms in the processing of words and word-like symbols. Ciba Found Symp. 1991;163:198–204. Discussion 204
43. Wise R, Hadar U, Howard D, Patterson K. Language activation studies with positron emission tomography. Ciba Found Symp. 1991;163:218–28. Discussion 228
44. Steinmetz H, Seitz RJ. Functional anatomy of language processing: neuroimaging and the problem of individual variability. Neuropsychologia. 1991;29:1149–61.
45. Petersen SE, Fiez JA. The processing of single words studied with positron emission tomography. Annu Rev Neurosci. 1993;16:509–30.
46. Binder JR, Frost JA, Hammeke TA, et al. Human brain language areas identified by functional magnetic resonance imaging. J Neurosci. 1997;17:353–62.
47. Demonet JF, Chollet F, Ramsay S, et al. The anatomy of phonological and semantic processing in normal subjects. Brain. 1992;115:1753–68.

48. Wise R, Chollet F, Hadar U, et al. Distribution of cortical neural networks involved in word comprehension and word retrieval. Brain. 1991;114:1803–17.
49. Price CJ. The anatomy of language: contributions from functional neuroimaging. J Anat. 2000;197(Pt 3):335–59.
50. Bookheimer SY. Functional MRI of language: new approaches to understanding the cortical organization of semantic processing. Ann Rev Neurosci. 2002;25:151–88.
51. Démonet JF, Thierry G, Cardebat D. Renewal of the neurophysiology of language: functional neuroimaging. Physiol Rev. 2005;85:49–95.
52. Vigneau M, Beaucousin V, Herve PY, et al. Meta-analyzing left hemisphere language areas: phonology, semantics, and sentence processing. Neuroimage. 2006;30:1414–32.
53. Poldrack RA, Wagner AD, Prull MW, et al. Functional specialization for semantic and phonological processing in the left inferior prefrontal cortex. Neuroimage. 1999;10:15–35.
54. Binder JR. The Wernicke area: modern evidence and a reinterpretation. Neurology. 2015;85:2170–5.
55. Wise RJ, Scott SK, Blank SC, et al. Separate neural subsystems within 'Wernicke's area. Brain. 2001;124:83–95.
56. Dronkers NF, Wilkins DP, Van Valin RD, et al. Lesion analysis of the brain areas involved in language comprehension. Cognition. 2004;92:145–77.
57. Hickok G, Poeppel D. The cortical organization of speech processing. Nat Rev Neurosci. 2007;8:393–402.
58. Gage N, Hickok G. Multiregional cell assemblies, temporal binding and the representation of conceptual knowledge in cortex: a modern theory by a "classical" neurologist, Carl Wernicke. Cortex. 2005;41:823–32.
59. Alamia A, Solopchuk O, D'Ausilio A, et al. Disruption of Broca's area alters higher-order chunking processing during perceptual sequence learning. J Cogn Neurosci. 2016;28:402–17.
60. Hamberger MJ, Goodman RR, Perrine K, Tamny T. Anatomic dissociation of auditory and visual naming in the lateral temporal cortex. Neurology. 2001;56:56–61.
61. Roux FE, Lubrano V, Lauwers-Cances V, et al. Intra-operative mapping of cortical areas involved in reading in mono- and bilingual patients. Brain. 2004;127:1796–810.
62. Lubrano V, Roux FE, Demonet JF. Writing-specific sites in frontal areas: a cortical stimulation study. J Neurosurg. 2004;101:787–98.
63. Giussani C, Roux FE, Lubrano V, et al. Review of language organisation in bilingual patients: what can we learn from direct brain mapping? Acta Neurochir. 2007;149:1109–16. Discussion 1116
64. Rutten GJ, Ramsey NF. Functional neuroimaging in neurosurgical practice. In: Duffau H, editor. Brain mapping: from neural basis of cognition to surgical applications. Berlin: Springer; 2011. p. 207–27.
65. Koechlin E, Jubault T. Broca's area and the hierarchical organization of human behavior. Neuron. 2006;50:963–74.
66. Fadiga L, Craighero L, D'Ausilio A. Broca's area in language, action, and music. Ann N Y Acad Sci. 2009;1169:448–58.
67. Kunert R, Willems RM, Casasanto D, et al. Music and language syntax interact in Broca's area: an fMRI study. PLoS One. 2015;10:e0141069.
68. Eickhoff SB, Heim S, Zilles K, Amunts K. A systems perspective on the effective connectivity of overt speech production. Philos Trans A Math Phys Eng Sci. 2009;367:2399–421.
69. Hirsch J, Moreno DR, Kim KH. Interconnected large-scale systems for three fundamental cognitive tasks revealed by functional MRI. J Cogn Neurosci. 2001;13:389–405.
70. Indefrey P, Levelt WJ. The spatial and temporal signatures of word production components. Cognition. 2004;92:101–44.
71. Indefrey P, WJML. The neural correlates of language processing. In: Gazzaniga M, editor. The new cognitive neurosciences. Cambridge: MIT Press; 2000.
72. Price CJ, Friston KJ. Cognitive conjunction: a new approach to brain activation experiments. Neuroimage. 1997;5:261–70.

73. Ramsey NF, Sommer IE, Rutten GJ, Kahn RS. Combined analysis of language tasks in fMRI improves assessment of hemispheric dominance for language functions in individual subjects. Neuroimage. 2001;13:719–33.
74. Mesulam MM. Large-scale neurocognitive networks and distributed processing for attention, language, and memory. Ann Neurol. 1990;28:597–613.
75. Bressler SL, Menon V. Large-scale brain networks in cognition: emerging methods and principles. Trends Cogn Sci. 2010;14:277–90.
76. Mesulam M. The evolving landscape of human cortical connectivity: facts and inferences. Neuroimage. 2012;62:2182–9.
77. Friederici AD. Towards a neural basis of auditory sentence processing. Trends Cogn Sci. 2002;6:78–84.
78. Poeppel D, Hickok G. Towards a new functional anatomy of language. Cognition. 2004;92:1–12.
79. Lubrano V, Draper L, Roux FE. What makes surgical tumor resection feasible in Broca's area? Insights into intraoperative brain mapping. Neurosurgery. 2010;66:868–75. Discussion 875
80. Nieto-Castanon A, Fedorenko E. Subject-specific functional localizers increase sensitivity and functional resolution of multi-subject analyses. Neuroimage. 2012;63:1646–69.
81. Rutten GJ, Ramsey NF, van Rijen PC, et al. Development of a functional MRI protocol for intraoperative localization of critical temporoparietal language areas. Ann Neurol. 2002;51:350–60.
82. Bennett CM, Miller MB. How reliable are the results from functional magnetic resonance imaging. Ann N Y Acad Sci. 2010;1191:133–55.
83. Rombouts SA, Barkhof F, Hoogenraad FG, et al. Within-subject reproducibility of visual activation patterns with functional magnetic resonance imaging using multislice echo planar imaging. Magn Reson Imaging. 1998;16:105–13.
84. Benson RR, FitzGerald DB, LeSueur LL, et al. Language dominance determined by whole brain functional MRI in patients with brain lesions. Neurology. 1999;52:798–809.
85. McGonigle DJ, Howseman AM, Athwal BS, et al. Variability in fMRI: an examination of intersession differences. Neuroimage. 2000;11:708–34.
86. Jansma JM, Ramsey NF, Rutten GJ. A comparison of brain activity associated with language production in brain tumor patients with left and right sided language laterality. J Neurosurg Sci. 2015;59(4):327–35.
87. Pujol J, Deus J, Losilla JM, Capdevila A. Cerebral lateralization of language in normal left-handed people studied by functional MRI. Neurology. 1999;52:1038–43.
88. Springer JA, Binder JR, Hammeke TA, et al. Language dominance in neurologically normal and epilepsy subjects: a functional MRI study. Brain. 1999;122:2033–46.
89. Janecek JK, Swanson SJ, Sabsevitz DS, et al. Language lateralization by fMRI and Wada testing in 229 patients with epilepsy: rates and predictors of discordance. Epilepsia. 2013;54:314–22.
90. Knecht S, Floel A, Drager B, et al. Degree of language lateralization determines susceptibility to unilateral brain lesions. Nat Neurosci. 2002;5:695–9.
91. Ross ED. Affective prosody and the aprosodias. In: Mesulam MM, editor. 2 Principles of behavioral and cognitive neurology (6). New York: Oxford University Press; 2000. p. 316–31.
92. Binder JR. Use of fMRI language lateralization for quantitative prediction of naming and verbal memory outcome in left temporal lobe epilepsy surgery. In: Ulmer S, Jansen O, editors. fMRI: basics and clinical applications. Berlin: Springer; 2010.
93. Thiel A, Habedank B, Winhuisen L, et al. Essential language function of the right hemisphere in brain tumor patients. Ann Neurol. 2005;57:128–31.
94. Saur D, Lange R, Baumgaertner A, et al. Dynamics of language reorganization after stroke. Brain. 2006;129:1371–84.
95. Seghier ML. Laterality index in functional MRI: methodological issues. Magn Reson Imaging. 2008;26:594–601.

96. Bauer PR, Reitsma JB, Houweling BM, et al. Can fMRI safely replace the Wada test for preoperative assessment of language lateralisation? A meta-analysis and systematic review. J Neurol Neurosurg Psychiatry. 2014;85:581–8.
97. Baxendale S, Thompson PJ, Duncan JS. The role of the Wada test in the surgical treatment of temporal lobe epilepsy: an international survey. [letter]. Epilepsia. 2008;49(4):715–20. Discussion 720
98. Rutten GJ, Ramsey NF, van Rijen PC, et al. fMRI-determined language lateralization in patients with unilateral or bilateral language dominance according to the Wada test. Neuroimage. 2002;17:447–60.
99. FitzGerald DB, Cosgrove GR, Ronner S, et al. Location of language in the cortex: a comparison between functional MR imaging and electrocortical stimulation. Am J Neuroradiol. 1997;18:1529–39.
100. Kuchcinski G, Mellerio C, Pallud J, et al. Three-tesla functional MR language mapping: Comparison with direct cortical stimulation in gliomas. Neurology. 2015;84(6):560–8.
101. Fernandez Coello A, Moritz-Gasser S, Martino J, et al. Selection of intraoperative tasks for awake mapping based on relationships between tumor location and functional networks. J Neurosurg. 2013;119:1380–94.
102. Giussani C, Roux FE, Ojemann J, et al. Is preoperative functional magnetic resonance imaging reliable for language areas mapping in brain tumor surgery? Review of language functional magnetic resonance imaging and direct cortical stimulation correlation studies. Neurosurgery. 2010;66:113–20.
103. Sabsevitz DS, Swanson SJ, Hammeke TA, et al. Use of preoperative functional neuroimaging to predict language deficits from epilepsy surgery. Neurology. 2003;60:1788–92.
104. Janecek JK, Swanson SJ, Sabsevitz DS, et al. Naming outcome prediction in patients with discordant Wada and fMRI language lateralization. Epilepsy Behav. 2013;27:399–403.
105. Hart MG, Price SJ, Suckling J. Connectome analysis for pre-operative brain mapping in neurosurgery. Br J Neurosurg. 2016;30:506–17.
106. Fornito A, Zalesky A, Breakspear M. The connectomics of brain disorders. Nat Rev Neurosci. 2015;16:159–72.
107. James W. The principles of psychology. Harvard UP, Cambridge, MA. 1890.

Recovery from Brain Damage

9

It has been known for ages that some form of spontaneous recovery usually follows a disabling injury. In the past, this was generally attributed to 'the healing power of nature' (Hippocrates' vis naturae medicatrix) or to supernatural forces [1]. Scientific inquires have revealed the nature of many of these recovery processes, and we have learned, to some extent at least, to understand and influence the course of injury and disease. In the case of a fractured bone, for example, physical and physiological processes have been elucidated quite accurately, and several treatments have been developed that facilitate healing of the fracture and recovery of a person's functionality. But clearly, in many other cases, in particular when there is damage to the brain, restoration of function is often incomplete or insufficient, and we are failing to grasp all the relevant factors that are involved in the process [2, 3]. This chapter reviews neural plasticity from a clinical point of view and specifically focuses on the brain's potential to reorganize the neural circuitry for language functions at the macroscopical level (i.e. in terms of brain areas and white matter pathways).

9.1 Historical Perspective on Restoration of Function

9.1.1 Momentum

Clinicians know from experience that lesions with an acute onset, for instance, due to trauma or stroke, produce far more symptoms than those that develop more slowly, such as brain tumours. This was already described more than two centuries ago by Morgagni (1761) and Hall (1841). Hall noted that slowly growing brain tumours may even exist without any symptoms [4]. A few decades later, Hughlings Jackson used the term 'momentum' to refer to the combined impact of the size and the rate of growth of lesions to understand the patient's symptomatology (i.e. mass × velocity). He disagreed with others, such as Flourens, who mainly considered the size of the lesion important and who believed in an almost complete

G.-J. Rutten, *The Broca-Wernicke Doctrine*, DOI 10.1007/978-3-319-54633-9_9

'equipotentiality' of the brain. His convictions later found support in experimental studies in animals (in the 1950–1960s), which showed that successive but partial surgical lesions generally produce far less disability than one large ablation [1, 5, 6]. A comparable multistage surgical approach has in recent times been advocated for brain tumour patients, whereby a remnant was deliberately left during a first operation (because of tumour ingrowth in brain areas that are still important for motor or language functions). When these patients were reoperated on several years later, the extent of resection could often be significantly increased with minimal morbidity [7, 8]. Apparently, further slow growth of the tumour remnant, possibly in combination with the effects of surgery, allowed further functional reshaping.

Scientific studies have confirmed what clinicians already learned a long time ago, namely, that the brain can in many cases compensate for the disabling effects of a lesion and that time is an important factor in this process. Stroke and brain tumour patients, when matched for the size and location of their lesion, clearly have different patterns of neuropsychological impairments (e.g. Anderson 1990) [9]. Tumour patients are generally least affected and may even perform completely normal on neuropsychological tests. These findings obviously confound any of the straightforward relationships between brain structure and function that were drawn in the past. Head, in 1926, was among those who were already well aware of this more complex situation. In his standard work on aphasia, he wrote:

> in all attempts to correlate the site of structural changes with defects of function it must never be forgotten that the severity and acuteness of the lesion exert an overwhelming effect on the manifestations.
>
> Thus a complete act of speech comes to be a wide-spread response of the organism to each fresh situation; this employs conscious, subconscious, automatic and purely physiological processes. The deeper the lesion the grosser and more definite the disorder. On the other hand, since the cortex is a more flexible organ with less rigid and preordained reactions, the disturbance of function due to injury of the surface is not so severe and permanent and is less easily determined [10].

9.1.2 Age

In addition to the spatial distribution and growth rate of a brain lesion, there is another variable that significantly influences functional outcome of the patient: age at the time of the brain insult. This phenomenon was also well recognized a long time ago. In 1868, just a few years after Broca's landmark publications, Cotard published a study where he analysed the autopsy reports of 42 patients [11]. He found damage to the left frontal lobe or even the entire hemisphere in seven patients with a right-sided hemiplegia since childhood, but with normal language and intellectual functions. Cotard thus demonstrated that left hemisphere damage that is acquired in early life does not necessarily lead to lasting language deficits. Steffen, in 1885, documented a complete recovery in 18 of 25 patients with acquired childhood aphasia and reported a good outcome in most of these cases [12]. It appeared that recovery in children was better and quicker than in older patients with similar deficits [13]. These authors, along with others such as Broca (1865), Clarus (1874), Wernicke (1874) and Bernhardt (1885), proposed that the right hemisphere could support language functions in the case of severe left hemisphere damage [12, 13].

More than a century later, these initial findings were confirmed in patients where a lesioned hemisphere was surgically removed [14]. In 1962, Basser reported on a series of 34 children where a hemispherectomy was performed because of intractable epileptic seizures. All but one patient developed normal language functions, and Basser concluded that for development of speech, 'the left and right hemisphere are equipotential' [15]. A similar conclusion was drawn by Wilson (1970) who found in his series of 50 patients that the side of the hemispherectomy did not influence subsequent language development [16]. These observations, according to Bates and Roe:

> led Lenneberg (1967) to his controversial notion that the brain is 'equi-potential' at birth, with lateralization determined gradually across the course of development. As a corollary, Lenneberg also argued that this period of equipotentiality and plasticity is brought to an end at puberty, providing the first systematic argument in favor of a 'critical period' for language [11].

To date, it remains unclear to what extent the left hemisphere is the default site for the 'organ of language' and what eventually determines language lateralization and language abilities [11, 17, 18]. Atypical (right hemisphere) language representation has been associated with developmental disorders, but is also present in the healthy population [19]. Laterality seems determined by multiple factors that are both environmental (acquired brain damage; exposure to language) and genetic (structural brain asymmetries; innate language abilities; gender) [19–21]. According to Bates and Roe, one of the most puzzling results in children with brain lesions is that 'most studies fail to find any significant difference in language outcomes when direct comparisons are made between children with left- vs. right hemisphere damage' (see also Chap. 6, Fig. 6.25) [11, 22]. These observations fit a model whereby language functions gradually develop from a bilateral to a more unilateral representation. There is also some evidence that language functions become more focally represented when infants improve their language skills. According to Bishop (2013), 'language-impaired children with poor phonological skills have more diffuse and bilateral processing of speech sounds than typically developing children' [19]. Other observations, whereby the type of aphasic disorders is age-specific, are also consistent with such theories:

> When Brown and Jaffe (1975) reviewed the clinical aphasia material, they noted an age specificity of certain aphasic disorders. For example, a lesion in Wernicke's area will produce motor aphasia in a child, conduction aphasia in middle age, and jargon aphasia in late life. This suggests a progressive differentiation or regional specification within the dominant hemisphere language zone. (…) In this process, expression may lateralize earlier than comprehension. The more diffuse representation of production mechanisms in younger patients accounts for the occurrence of nonfluent aphasia with more widely distributed lesions. The relative preservation of comprehension in these patients may reflect the contribution of the right hemisphere or intact portions of the left [23].

9.1.3 Experimental Studies in Animals

In the nineteenth century, experimental studies began to investigate the brain's potential to compensate for damage. In 1824, Flourens described recovery of function after surgical damage to parts of the brains of birds and mammals. As was described in Chap. 1, these studies were rather primitive and crude [24]. Fifty years later, in the era of Fritsch and Hitzig and Ferrier, the precision and reliability of experimental studies had significantly improved (see Chap. 6). In 1876, for instance, Soltmann studied the cortex of dogs and rabbits, which he believed to play an important role in willed movements [25]. He made lesions to cortical motor regions and found remarkable functional preservation, in particular in younger animals:

> In one puppy, who could walk normally following such a lesion, he showed that [electrical] stimulation of the undamaged cortex elicited bilateral movements. This was not seen in dogs lesioned as adults. He postulated that the undamaged motor cortex had taken over the function of the damaged cortex [26].

Around the turn of the nineteenth century, several studies had demonstrated that surgical lesions to the motor cortex of monkeys frequently resulted in very significant recovery of function, even when contralesional motor areas were subsequently removed in a second or third surgical session. In the same year that Leyton and Sherrington published their landmark paper on the electrically excitable motor cortex of apes and monkeys (1917), the potential importance of postlesional rehabilitation was indicated in a study by Ogden and Franz [27]. In their experiments a large part of the motor cortex in monkeys was lesioned, resulting—as expected—in a flaccid paralysis of the contralateral limbs (Fig. 9.1). In some of the animals, they constrained the contralesional upper limb immediately after surgery and also instituted daily movement therapy for the affected upper and lower limbs. What they found was that these animals regained, in their view, full functionality of their limbs, whereby much of this recovery was seen in the first 2–3 weeks. What was even more remarkable was that the animals who did *not* receive the limb constrainment and movement therapy remained greatly impaired. These, and several other studies, highlighted the enormous potential of the nervous system for recovery, even after it had suffered extensive damage. However, these results seemed largely forgotten for many decades, until constrained-induced movement therapy was reintroduced for

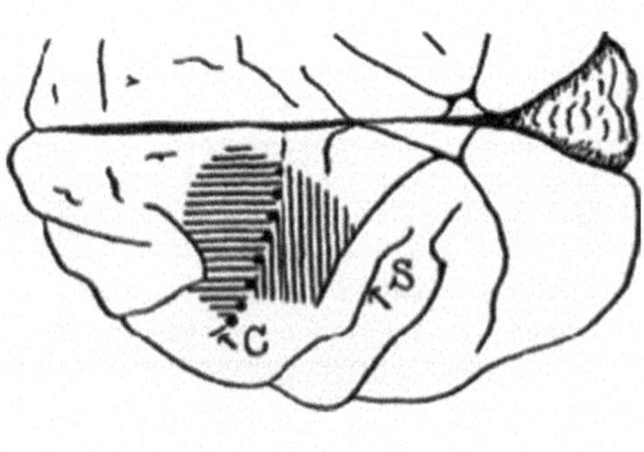

Fig. 9.1 Top view on a schematic drawing of the left hemisphere. Horizontal lines indicate the classic lesion site in the left precentral motor cortex of monkeys in the study of Ogden and Franz (1917) [27]

hemiplegic stroke patients by Edward Taub (1931) [28]. Taub's work, which originated in the 1960s, was predominantly based on surgically induced spinal cord lesions that left the monkeys' limbs deafferentiated. His idea was that a monkey, or for that matter a human, did not use a paretic extremity because it had not learned to reuse it properly immediately after the injury. Under normal circumstances, an animal gets positive feedback from successful limb movements. Taub hypothesized that under pathological conditions, when movements cannot be adequately executed, this reinforcement becomes impaired. As a result, the animal gives up and starts to use other limbs to compensate for the behavioural deficit ('learned non-use'). This will lead to deterioration of the motor programmes of the affected limb, further diminishing the chances of functional recovery:

> Taub believed that even after stroke, there was good chance that the motor programs for movement were present in the nervous system. Thus the way to unmask motor capacity was to do to human beings what he did to monkeys: constrain the use of the good limb and force the affected one to begin moving.[a]

Patient studies with 'constraint-induced movement therapy' were successful and demonstrated relevant improvements in the use of the affected arm [30]. Today, the therapy is widely used in the rehabilitation of stroke patients.

In the period 1980–1990, experiments that used neuronal microrecordings in monkeys confirmed that cortical maps were plastic, in the sense that they could increase or decrease in size when afferent input was modified (see for a review Kaas 1991) [31]. For instance, when a digit is amputated or a peripheral nerve transected, the cortical region for skin sensation becomes 'invaded' by neighbouring regions whose sensory functions have remained intact [32]. Effectively, the area is now involved in these new functions, as was demonstrated by Michael Merzenich (1942) and colleagues in a series of famous experiments [32, 33]. His group showed that within 2 months after amputation of a digit, the corresponding cortical area begins to respond to sensory stimuli of the adjacent fingers. Initially, it was estimated from these experiments that the expansion of neighbouring cortical representations did not extend beyond a distance of 1–2 mm. Subsequent experiments have suggested that the upper limit of cortical reorganization was at least a magnitude larger (i.e. up to 1 cm) [34]. Evidence for local adaptation of cortical maps has meanwhile been demonstrated for other modalities (auditory, motor, visual) in many other studies and also in humans (among others, in braille readers and in patients with phantom limb experiences) [35–38].

Even during ontogenesis, cortical specialization seems predominantly dependent on the type of afferent input that a particular area receives and not so much on the genetic predisposition. The location of an area therefore does not necessarily determine its functions. When retinal input is surgically fed into the auditory pathways of young ferrets, for example, the temporal cortex (i.e. the classic location of primary auditory cortex) will take on a systematic representation of visual space [39].

[a]Quote taken from *The brain that changes itself* by Norman Doidge (2007) [29]

And if cortical areas are surgically relocated in young animals, functionality will change accordingly. O'Leary and Stanfield (1989) transplanted parts of the foetal cortex from one sensory domain to another (e.g. from sensorimotor to visual areas) and found that these will now take on the function that is appropriate to the input they receive at their new location [40]. Following similar rules, newborn kittens become cortically blind if one eye is deprived from normal visual experience (e.g. by suturing one eye closed). Without the appropriate visual input, this part of the visual cortex will not develop normally, and after a certain critical period, the anatomical and physiological changes become irreversible. Most cells will now have lost the ability to respond to the eye that was closed. If similar experiments are performed on adult cats, these deprivation-derived changes do not occur [41].

9.1.4 General Observations on Brain-Injured Patients

The idea that changes in behaviour are somehow anchored in brain structures originated in the ideas of James, Cajal and Hughlings Jackson, well over a century ago. Hughlings Jackson reasoned that if representations in the brain were static, then the degree to which undamaged areas could compensate would also have to be fixed. This, however, conflicted with his many observations of recovery after a focal brain lesion and led him to formulate his *Principle of Compensation*: functions had a dynamic representation in the brain [42]. At the end of the nineteenth century, there were several other investigators that argued against the classic and static localistic theories that dominated that era. Friedrich Goltz (1834–1902) and Charles-Édouard Brown-Sequard (1817–1894), for instance, wrote about the non-local effects that brain lesions had on the remaining healthy parts of the brain, thereby raising questions about strict localization of functions [43, 44]. Although not the first, Constantin von Monakow (1853–1930) would become the proponent of these distant-effect theories and introduced the concept of *diaschisis* [43]. A similar concept would decades later be formulated by Geschwind, who referred to it as a *disconnection syndrome*. As we have seen in Chap. 7, Geschwind's ideas relate to those of Wernicke and his school of associationism. Today, the concept of diaschisis seems again at the heart of our understanding of brain functions and has clearly resurfaced along with new MRI techniques that are able to visualize and quantify connectivity within the brain. A good overview is provided in the paper by Carrera and Tonini (2014) [44].

von Monakow observed that the functional impairments that resulted from brain lesions were often out of proportion to the amount of brain tissue that was damaged or were better explained by a dysfunction of remote brain areas. To him, the sole location of a lesion was not a very reliable indicator of its function, except maybe for elementary sensory and motor functions. This led him to propose that any lesion in the brain causes a functional standstill (German: Stillstand) of distant areas due to a disruption of anatomical connections (Fig. 9.2). This inactivation was to a certain extent a temporary process, so he hypothesized, explaining the well-known observations, that brain damage is commonly followed by functional recovery. von Monakow did not favour the idea that new brain areas compensated for the loss of

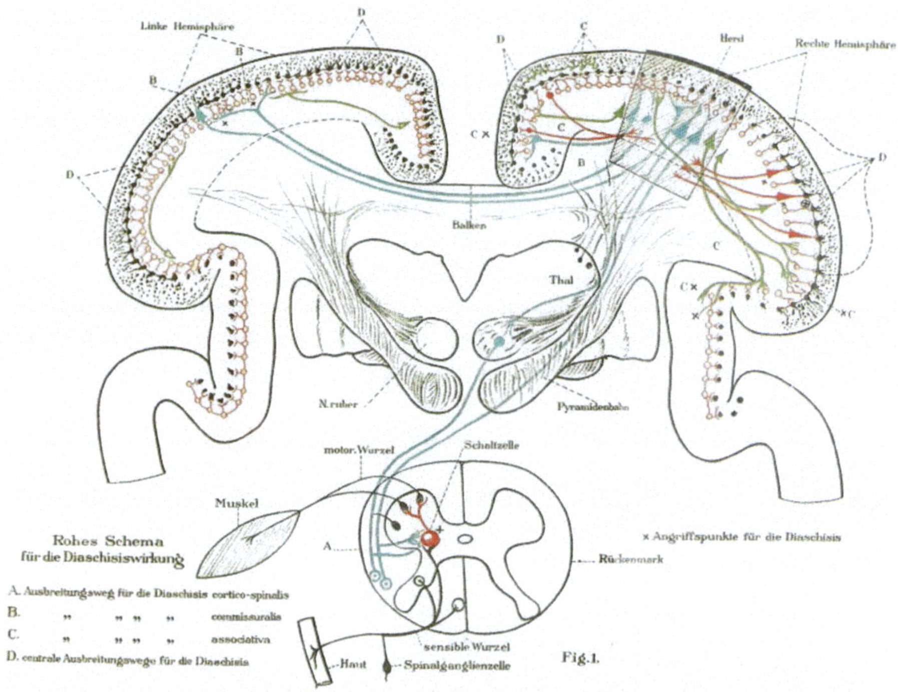

Fig. 9.2 von Monakow's illustration (1914) of the remote effects of a focal lesion (German: *Herd*) in the right hemisphere. Shown are fibres of the corticospinal tract (*A*) that are descending to the spinal cord and cause local dysfunction. Diaschisis (i.e. 'suspension of activity' via collateral and associative fibres) causes dysfunction in several remote ipsilateral and contralateral hemispheric areas (*B–D*). von Monakow often observed recovery of function after focal lesions, a process that could already begin in the first hours after lesion onset and would usually continue in the following days and weeks. He attributed recovery to (partial) resolution of the diaschisis (Figure from von Monakow's book *Die Lokalisation im Grosshirn und der Abbau der Funktion durch Kortikale Herde* (1914) [44])

functionality. Instead, he argued that the 'inhibition' of uninjured areas gradually resolved and that this led the patient to regain part of his functionality [45]:

> The key-note of his [von Monakow's] position is given by the universally recognized fact that the focal symptoms, which immediately follow a non-progressive lesion, are severer, more extensive and often less sharply determined than is the case after expiration of some days or weeks. Moreover, these initial manifestations may differ profoundly in character from the residual consequences of irreparable anatomical destruction. Thus, an operation upon the cortex and sub-cortical tissues may be immediately followed by a total flaccid hemiplegia accompanied by loss of sensation so gross that it would appear to be the result of a mid-brain lesion. After many weeks to months these temporary signs pass away and the clinical phenomena come to correspond more nearly in form to those consonant with injury to higher cerebral centers.[b]

[b] Quotation taken from *Aphasia and Kindred Disorders of Speech* by Head (1926) [10]

von Monakow's ideas resulted from his many years of clinical, anatomical and experimental studies. He eventually became head of his own research institute in Zurich (*Hirnanatomisches Institut*) and famously described several anatomical structures that were later given his name. The lateral cuneate nucleus and the rubrospinal fasciculus are nowadays known as von Monakow's nucleus and von Monakow's bundle, respectively. He was also the first to identify the arcuate fasciculus as the tract connecting the Broca and Wernicke areas, a view later adopted by Wernicke himself in 1908 [46]. von Monakow was particularly interested in the timed sequence in which cortical and subcortical components participated in brain functions, a principle that he referred to as 'chronogenic localization'. He was convinced that the spatial representation of functions can vary over time and, for instance, emphasized that aphasia can be absent in the case of slowly growing tumours in Broca's area.

von Monakow's ideas can be seen as a bridge between localizable and non-localizable concepts of functional organization. His work was acknowledged by scientist such as Kurt Goldstein (1878–1965) and Aleksandr Luria (1902–1977), who themselves had moved away from classic localistic views. As Goldstein wrote in 1934 in his book *The Organism*:

> In summarizing our discussion of the possibility of a correct anatomical evaluation, as a basis for localization, one thing seems certain: it renders rather unsatisfactory service to determine simply the location of a lesion. Whether a symptom will appear on account of a local injury, especially whether it will become a permanent symptom, certainly depends on many factors: on the nature of the disease process, on the condition of the rest of the brain, on the state of the circulation, and on the psycho-physical constitution of the patient. It also depends on the 'difficulty' of that performance, the disturbance of which represents the symptom, and, finally, on the reaction of the entire organism to the defect [47].

Goldstein rejected classic neurological theories on several grounds and provided alternative views not only for the representation of functions, but in particular for the treatment and rehabilitation of brain-damaged patients. He studied and treated the injured patient as a 'whole' and focused on altered behaviour and not only on focal neurological symptoms such as dysphasia or motor disturbances. The complex neurological and behavioural disturbances that he had seen in the many World War I casualties had broadened his scientific and clinical view. He did not merely see symptoms as the consequence of a single damaged brain area, but rather as an 'attempted solution' of an organism that dealt with disease [47]. Certain symptoms, so he stated, were the direct consequence of damage to a particular part of the brain. However, he considered many other symptoms 'the expression of the struggle of the changed organism to cope with the defect, and to meet the demands of a milieu with which it is no longer equipped to deal' [48].

Oliver Sacks (1933–2015) wrote in a foreword to one of Goldstein's reprinted books that he was 'one of the most important, most contradictory, and now most forgotten figures in the history of neurology and psychiatry' [47]. To Sacks, when he read the works in the 1950s as a medical student, Goldstein's work seemed to have:

> a vigor, a vitality, a largeness of vision, that radically contrasted with the tight atmosphere of classical neurology in which we were (...) being educated. (...) He talked about 'reactions' to illness, about 'adaptation', 'compensation', 'coming to terms', reactions that we could see in our patients all the time and that were crucial to understand if any rehabilitation was to be achieved, but ones that our textbooks completely ignored [47].

Although Goldstein acknowledged that sensorimotor functions or some of the motor abilities in speech could be linked to certain specific areas, he held that impairment of cortical functions was always more or less related to global changes in cortical organization. As he wrote in 1942:

> Localization of a performance means to me not an excitation in a certain place, but a dynamic process which occurs in the entire nervous system, even in the whole organism, and which has a definite configuration for each performance. (...) A specific location is characterized by the influence which a particular structure of that area exerts on the total process (...). However, that does not exclude the fact that we consider some symptoms more closely related to some areas and can make our decision, for practical purposes, on this basis [48].

Goldstein, and later Luria, thus strongly believed that no form of mental activity could be related to a localized and limited group of nerve cells or area [49].[c] Cortical and subcortical areas operated in concert, and each made a specific contribution to the complex 'dynamic structure' that constitutes a function. Luria (1947, 1970) stressed that we should not speak of 'localization of function'—as that is never possible—but rather of 'localization of symptoms':

> This position which has been so strongly defended by Goldstein was originally formulated by Jackson. According to his conception, a focal lesion of the brain which disturbs a process necessary for the performance of a certain function produces a symptom. In this sense the symptom is correlated with the site of the damage, i.e. 'localized' to a certain area of the cortex [49].

Luria dedicated a great deal of his professional life to the study of brain-injured patients. He was one of the first researchers who realized that analysis of the functional impairments of focal lesions had important implications for the study of normal (and not only abnormal) functional systems. Before him, researchers had mainly tried to establish a correlation between function and the structure of the damaged region, thereby largely ignoring non-local effects or the recovery potential of the undamaged part of the brain. Lucia's dynamic functional system was in line with previous work by Goldstein and in particular Hughlings Jackson, who he often cited and was keen to acknowledge. These authors spoke of 'disintegration' or 'impairment' of functions, and not so much of a 'loss' of function. A function, says Luria, is seldom completely lost in patients with spatially restricted brain lesions.

Luria was an exceptionally talented observer, who had learned from the large number of war casualties that traumatic speech deficits could be reduced or overcome by suitable therapy. He also deduced from his many cases that functional

[c] See for other pioneers Tesak and Code's *Milestones in the History of Aphasia* [24].

deficits had no simple relationship to the area that was damaged. There were simply too many different functional impairments that could develop when a single area of cortex was damaged:

> If (…) damage to the posterior-superior parts of the left temporal area disrupts normal activity in the system of structures responsible for acoustic analysis and synthesis, then it cannot but produce a disturbance of auditory speech (phonemic hearing). In doing this it must affect all forms of activity in which phonemic hearing and the connections related to it play a role. With the disturbance of structured hearing it is impossible to maintain precise and differentiated speech, to find necessary words, or to perform sound analysis and synthesis necessary for writing. When this basic disturbance is present, secondary symptoms inevitably develop as indirect or compensatory effects. Literal and verbal paraphasias and disturbances in the grammatical structure of speech appear, but such functions as object recognition, orientation in space, and calculation involving elementary written numbers remain unimpaired [49].

In line with von Monakow's theories on diaschisis, Luria proposed that functional disturbances are caused by two major mechanisms. First, there is functional loss due to damage of a particular cortical area. Second, there is a functional disturbance of several distant cortical areas, 'very likely as the result of loss of the normal conduction of excitation in the areas directly involved' [49]. As Luria considered the local damage to a large extent irreversible, he proposed that recovery of function must either be associated with a resolution of diaschisis or with a major reorganization of cortical processes [49]. To him, the latter was the major mode of recovery in most cases of focal brain damage. In this aspect, his opinion differed from that of von Monakow.

9.1.5 Mechanisms of Neural Plasticity

Jerzy Konorski (1903–1973), a Polish neuroscientist, gave one of the first modern definitions of neuroplasticity, emphasizing the role of structural changes in already existing pathways.[d] In Konorski's view, neurons can immediately react to incoming changes ('excitability'), leading to a 'permanent transition' of a system of neurons which he called plasticity [37]. Over time, there have been many definitions, basically all revolving around the concept of change. Plasticity is thereby considered a natural and lifelong property of the brain, enabling it to change its structure in response to experience. It is driven primarily by an imbalance between functional performance and environmental demands [52]. It is the key principle of learning and memory, but is also acting when the brain is aging or affected by disease. After

[d] Konorski published similar ideas on synaptic plasticity to Donald Hebb, who is usually credited for his rules on learning and the concept of distributed memory [50]. Konorski's research was suppressed for political reasons, and the impact of his work in the West was therefore probably less than it should have been. According to Markram (2011), his proposals were nevertheless appreciated at an early stage by well-known researchers such as Hebb, Adrian and Eccles [51]. 'Some researchers prefer to speak of Hebb–Konorski plasticity (…), although the concept of Hebbian plasticity is clearly in wider use' [51].

damage to the nervous system, due to for instance a tumour or stroke, it aims to restore the functional homeostasis of the nervous system.

Animal experiments have revealed several underlying microscopic processes that allow the brain to change its functionality. The reader is referred to the extensive literature on this subject. In short summary these mechanisms are axonal sprouting (allowing for new connections), unmasking of connections, neurogenesis, neuron–glia interaction, modification of synaptic weights, formation of new synaptic connections, and cell death and synaptic pruning (for fine-tuning of connectivity) [53–56]. At the macroscopic level, plastic changes involve restoration or redistribution of function within networks, as well as development of new compensatory cognitive strategies. These processes are still not much understood, but are increasingly studied with functional neuroimaging techniques (fMRI, PET, TMS). Although it is sometimes stated that neuronal plasticity aims to optimize the organism's behaviour, or that the brain can even 'change itself', it should be remembered that there probably exists no intrinsic frame of reference against which the brain can compare its results (in a similar manner that evolutionary processes do not strive for a particular goal, but are merely the result of competition). Effects of plasticity can be positive or negative, but that is ultimately a subjective judgement.[e]

9.1.6 Redundancy

Alternatively, individual brain functions may be represented by more than one neural representation to begin with. In this view, that is known as multiple realizability, there is not a one-to-one but a one-to-many relationship between functional and physical brain states (see, for instance, Overgaard and Mogensen (2011) [59]). The notion that functional information is stored in multiple different anatomical configurations would explain the brain's redundancy to (sub)cortical damage, as well as its capacity to regain functionality.

Redundancy to brain damage is a common observation in clinical practice. Neurosurgeons, for example, frequently place ventricular catheters or electrodes via an entrance that requires a small cortical lesion, or remove a tumour via a larger cortical opening without the introduction of new neurological deficits. More 'natural' causes also do damage to the brain without acute behavioural consequences. Aging, for instance, leads to an accumulation of small white matter lesions. Of course, all of these damaged areas were never afunctional before, but part of a larger brain network that apparently almost instantaneously compensated for the loss of a part of it. This seems only possible if there is enough redundant capacity within the network to maintain its performance. From a theoretical perspective, at least, any lesion would have to come at some cost (e.g. speed of informational processing or accuracy of behaviour). In practice, these effects are often subclinical and not measurable in neurological or behavioural terms.

[e] Maladaptive effects of neural plasticity are, for example, the experience of a phantom limb [35] or dystonia and motor hand weakness in professional musicians [57, 58].

Redundancy, in one form or another, has a long history in the discussion on cerebral localization of function. Already, in 1888, Golz pointed out that small cortical lesions in animals had absolutely no effect on their behaviour and that remaining regions were equally capable of supporting these functions [60]. Even larger lesions, so he concluded, produced surprisingly small effects due to this mechanism [1]. Lashley, in the 1930s, hypothesized an extreme equipotentiality for all cortical areas that were involved in sensory and perceptual processes, based on his extensive brain-lesioning experiments in rats [61]. He concluded that in these animals, behavioural impairments were simply related to the amount of tissue that was removed.

9.2 The Lateral Shift Hypothesis

Already at the end of the nineteenth century, the idea was discussed that not only left hemispheric areas could potentially play a role in the recovery of language functions. Many scholars, including Broca and Wernicke, considered the right hemisphere a 'backup' in case of damage to the left hemisphere. In fact, this was not so much of an issue at the time, perhaps because earlier ideas on brain functions—before hemispheric specialization became the dominant view—considered both sides of the brain functionally equivalent. As Head (1926) summarized:

> Normally in right-handed persons the right hemisphere is a vast uncultivated field, which plays a subservient or secondary part in the mechanism of speech. Impressions received by the sensory surfaces are worked up by centers of the left hemisphere into factors, which underlie the highest forms of logical thinking. These are transferred to the opposite site, to be stored up as unconscious memories in this special seat of latent and automatic psychical activities. Under normal conditions centers is the right hemisphere responsible at most for affective or interjectional speech; but as the result of suitable education and training they may play a material role in the re-acquisition of power to employ language [10].

In 1887, William Gowers (1845–1915) was among the first to provide evidence that the right hemisphere might indeed contribute to recovery of language functions [62]. He noted that in patients that had recovered from aphasia due to a stroke in the left hemisphere, a second lesion in the right hemisphere sometimes induced worsening of language functions. Gowers argued that this was powerful evidence for a role of the right hemisphere in the initial recovery from aphasia after the first left hemisphere stroke:

> Loss of speech due to permanent destruction of the speech region in the left hemisphere has been recovered from, and that this recovery was due to supplemental action of the corresponding right hemisphere is proved by the fact that in some cases, speech has been again lost when a fresh lesion occurred in this part of the right hemisphere [24].

At the time, Thomas Barlow (1845–1945) presented a case that was subsequently cited by many of his contemporaries as strong evidence for a functional takeover of speech functions by the right hemisphere [63]. However, the case lends itself to different interpretations. Interestingly, Barlow himself never discussed reorganization of speech functions. The case is shortly described in Fig. 9.3 and was critically reviewed before by Finger (2003) and Hellal (2007) [64, 65].

Others pleaded for a language role of the right hemisphere already in the normal, healthy situation. Hughlings Jackson was one of the first who considered both hemispheres to be important for language functions. Although he agreed with Broca's statement that a loss of speech was associated with damage to the left hemisphere, he believed that comprehension was a bilateral function. As Finger (2001) wrote:

> He theorized the right hemisphere could still learn the meaning of words as a result of associative laws (hearing of the word "horse" while also seeing a horse) and that it possessed a consciousness for place and things. Yet, it seemed clear to him that only the left hemisphere could truly become "conscious in words". (...) Jackson (1868) accepted the belief that the left frontal lobe grew faster than the right frontal lobe and took the lead in voluntary speech. He also believed the right posterior lobe grew in advance of that on the left, and from this he deduced that the posterior right hemisphere was the leading part for perception and imagination [66].

9.2.1 Broca

Broca's view of the right hemisphere's role in language is surprisingly different to his legacy of strict left hemisphere language dominance, and resembles some of the contemporary thoughts about this issue [11]. For that reason, I will discuss it at some length here. According to Broca, 'articulated speech' is normally (i.e. in healthy conditions) the privilege of the left hemisphere. To explain this organic predisposition, he referred to the works of Gratiolet and Bertillon, who had provided evidence that left hemisphere development starts earlier than that of the right hemisphere. As it is thus more precocious at birth, it best fits to encompass language functions at a young age. 'The tendency to speak with the left hemisphere is born', so wrote Broca in his 1865 paper [67]. But Broca did not link language functions exclusively to the left hemisphere. In that same paper he also explained that the 'connection' between ideas and words is a capacity that belongs to both hemispheres. Proof of this, according to Broca, is seen in patients with deep and extensive lesions in the left hemisphere. These patients are disabled in their speech, but continue to understand what is spoken to them. Each hemisphere can therefore, in case of a lesion or disease, 'reciprocally substitute for each other; however, the faculty to express them by means of coordinated movements, in which the practice requires a very long period of training, appears to belong to but one hemisphere, which is almost always the left one' [67].

In Broca's view, both hemispheres are thus capable of language comprehension, hence the possibility of functional recovery after brain damage (note that a left temporal language centre was only to be discovered a decade later). In order for people to become *right-brained*, Broca assumed that the left third frontal convolution was involved in an injury very early in life. Under those circumstances, the right homologue area would substitute and have the ability to learn to speak. He compared this process to a child that is born without a right hand, but still becomes as skilful with his remaining left hand as he would have ordinarily been with the other. Broca stressed that the mastery of articulated speech is complex and difficult, something that a child only manages to succeed after long and tentative efforts 'of the most complicated degree' [67]. Broca was convinced that training would be very beneficial for functional recovery in patients, but it should be given with the same intensity and duration as a child that learns to speak. These patients, so he concluded, should be treated with the 'tireless patience of a mother who teaches her son how to speak' [67]. But obviously that was not done at that time.

Broca's ideas were supported by one of his cases from the Salpêtrière Hospital. During the autopsy of a 47-year-old patient, he noted that the third left frontal convolution was lacking, along with the inferior parietal convolution and the superior temporal convolution [67]. In other words, the *convolution d'enceinte* was missing

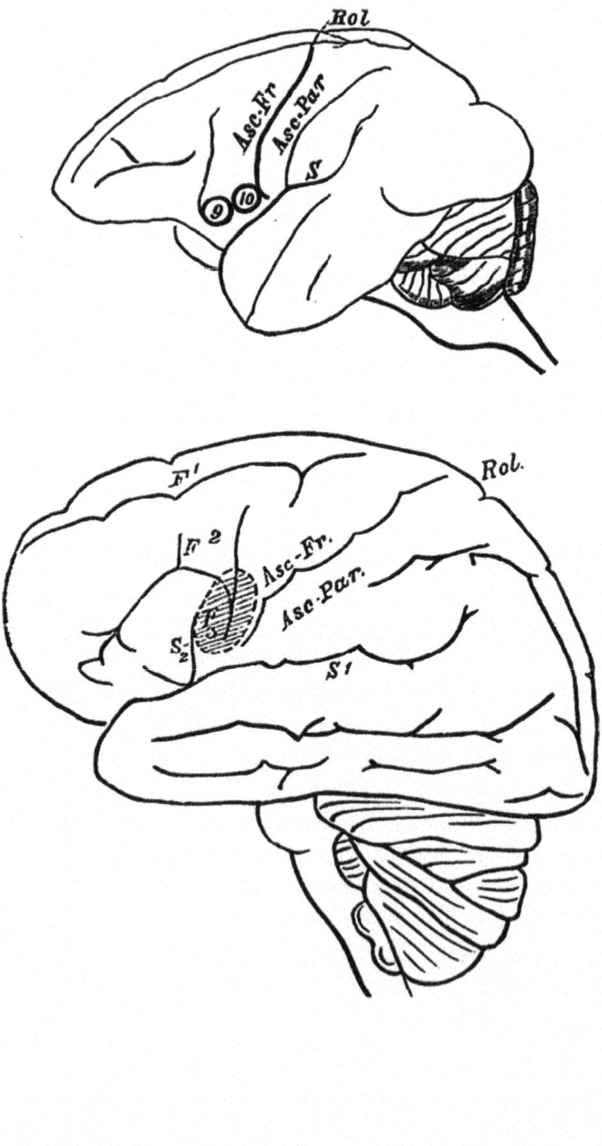

(see Fig. 1.4 in Chap. 1). In addition to this, several other areas in the left hemisphere were underdeveloped. Broca concluded that there was a congenital atrophy of the left side of the brain and speculated that this was due to congenital absence of the Sylvian artery that he was unable to find during the autopsy. The patient had seizures from a very early age and had atrophic and paretic right extremities. But there were no obvious language problems. Broca concluded that in this case there was no doubt that the right inferior frontal convolution was involved in language functions. Immediately, he asked himself a much broader question: why does such a functional compensation not occur in all cases of *aphémie*? If both hemispheres can indeed contribute to language functions, then a lesion in only one hemisphere should not be enough to cause lasting language impairments. But why was this not the case in everyday practice? To explain this, Broca assumed that larger lesions generally affected the intellect of most patients. Consequently, he stated that their 'weakened mind' prevented them from learning to speak exclusively with the right hemisphere, 'which up to now had only played an accessory role in the function of expression by means of articulated speech'. So it is not so much that the right

Fig. 9.3 Barlow's case (1877) of a patient who presented with two brain infarcts, first on the left side and later on the right side [63]. The case report was seen by many of Barlow's contemporaries as strong evidence for the ability of the right hemisphere to restore speech functions after damage. Barlow described the medical history of a 10-year-old boy who suddenly developed a right-sided hemiplegia, paralysis of the tongue and severe difficulties with speech and swallowing. In the weeks afterwards, the patient made a near complete recovery. Unfortunately, 4 months after the first insult, a hemiplegia on the other (left) side occurred, and speech again deteriorated. After initial recovery of function, the boy died 2 months later. Autopsy revealed that he had died from cardiac disease that had caused embolic infarctions in several organs including the brain (first on the left side and later on the right). Fairly small but nearly symmetrical lesions were noted on both sides of the brain (bottom figure in Fig. 9.3 is an Ecker figure that shows the lesion in the left hemisphere): 'on each side, these regions consisted of the lower end of the ascending frontal and the hinder end of the middle and inferior frontal convolutions. These areas were pale, buff coloured, slightly depressed and slightly softer than the surrounding brain substance. Reckoning from the surface, they were less than a quarter of an inch deep, i.e. they involved the cortical and a little of the subjacent white substance [63]'. Remarkably, Barlow himself never suggested that the right hemisphere had taken over speech functions from Broca's area. He attributed the speech disorder to damage to the face area of the motor cortex and referred to the experiments of Ferrier, who had shown that stimulation of inferior frontal areas in monkeys—in a homologue location to those of his patient—results in movement of mouth and tongue (top figure in Fig. 9.3 from Ferrier and modified by Barlow for his publication). 'The bilateral muscles, which act together, are represented on the two sides of the brain. After the first attack of hemiplegia, although this region on the left side was probably permanently damaged, yet still the right side remained intact. But, after the second attack of hemiplegia, the corresponding region on the right side became damaged; and henceforth, as far as voluntary movements of the mouth and tongue were concerned, the boy was irretrievably deficient'. As Finger (2003) concluded in his analysis of the case [64]: 'thus, from Barlow's perspective, the quite capable and partially redundant right motor cortex was simply left with sole control of the bilateral mouth musculature after the boy's left hemisphere was severely damaged and the remote effects of the insult subsided. But after the right hemisphere subsequently suffered ischemic damage, neither side of the brain still possessed the circuitry needed to control these muscles voluntarily. This was why the boy was "irretrievably deficient" when he now tried to converse [64]' (Figures and text taken from Barlow (1877) [63]) hemisphere is not capable of language functions, but that these patients lack the cognitive abilities to learn to speak again and to complete the intensive training that is needed for successful rehabilitation.

Broca was one of the first to consider a 'takeover' of language functions by the right hemisphere [68]. Since the latter half of the nineteenth century, this explanation has been frequently brought forward in particular in patients with a damaged left hemisphere but only minor language impairments. This lateral shift hypothesis is certainly very intuitive, something that undoubtedly contributed to its remaining popularity. However, the evidence for it is rather meagre, and cohesive theories are still lacking [69]. Code, in 1987, wrote that it is predominantly based on 'a loose collection of clinical anecdotes and speculation' [68]. Even with the advent of modern neuroimaging techniques, the issue has not been settled. Careful interpretation of results, though, indicates that the left hemisphere very often remains critically involved in language functions, as will be explained in the next sections.

9.2.2 The Confusing Results of Modern Imaging Methods

Non-invasive structural and functional neuroimaging techniques, which have evolved over the last few decades, have dramatically improved the ability of researchers to map the brain. New possibilities arose to examine the brain's reactions to damage and to identify more accurately the areas that are involved in postlesional language organization and recovery. Not surprisingly, the number of brain mapping studies in the field of neural plasticity has grown exponentially. Functional imaging techniques, however, are known to have drawbacks and limitations, in particular in a clinical setting.[f] There is no broad consensus on how to compare, quantitatively, one brain map to another in order to measure the presence of plasticity or to identify a language-dominant hemisphere. Researchers and clinicians do this most of the time by eyeballing the fMRI maps of a patient and comparing these to their own ideas of a 'normal' pattern of language areas or—if they are lucky enough to work at an experienced clinical research institute—use home-grown algorithms to compare results against own historical series and clinical experience [72]. But even when quantitative measures are used by these groups, for example, to express the degree of language lateralization, inconsistencies remain compared to the results from clinical techniques.

---

[f]Possibilities and pitfalls of functional neuroimaging techniques are extensively discussed in Chap. 8. In short, there are two main issues that put limitations on the use of fMRI: (1) averaging individual results improves statistical power, but will decrease spatial resolution of the group results to 1–2 cm. (2) Functional imaging techniques cannot differentiate between *critical* and *involved* language areas. When an area shows up on a brain map, this does not necessarily mean that it is crucially important for language nor that it is even involved in language functions per se. Consequently, the presence of right hemisphere activation does not automatically imply that there is atypical or 'abnormal' language organization. This is simply demonstrated by the fact that fMRI maps of healthy subjects invariably show bilateral language-related activation, obviously challenging the clinical dogma that most people have a language-dominant left hemisphere [70, 71].

As was explained in Chap. 8, the most common approach to studying the functional reshaping of brain areas is to measure brain responses in groups of patients and to compare them to those of healthy subjects. One should be cautious, though, as potentially relevant individual differences may get lost when brain maps are averaged to create group maps.[g] Conclusions drawn from group studies therefore do not necessarily hold for individual brain-injured subjects. To minimize this bias, groups should be matched as closely as possible for variables that are supposed to influence brain plasticity (such as age, educational level, and chronicity and spatial distribution of the lesion), except for the variable that is of interest. The problem is, of course, that not all potential confounding effects are known a priori. As Halai and colleagues (2016) state it:

> The generation of stable, reliable models of normal and impaired function relies on the ability to understand the nature and sources of individual differences across patients. This has always been a key challenge for the field, leading to numerous debates and discussions, and arguably still is. The kernel of the problem relates to the fact that there are multiple sources that underlie variable neuropsychological results [73].

So far, the results from functional imaging studies have been very diverse and led to contradicting views about the way the brain remaps functions at the macroscopical level. Some of the first functional imaging studies indeed proposed that right hemisphere areas facilitate recovery of impaired language functions and described increased activation in the contralesional hemisphere of patients with a left-sided stroke or brain tumour [74–79].[h] However, subsequent PET and fMRI studies provided convincing arguments for exactly the opposite view and argued that right hemisphere activation is usually an impediment to language recovery and that restoration of language functions largely depends on the repair or recruitment of left hemisphere areas [81, 82]. The compensatory potential of the right hemisphere, so it seems, is lower than the recovery potential of the left [83–85]. Although these contradicting views are not mutually exclusive—as there are likely individual differences in which the brain reacts to damage—it is reasonable to assume that functional neuroimaging studies in general tend to overestimate the role of the right hemisphere in core language functions [86]. What is often interpreted as a shift in laterality, or even as a recruitment of 'new' parts of the brain after recovery of aphasia, could well be the expression of a pre-existent and large-scale bilateral language network [54, 70, 75, 77].

---

[g] Alternatively, brain plasticity can be studied in single-case studies. Such an approach, however, does not allow for generalization of results or testing of scientific hypotheses and also suffers from the limitations of single-subject fMRI that are mentioned in Chap. 8 [73].

[h] The explanation frequently given is that a lesion—via the transcallosal pathways—causes disinhibition of homologue areas in the other hemisphere. This assumes that under normal circumstances, the language-dominant left hemisphere inhibits right hemisphere homologue areas via the transcallosal pathways. Due to a lesion, this inhibition is lost, and there is a reactive increase in the activation of the contralesional homologue area. There is some experimental evidence for this phenomenon (see Fig. 9.4) [80].

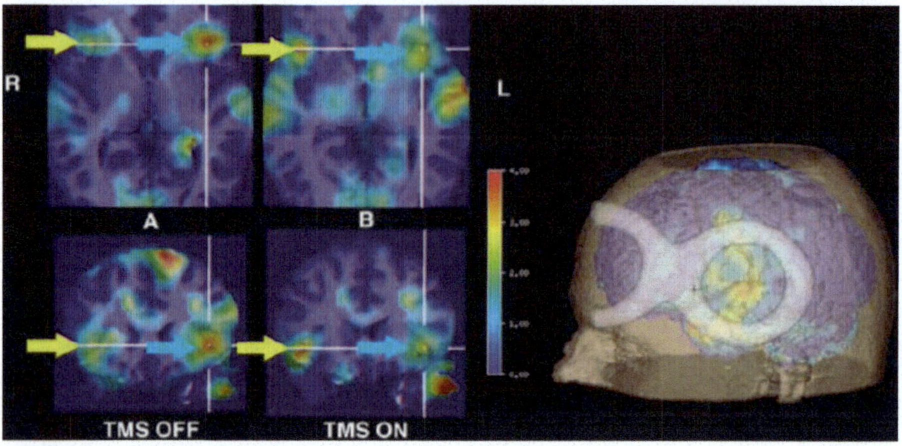

Fig. 9.4 In this experiment by Thiel and colleagues (2006), six healthy subjects were asked to perform a verb-generation task. Exemplary results of one subject are shown in image A. As expected, PET images show more activation in the left inferior frontal cortex (*blue arrow*) than the right (*yellow arrow*). Next, TMS was applied over the left inferior frontal region. With the coil, seen in the figure on the right, electromagnetic pulses are applied over a targeted area in Broca's region. Stimulation lowered the performance on the language task, indicating that the function of the stimulated area is disturbed. Also, the activation pattern as simultaneously measured with PET changed: activation decreased in the left hemisphere (*blue arrow*) and increased in the right hemisphere (*yellow arrow*). The authors suggested that the right hemisphere activation that they had previously observed in left hemisphere tumour patients was caused by a similar mechanism of transcallosal disinhibition and not so much by increased language dominance of the right hemisphere (Figure taken from Thiel and colleagues, 2006 [80])

Importantly, brain maps of language-impaired patients are—by definition—confounded by altered performance. But even when performance on a certain fMRI language task is within normal limits, patients may have used a different cognitive strategy to perform the task, performed the task less efficiently, or required more mental effort. The exact influence of these effects on brain maps is not well known, but it is likely that they contribute in some form to the right hemisphere activity that is seen in fMRI studies. Some researchers have used transcranial magnetic stimulation (TMS) to investigate, in a more direct manner, the functionality of these areas (see Fig. 9.4) [80, 87]. Language mapping with TMS, however, is challenging, and only a few studies have been published yet. There are other reasons why the functionality of right hemisphere areas can get misjudged or overrated, namely, when left hemisphere areas go undetected due to flaws in the experimental design or the effects of pathology. Pathophysiological factors such as tumour infiltration, neovascularity or changed haemodynamics have been reported to disturb the BOLD signal, unintentionally lowering the sensitivity for task-related signal changes [88]. This causes a shift of language-related activation towards the contralesional hemisphere that is not so much caused by an actual increase of activation in the right hemisphere, but due to reduced activation in the lesional left hemisphere [88, 89]. This phenomenon has been named 'pseudo-dominance'.

In conclusion, modern brain mapping techniques are adding new and valuable parts to the puzzle of brain plasticity, but proper use requires thorough knowledge of their methodology as well as an unbiased scientific perspective; otherwise they will lead to misinterpretations [73, 90, 91]. One thing is clear: neural reorganization is not simply a transfer of functions to contralesional homologue areas. The next two sections summarize in somewhat more detail findings in patients with acute-onset and slowly growing lesions, respectively.

9.2.3 Acute-Onset Lesions

In adult patients that become aphasic after a stroke, most of the functional recovery occurs within the first few weeks to months, depending on the severity of the initial language disturbances (Fig. 9.5) [92]. After a month, speech has normalized in approximately one-third of patients [92, 93]. One year after stroke, most patients have made a substantial recovery, with an estimated one-fifth of patients that are left with severe language impairments [94]. Different mechanisms have been postulated to play a role during functional recovery. In the immediate phase after the stroke, outcome strongly depends on reperfusion of perilesional tissue (i.e. restoration of the penumbra) and not so much on relocation of functions [95]. In the weeks afterwards, right hemisphere areas seem to contribute to the recovery process [96]. After a longer period of at least one year, however, there is substantial evidence that many of the recovered patients again rely on areas in the left hemisphere for their language functions [81]. Data therefore suggests that the most effective manner to regain language functions is to restore left hemisphere networks and to successfully

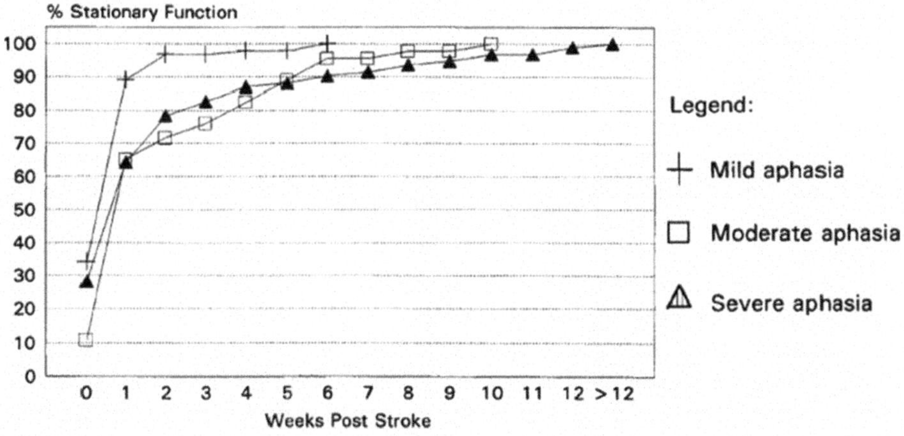

Fig. 9.5 Time course of recovery from aphasia after stroke, as measured with clinical language tests in a cohort of 881 stroke patients. The graph shows the cumulative percentage of patients that have reached stationary language functions in relation to the initial severity of their aphasia. Note that most improvement occurs within the first 2 months and in case of initial mild aphasia the first 2 weeks (Figure from Pedersen (1995) [92])

integrate available ipsilesional areas [54, 96–98]. A permanent shift of language-related fMRI activity to the non-dominant hemisphere is predominantly seen in patients that keep their language impairments, and somehow reflects inefficient language processing [82]. Nonetheless, a number of studies have associated language recovery with functional and structural changes in the right hemisphere [71, 76, 99].

The situation in infants and young children is different, in the sense that they have a much better functional perspective than adults after comparable brain lesions. It is commonly accepted that the immature brain can handle damage much better and that even larger lesions only infrequently lead to significant (language) deficits. This is best understood by the fact that the brain of a child is not fully matured and thus has not established a definite language network yet [11]. The question of *how* exactly this is accomplished is however still open for discussion. Some have proposed that there is a certain degree of equipotentiality for language functions at birth (as was already discussed in Sect. 9.1.2). As children with a damaged left hemisphere may not undergo the usual left-lateralization process, this would leave them with a bilateral or even right-sided language presentation [100, 101]. Such a process is in line with the long-standing belief that language development after early left hemisphere injury takes place by compensatory activity in the right hemisphere [102]. Behavioural studies have indeed generally found more involvement of the right hemisphere in early-lesion patients than in healthy controls, although findings in the literature—and in particular their interpretations—are heterogenous [100, 103, 104]. Patients in whom the lesional left hemisphere was surgically removed (leaving no room for discussion on the functionality of the remaining right hemisphere) can indeed do remarkably well, but this seems to be the exception rather than the rule [14]. Most of these patients have at least moderate language impairments when carefully examined (note that only small series of these patients exist in the literature) [105, 106].

Functional neuroimaging studies, at first impression, support the hypothesis that the right hemisphere takes over language functions after early left hemisphere injury. However, on closer inspection, it appears that not all of the right hemisphere activation is effectively contributing to language—a situation that is also observed in patients that acquired their lesion in adult life. When the relationship is examined between language performance and the degree of lateralization, there is clear evidence that the left hemisphere, and in particular frontal areas, remain playing a critical role, even in the case of extensive left hemisphere damage [100]. Such studies argue that there is an early bias for certain aspects of language organization in the left hemisphere and plead against the concept of equipotentiality.

In summary, there are clear age-dependent differences in the recovery potential of the brain after acute-onset lesions. It remains unclear, however, whether these differences are also associated with different mechanisms of neural reorganization and what factors determine the individual level of recovery (other than age) [104]. In further search of explanatory and predictive models, we should probably best adopt a network approach. Although such 'connectome analysis' is still in its infancy, it may help characterizing the dynamics of language functions in the presence of a lesion [107–109].

9.2.4 Lesions That Tend to Grow Slowly or Are Congenital

What about the functional impact of slowly growing lesions? Unlike stroke patients, most of these patients—even with lesions that are located within classic language territories—have no or only subtle language deficits, something that is commonly 'explained' by neural plasticity [110, 111]. Good examples can be found in patients with arteriovenous malformations (AVMs) and low-grade gliomas.

AVMs are vascular anomalies that occur during foetal development or at least very early in life. As a result, these lesions are part of the brain before the language system is fully matured. AVMs are typically detected between the age of 20 and 40 years, either after a bleeding or a seizure, or by coincidence (when a scan was ordered for other reasons). The fact that an AVM within a classic language region does not cause language impairments is frequently taken as evidence that language functions have relocated to other areas [110]. In the largest series to date by Deng (2015), 63 patients with unbled AVMs in the left hemisphere were studied with fMRI to assess language representation [112]. In 23 of these patients (37%), lateralization to right hemisphere homologue areas was described, suggesting reorganization of language functions (Fig. 9.6). The authors conclude that their 'results seem to indicate that in patients with an AVM, a nidus near the Broca area mainly leads to right-sided activation of the Broca area, whereas a nidus near the Wernicke area mainly leads to right-sided activation of the Wernicke area' [112]. Importantly, however, in a follow-up study of these left hemisphere AVMs after surgery, the authors did not find evidence that the patients with right-sided fMRI lateralization did better than those that were left-lateralized [113]. The incidence of postoperative aphasia in both groups was the same (respectively 39% and 44% immediately after surgery and 17% and 16% several months to years after surgery). Apparently, language lateralization was not predictive of hemispheric language dominance, and important language structures had still remained in the vicinity of the lesion in the left hemisphere.[i]

The same study of Deng also included 38 patients with left hemisphere gliomas in or near the classic language regions. Gliomas, in contrast to AVMs, are not of congenital origin, but develop in adult life. The glioma patients, who had a mix of low-grade and high-grade gliomas,[j] showed significantly less lateralization to the right hemisphere than the patients with AVMs and suggested relocation in only 6 of 38 patients (16%) [112]. This result is in accordance with other studies that yielded

[i] The abnormal blood vessels of AVMs may disturb the local haemodynamic response and reduce fMRI signals. There is some concern in the literature that this may significantly affect the clinical interpretation of fMRI maps, although studies are not conclusive at this point [88, 114]. In the series of Deng, these local disturbances may have falsely exaggerated right-sided lateralization (pseudo-dominance), but obviously cannot explain the increased fMRI activity in the contralesional hemisphere.

[j] Low-grade gliomas are typically present for more than a decade before they become clinically manifest and in the majority of cases debut with a seizure [115]. High-grade gliomas grow much faster, presumably in the order of months, and frequently cause neurological and cognitive deficits.

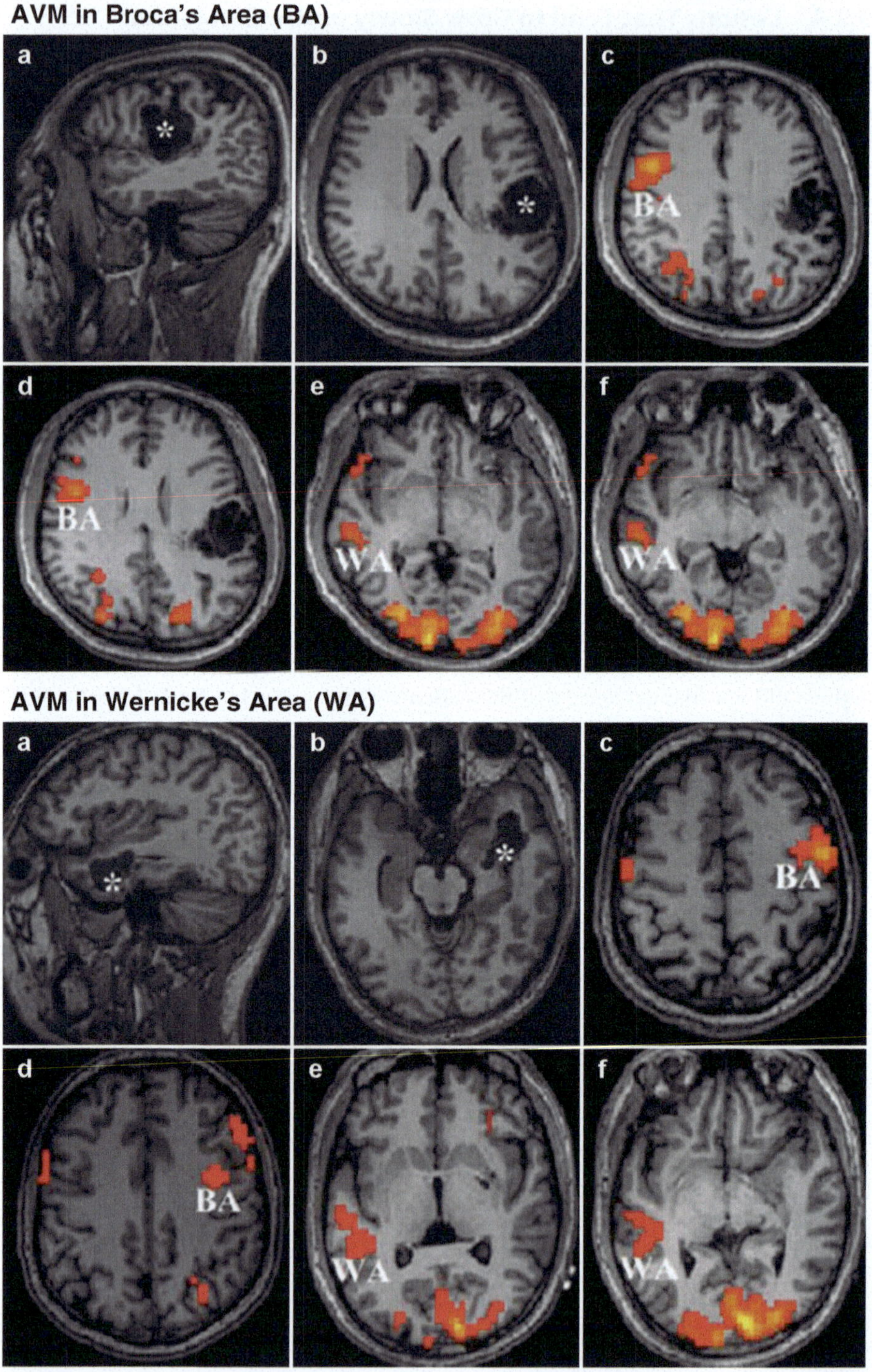
AVM in Broca's Area (BA)
a
b
c
BA
d
BA
e
WA
f
WA
AVM in Wernicke's Area (WA)
a
b
c
BA
d
BA
e
WA
f
WA

low estimates of right hemisphere language dominance in brain tumour patients. Thiel and colleagues (2001, 2005), for example, found that 63% of left hemisphere brain tumour patients had activation of the right inferior frontal cortex, but—after calculation of a lateralization index—classified only 18% of these patients as having truly 'reversed dominance' [77, 116]. Others found no evidence at all for contralesional reorganization to homologue areas in patients with a tumour in the left hemisphere [117]. The fact that the left hemisphere remains critical for language in glioma patients finds strong support from awake surgical procedures [86, 117]. In almost all cases, even when the glioma has infiltrated classic language territory, the surrounding (sub)cortex is still involved in language functions, as demonstrated by direct electrical stimulation mapping and the high incidence of temporary language disturbances after surgery (see for case examples Figs. 4.9, 6.20 and 8.16 in respectively Chaps. 4, 6 and 8) [86, 118].

These studies, and many others in the literature, strongly suggest that exclusive right hemisphere language representation is rare and that completely normal language functioning without support of the left hemisphere is unlikely. Although there is certainly evidence that patients with brain lesions more often have bilateral language representation than healthy subjects, it remains unclear how this relates to their language abilities [87]. In this respect, we are still far from understanding the neuroplasticity of language networks.

9.3 The Efficacy of Aphasia Treatment

Until a few decades ago, it was widely believed that the structure of the brain was largely immutable after childhood and that there was nothing much one could do to help patients recover from serious brain damage. There was room for some small changes, given the brain's ability to learn and adapt, but more radical changes were

Fig. 9.6 Deng and colleagues (2015) studied the language representation of 63 patients with an unruptured arteriovascular malformation (AVM) in the left hemisphere. fMRI was used with a silent reading task of Chinese characters. AVMs were selected that were either located in or near Broca's area (BA) or Wernicke's area (WA). Based on the amount of voxels in both hemispheres, laterality indices were calculated for both language regions. Right-sided lateralization was found in 37% of the AVM patients and 16% of the glioma patients. Interestingly, in the glioma patients, these were all Wernicke's area homologues, even with left frontal tumours. *Top images* (**a–f**) show an example of an AVM (*asterisk*) near Broca's area, with fMRI activation lateralized to the right-sided homologues of Broca's and Wernicke's area (note that the AVM is predominantly located in the inferior part of the precentral gyrus and not so much in the inferior frontal gyrus). The *bottom images* (**a–f**) show an AVM in the anterior part of the temporal lobe, with right-sided lateralization of Wernicke's area and left-sided lateralization of Broca's area. Opposite lateralization for frontal and temporal language areas has been described before in other fMRI studies [141] held to be impossible. This was reflected in the attitude towards patients with brain lesions, as paraphrased already in 1948 by Goldstein [119]:

> Treatment of aphasic disturbances was not and is not very popular among neurologists. Usually the attitude was one of pessimism as to whether one can help these patients by systematic training, and it was said: Either the condition improves spontaneously or it remains essentially unchanged in spite of all attempts at retraining.

Neurologists and neurosurgeons have never been much involved in rehabilitation. Today, once diagnostics and acute medical treatment have finished, further support and treatment for recovery is provided by speech therapists, physiotherapists and neuropsychologists. Interventions are predominantly centred on functional compensation, whereby behavioural changes after brain damage are usually interpreted as the result of new strategies that are created from an old functional repertoire.[k]

In the literature, language recovery is either attributed to 'spontaneous' or 'reactive' mechanisms, or to the effects of learning or training [53, 121]. Learning and training are generally seen as a different, second form of plasticity. These processes are engaged by therapists but also by friends and family. They are effective for a much longer time—probably months to years—and have been termed 'experience-dependent' or 'learning-induced' plasticity. Examples are the genesis or modification of synapses [121, 122].

For patients, the relevant question is not so much whether the brain is amenable to change, or what the exact mechanisms of recovery are; everyone will agree that plasticity is a fundamental property of the nervous system [123]. The question is to what extent impaired behaviour can be restored, and whether this process can be enhanced by training or other measures. The rest of this section will briefly outline the field of neurorehabilitation and some of its current options for patients with speech or language disturbances. As will be seen, there is a lack of highly effective therapies.

9.3.1 Experience-Induced Plasticity

In healthy subjects, there is substantial evidence that training can improve a variety of sensorimotor and cognitive skills. Modern technologies, such as action video games, seem very suitable to providing the complex learning environments that are needed for this purpose [57]. To cite Bavelier, in Gazzaniga's *The Cognitive Neurosciences* (2009):

[k] The works of Goldberg, Luria and many others emphasized that there is more to functional recovery than changes in neural architecture and spatiotemporal reorganization of functions. Brain-damaged organisms also retrieve their goals 'through the employment of novel tactics or unusual behaviors' [1, 120]. Social support, motivation and a positive rehabilitation environment play an important role.

> the variety of different skills and the degree to which they are modified in video game players appears remarkable. These include improved hand-eye coordination, increased processing in the periphery, enhanced mental rotation skills, greater divided attention abilities, faster reaction times, and even job-specific skills such as laparoscopic manipulation and airplane piloting procedures [57].

However, to keep things in perspective, exercise and socialization also have a beneficial effect on sensorimotor and cognitive functions [124].

Since the second half of the last century, the efficacy of speech-language therapy for patients with aphasia has been studied with varying degrees of sophistication and with varying outcomes.[l] Progress, however, has been slow. Although language rehabilitation after stroke is often recommended by physicians [126], there are still no therapies or methods that have a significant impact (taking into account the effects of spontaneous recovery and normal social contacts) [127, 128]. Some aspects of language and communication can be improved by treatment, but it is unclear how long these effects last and what the best treatment is [82, 125]. One of the problems is that therapeutic interventions often remain limited to that which has been trained. Results, in other words, do not always generalize to untrained stimuli. A good example is lexical knowledge in patients with naming difficulties. Relearning words is labour intensive and not very effective, because the relationship between word meaning and word sound is largely arbitrary. In the words of Nadeau (2015): 'If one has learned the names of 100 objects, this provides no assistance in naming the 101st' [121]. There are some newly developed methods that aim to generalize training results to a broader context, but these are pending further scientific evaluation (see for an overview Nadeau, 2015) [121].

Fortunately, some therapies have proven to be effective. One famous example is a computer-based training for developmental dyslexia (a reading and spelling disorder) that is based on the ideas of brain plasticity. Paula Tallal and Steve Miller discovered that dyslexia is often an auditory speech-processing disorder, caused by a deficit in the identification of rapid speech sounds (e.g. with phonemes such as *b*, *p*, *g* or *d*).[m] As most of these children have no detectable difficulties with speech, this finding maybe somewhat counterintuitive at first. However, the deficiencies in phonological processing obstruct a good match between the sound of a word and its written form. This will cause problems for children who are learning to read. When Tallal and Miller met Merzenich and others, they started to use the principles of neuroplasticity to train dyslectic children. Their story is a fascinating read.[n]

> Might reading be improved in dyslexics, he [Merzenich] wondered, if their ability to process rapid phonemes were improved? And could that be done by harnassing the power of

[l] Speech-language therapy is defined as a formal intervention that aims to improve language and communication abilities, activity and participation [125]. Studies have predominantly been conducted with stroke patients.

[m] Others have argued against this theory; see, for example, Studdert-Kennedy and Mody (1995) [129] or Snowling (2001) [130].

[n] See the book of Schwartz and Begley, *The Mind and the Brain* (2002) [131].

> neuroplasticity? Just as his monkeys' digits became more sensitive through repeated manipulation of little tokens, Merzenich thought, so dyslexics might become more sensitive to phonemes through repeated exposure to auditory stimuli. But they would have to be acoustically modified stimuli: if the basis of dyslexia is that the auditory cortex failed to form dedicated circuits for explosive, staccato phonemes, then the missing circuits would have to be created [131].

Eventually, a computer program was developed that slows down the various speech sounds to a level where the child is able to parse them again in a correct manner. Once this modified speech tempo has been mastered, the interval between the sounds is gradually shortened until that of (almost) normal speech.[o] This approach proved to be effective and helps to overcome the inborn temporal processing deficits in dyslexic children [132–134].

9.3.2 Techniques to Enhance Plasticity

There are currently no devices that can speed up or aid language rehabilitation via a direct interaction with the brain. There are only few studies that have used electromagnetic fields for this purpose. Brain stimulation in patients with aphasia has either tried to enhance activity in brain areas that are thought to support language functions or—more commonly—tried to suppress brain activity in areas that are thought to interfere with recovery [82].

Naeser and colleagues were the first to provide evidence that non-invasive brain stimulation could be used to improve language functions [135]. In their initial studies with TMS in stroke patients, stimulation of the right hemisphere homologue of Broca's area led to better performance in language tasks, supporting theories that right hemisphere activation is in fact counterproductive and withholds recovery [135, 136]. Most other studies confirmed that TMS is apparently able to reduce the negative effects of presumed 'overactivated' right hemisphere areas [135]. These results should be seen as preliminary, as study populations are small and long-term impact is unknown.[p]

Electrical pulses have occasionally also been applied invasively, in the hope to enhance functional reorganization. Barcia and colleagues operated on a patient with a tumour in Broca's area and had to leave a significant remnant that seemed still involved in language functions [138]. They subsequently implanted subdural electrodes to allow for chronic electrical stimulation of the functional areas within the tumour. They hypothesized that 'through the artificial induction of a progressive virtual lesion in an eloquent area, we could promote plasticity by mimicking the ability of the tumour to evoke a reorganization of function to ipsilateral or contralateral areas' [138]. Stimulation was set to induce a 'mild speech disturbance' and lasted for 25 days, after which the patient was again operated under awake

[o] Fast ForWord is a commercially available program; see http://www.scilearnglobal.com/.

[p] See for reviews Ren (2014) [137] and Turkeltaub (2015) [82].

conditions. The number of cortical sites where language areas were detected had significantly lowered (from 12 to 2), and the surgical team was able to remove the remaining cortical part of the tumour without new impairments.[q]

Although these approaches are innovative and perhaps even promising, they should be met with caution, as many questions remain to be answered. Importantly, all of these procedures have been developed despite a clear understanding of how brain plasticity works or how language functions reorganize after a brain lesion. It is important that we identify and understand these mechanisms, as we are otherwise unable to extract the essential parts of these procedures and improve our therapies [54].

References

1. Levin HS, Grafman J. Cerebral reorganization of function after brain damage. Oxford: Oxford University Press; 2000.
2. Finniss DG, Kaptchuk TJ, Miller F, Benedetti F. Biological, clinical, and ethical advances of placebo effects. Lancet. 2010;375:686–95.
3. Logan AC, Selhub EM. Vis Medicatrix naturae: does nature "minister to the mind"? Biopsychosoc Med. 2012;6:11.
4. Finger S, editor. Recovery from brain damage. Research and theory. New York: Plenum Press; 1978.
5. Ades HW, Raab DH. Recovery of motor function after two-stage extirpation of area 4 in monkeys. J Neurophysiol. 1946;9:55–60.
6. Travis AM, Woolsey CN. Motor performance of monkeys after bilateral partial and total cerebral decortications. Am J Phys Med. 1956;35:273–310.
7. Dufau H. Plasticity of cognition in brain gliomas. In: Tracy JI, Hampstead BM, Sathian K, editors. Cognitive plasticity in neurologic disorders. New York: Oxford University Press; 2015. p. 125–51.
8. Southwell DG, Hervey-Jumper SL, Perry DW, Berger MS. Intraoperative mapping during repeat awake craniotomy reveals the functional plasticity of adult cortex. J Neurosurg. 2016;124:1460–9.
9. Anderson SW, Damasio H, Tranel D. Neuropsychological impairments associated with lesions caused by tumor or stroke. Arch Neurol. 1990;47:397–405.
10. Head H. Aphasia and kindred disorders of speech. Cambridge: Cambridge University Press; 1926.
11. Bates E, Roe K. Language development in children with unilateral brain injury. In: Nelson CA, Luciana M, editors. Handbook of developmental cognitive neuroscience. Cambridge: MIT Press; 2001. p. 281.
12. Coppens P, Lebrun Y, Basso A. Aphasia in atypical populations. New York: Psychology Press; 1998.
13. Paquier PF, Van Dongen HR. Is acquired childhood aphasia atypical. In: Coppens P, Lebrun Y, Basso A, editors. Aphasia in atypical populations. Cambridge: Cambridge University Press; 1998. p. 67–117.
14. Vanlancker-Sidtis D. When only the right hemisphere is left: studies in language and communication. Brain Lang. 2004;91:199–211.

[q] Relocation of functional areas has, incidentally, also been reported in patients with a tumour in the primary motor cortex after multistage surgery (but without chronic electrical stimulation) [139, 140].

15. Basser LS. Hemiplegia of early onset and the faculty of speech with special reference to the effects of hemispherectomy. Brain. 1962;85:427–60.
16. Wilson PJ. Cerebral hemispherectomy for infantile hemiplegia. A report of 50 cases. Brain. 1970;93:147–80.
17. Deacon W. The symbolic species: the co-evolution of language and the brain. New York: WW Norton & Company; 1997.
18. Chilosi AM, Cipriani P, Pecini C, et al. Acquired focal brain lesions in childhood: effects on development and reorganization of language. Brain Lang. 2008;106:211–25.
19. Bishop DV. Cerebral asymmetry and language development: cause, correlate, or consequence? Science. 2013;340:1230531.
20. Minagawa-Kawai Y, Cristia A, Dupoux E. Cerebral lateralization and early speech acquisition: a developmental scenario. Dev Cogn Neurosci. 2011;1:217–32.
21. Annett M. Left, right, hand and brain: the right shift theory. New York: Psychology Press; 1985.
22. Ballantyne AO, Spilkin AM, Trauner DA. Language outcome after perinatal stroke: does side matter. Child Neuropsychol. 2007;13:494–509.
23. Brown JW. The life of the mind. Hillsdale: Lawrence Erlbaum Associates; 1988.
24. Tesak J, Code C. Milestones in the history of aphasia: theories and protagonists. Hove: Psychology Press; 2008.
25. Finger S, Beyer T, Koehler PJ. Dr. Otto Soltmann (1876) on development of the motor cortex and recovery after its removal in infancy. Brain Res Bull. 2000;53:133–40.
26. Connolly KJ, Forssberg H. Neurophysiology and neuropsychology of motor development. New York: Cambridge University Press; 1997.
27. Ogden R, Franz SI. On cerebral motor control: the recovery from experimentally produced hemiplegia. Psychobiology. 1917;1:33.
28. Darling WG, Pizzimenti MA, Morecraft RJ. Functional recovery following motor cortex lesions in non human primates: experimental implications for human stroke patients. J Integr Neurosci. 2011;10:353–84.
29. Doidge N. The brain that changes itself: stories of personal triumph from the frontiers of brain science. Penguin; 2007.
30. Taub E, Uswatte G, Elbert T. New treatments in neurorehabilitation founded on basic research. Nat Rev Neurosci. 2002;3:228–36.
31. Kaas JH. Plasticity of sensory and motor maps in adult mammals. Annu Rev Neurosci. 1991;14:137–67.
32. Merzenich MM, Kaas JH, Wall J, et al. Topographic reorganization of somatosensory cortical areas 3b and 1 in adult monkeys following restricted deafferentation. Neuroscience. 1983;8:33–55.
33. Merzenich MM, Nelson RJ, Stryker MP, et al. Somatosensory cortical map changes following digit amputation in adult monkeys. J Comp Neurol. 1984;224:591–605.
34. Pons TP, Garraghty PE, Ommaya AK, et al. Massive cortical reorganization after sensory deafferentation in adult macaques. Science. 1991;252:1857–60.
35. Ramachandran VS, Hirstein W. The perception of phantom limbs. The D. O. Hebb lecture. Brain. 1998;121:1603–30.
36. Pascual-Leone A, Cammarota A, Wassermann EM, et al. Modulation of motor cortical outputs to the reading hand of braille readers. Ann Neurol. 1993;34:33–7.
37. Gilbert CD, Sigman M, Crist RE. The neural basis of perceptual learning. Neuron. 2001;31:681–97.
38. Xerri C. Experience-dependent reorganization of somatosensory and motor cortical areas: towards a neurobiology of rehabilitation. In: Duffau H, editor. Brain mapping: from neural basis of cognition to surgical applications. Wien: Springer; 2011. p. 111–29.
39. Roe AW, Pallas SL, Hahm JO, Sur M. A map of visual space induced in primary auditory cortex. Science. 1990;250:818–20.
40. O'Leary DD, Stanfield BB. Selective elimination of axons extended by developing cortical neurons is dependent on regional locale: experiments utilizing fetal cortical transplants. J Neurosci. 1989;9:2230–46.

41. Hubel DH, Wiesel TN. Ferrier lecture. Functional architecture of macaque monkey visual cortex. Proc R Soc Lond B Biol Sci. 1977;198:1–59.
42. York GK, Steinberg DA. Hughlings Jackson's theory of recovery. Neurology. 1995;45:834–8.
43. Finger S, Koehler PJ, Jagella C. The Monakow concept of diaschisis: origins and perspectives. Arch Neurol. 2004;61:283–8.
44. von Monakow C. Die Lokalisation im Grosshirn und der Abbau der Funktion durch Kortikale Herde. Wiesbaden, Germany: JF Bergmann; 1914.
45. York GK. Localization of language function in the twentieth century. J Hist Neurosci. 2009;18:283–90.
46. Geschwind N. Wernicke's contribution to the study of aphasia. Cortex. 1967;3:449–63.
47. Goldstein K. The organism: a holistic approach to biology derived from pathological data in man; 1934.
48. Goldstein K. Aftereffects of brain injuries in war: their evaluation and treatment. The application of psychologic methods in the clinic; 1942.
49. Luria AR. Traumatic aphasia: Its syndromes, psychology and treatment. Berlin: Walter De Gruyter; 1970.
50. Hebb DO. The organization of behavior: a neuropsychological theory. New York: Wiley; 1949.
51. Markram H, Gerstner W, Sjöström PJ. A history of spike-timing-dependent plasticity. Front Synaptic Neurosci. 2011;3:4.
52. Lövdén M, Bäckman L, Lindenberger U, et al. A theoretical framework for the study of adult cognitive plasticity. Psychol Bull. 2010;136:659–76.
53. Grafman J, Christen Y. Neuronal plasticity: building a bridge from the laboratory to the clinic. Berlin: Springer; 1999.
54. Rijntjes M, Weiller C. Recovery of motor and language abilities after stroke: the contribution of functional imaging. Prog Neurobiol. 2002;66:109–22.
55. Fields RD, Stevens-Graham B. New insights into neuron-glia communication. Science. 2002;298:556–62.
56. Cicchetti D, Curtis JW, Mikl M, et al. The developing brain and neural plasticity: implications for normality, psychopathology, vol. 2. New York: Wiley; 2006. p. 1.
57. Bavelier D, Green CS, Dye MWG. Exercising your brain: training-related brain plasticity. In: Gazzaniga MS, editor. The cognitive neurosciences. Cambridge: MIT Press; 2009. p. 153–64.
58. Classen J. Focal hand dystonia—a disorder of neuroplasticity. Brain. 2003;126:2571–2.
59. Overgaard M, Mogensen J. A framework for the study of multiple realizations: the importance of levels of analysis. Front Psychol. 2011;2
60. Goltz F. Über die verrichtungen des grosshirns. Pfluegers Archiv fuer die Gesamte Physiologie. 1888;42:419–67.
61. Lashley KS. Brain mechanisms and intelligence. A quantitative study of injuries to the brain. Chicago: University of Chicago Press; 1929.
62. Gowers WR. Lectures in the diagnosis of diseases of the brain. Philadelphia: Blakiston; 1887.
63. Barlow T. On a case of double hemiplegia, with cerebral symmetrical lesions. Br Med J. 1877;2:103–4.
64. Finger S, Buckner RL, Buckingham H. Does the right hemisphere take over after damage to Broca's area? the Barlow case of 1877 and its history. Brain Lang. 2003;85:385–95.
65. Hellal P, Lorch MP. The validity of Barlow's 1877 case of acquired childhood aphasia: case notes versus published reports. J Hist Neurosci. 2007;16:378–94.
66. Finger S. Origins of neuroscience: a history of explorations into brain function. New York: Oxford University Press; 2001.
67. Broca P. Sur le siège de la faculté du langage articulé. Bull Soc Anthropol. 1865;6:337–93.
68. Code C. Language, aphasia and the right hemisphere. New York: Wiley; 1987.
69. Basso A, Gardelli M, Grassi MP, Mariotti M. The role of the right hemisphere in recovery from aphasia. Two case studies. Cortex. 1989;25:555–66.
70. Springer JA, Binder JR, Hammeke TA, et al. Language dominance in neurologically normal and epilepsy subjects: a functional MRI study. Brain. 1999;122:2033–46.

71. Baron JC, Cohen LG, Cramer SC, et al. Neuroimaging in stroke recovery: a position paper from the First International Workshop on Neuroimaging and Stroke Recovery. Cerebrovasc Dis. 2004;18:260–7.
72. Janecek JK, Swanson SJ, Sabsevitz DS, et al. Naming outcome prediction in patients with discordant Wada and fMRI language lateralization. Epilepsy Behav. 2013;27:399–403.
73. Halai AD, Woollams AM, Lambon Ralph MA. Using principal component analysis to capture individual differences within a unified neuropsychological model of chronic post-stroke aphasia: revealing the unique neural correlates of speech fluency, phonology and semantics. Cortex. 2017;86:275–89.
74. Thompson CK, den Ouden DB. Neuroimaging and recovery of language in aphasia. Curr Neurol Neurosci Rep. 2008;8:475–83.
75. Weiller C, Isensee C, Rijntjes M, et al. Recovery from Wernicke's aphasia: a positron emission tomographic study. Ann Neurol. 1995;37:723–32.
76. Thulborn KR, Carpenter PA, Just MA. Plasticity of language-related brain function during recovery from stroke. Stroke. 1999;30:749–54.
77. Thiel A, Herholz K, Koyuncu A, et al. Plasticity of language networks in patients with brain tumors: a positron emission tomography activation study. Ann Neurol. 2001;50:629.
78. Perani D, Cappa SF, Tettamanti M, et al. A fMRI study of word retrieval in aphasia. Brain Lang. 2003;85:357–68.
79. Cao Y, Vikingstad EM, George KP, et al. Cortical language activation in stroke patients recovering from aphasia with functional MRI. Stroke. 1999;30:2331–40.
80. Thiel A, Schumacher B, Wienhard K, et al. Direct demonstration of transcallosal disinhibition in language networks. J Cereb Blood Flow Metab. 2006;26:1122–7.
81. Szaflarski JP, Allendorfer JB, Banks C, et al. Recovered vs. not-recovered from post-stroke aphasia: the contributions from the dominant and non-dominant hemispheres. Restor Neurol Neurosci. 2013;31:347–60.
82. Turkeltaub PE. Brain stimulation and the role of the right hemisphere in aphasia recovery. Curr Neurol Neurosci Rep. 2015;15:72.
83. Winhuisen L, Thiel A, Schumacher B, et al. Role of the contralateral inferior frontal gyrus in recovery of language function in poststroke aphasia: a combined repetitive transcranial magnetic stimulation and positron emission tomography study. Stroke. 2005;36:1759–63.
84. Winhuisen L, Thiel A, Schumacher B, et al. The right inferior frontal gyrus and poststroke aphasia: a follow-up investigation. Stroke. 2007;38:1286–92.
85. Szaflarski JP, Allendorfer JB, Byars AW, et al. Age at stroke determines post-stroke language lateralization. Restor Neurol Neurosci. 2014;32:733–42.
86. Duffau H. Brain plasticity and tumors. Adv Tech Stand Neurosurg. 2008;33:3–33.
87. Thiel A, Habedank B, Herholz K, et al. From the left to the right: how the brain compensates progressive loss of language function. Brain Lang. 2006;98:57–65.
88. Ulmer JL, Hacein-Bey L, Mathews VP, et al. Lesion-induced pseudo-dominance at functional magnetic resonance imaging: implications for preoperative assessments. Neurosurgery. 2004;55:569–81.
89. Lehericy S, Biondi A, Sourour N, et al. Arteriovenous brain malformations: is functional MR imaging reliable for studying language reorganization in patients? Initial observations. Radiology. 2002;223:672–82.
90. Mah YH, Husain M, Rees G, Nachev P. Human brain lesion-deficit inference remapped. Brain. 2014;137(Pt 9):2522–31.
91. Rutten GJ, Ramsey NF. The role of functional magnetic resonance imaging in brain surgery. Neurosurg Focus. 2010;28:E4.
92. Pedersen PM, Jorgensen HS, Nakayama H, et al. Aphasia in acute stroke: incidence, determinants, and recovery. Ann Neurol. 1995;38:659–66.
93. Laska AC, Kahan T, Hellblom A, et al. A randomized controlled trial on very early speech and language therapy in acute stroke patients with aphasia. Cerebrovasc Dis Extra. 2011;1:66–74.
94. Dijkerman HC, Wood VA, Hewer RL. Long-term outcome after discharge from a stroke rehabilitation unit. J R Coll Physicians Lond. 1996;30:538–46.

95. Furlan M, Marchal G, Viader F, et al. Spontaneous neurological recovery after stroke and the fate of the ischemic penumbra. Ann Neurol. 1996;40:216–26.
96. Saur D, Lange R, Baumgaertner A, et al. Dynamics of language reorganization after stroke. Brain. 2006;129:1371–84.
97. Warburton E, Price CJ, Swinburn K, Wise RJ. Mechanisms of recovery from aphasia: evidence from positron emission tomography studies. J Neurol Neurosurg Psychiatry. 1999;66:155–61.
98. Robson H, Zahn R, Keidel JL, et al. The anterior temporal lobes support residual comprehension in Wernicke's aphasia. Brain. 2014;137:931–43.
99. Xing S, Lacey EH, Skipper-Kallal LM, et al. Right hemisphere grey matter structure and language outcomes in chronic left hemisphere stroke. Brain. 2016;139:227–41.
100. Raja Beharelle A, Dick AS, Josse G, et al. Left hemisphere regions are critical for language in the face of early left focal brain injury. Brain. 2010;133:1707–16.
101. Szaflarski JP, Holland SK, Schmithorst VJ, Byars AW. fMRI study of language lateralization in children and adults. Hum Brain Mapp. 2006;27:202–12.
102. Lidzba K, Staudt M. Development and (re)organization of language after early brain lesions: capacities and limitation of early brain plasticity. Brain Lang. 2008;106:165–6.
103. Tillema JM, Byars AW, Jacola LM, et al. Cortical reorganization of language functioning following perinatal left MCA stroke. Brain Lang. 2008;105:99–111.
104. Liegois F, Connelly A, Helen CJ, et al. Language reorganization in children with early onset lesions of the left hemisphere: an fMRI study. Brain. 2004;127:1229–36.
105. Vargha-Khadem F, Isaacs EB, Papaleloudi H, et al. Development of language in six hemispherectomized patients. Brain. 1991;114:473–95.
106. Trudeau N, Colozzo P, Sylvestre V, Ska B. Language following functional left hemispherectomy in a bilingual teenager. Brain Cogn. 2003;53:384–8.
107. Forkel SJ, Thiebaut de Schotten M, Dell'Acqua F, et al. Anatomical predictors of aphasia recovery: a tractography study of bilateral perisylvian language networks. Brain. 2014;137:2027–39.
108. Pani E, Zheng X, Wang J, et al. Right hemisphere structures predict poststroke speech fluency. Neurology. 2016;86(17):1574–81.
109. Hart MG, Price SJ, Suckling J. Connectome analysis for pre-operative brain mapping in neurosurgery. Br J Neurosurg. 2016;30:506–17.
110. Ding D, Starke RM, Liu KC, Crowley RW. Cortical plasticity in patients with cerebral arteriovenous malformations. J Clin Neurosci. 2015;22:1857–61.
111. Duffau H. Lessons from brain mapping in surgery for low-grade glioma: insights into associations between tumour and brain plasticity. Lancet Neurol. 2005;4:476–86.
112. Deng X, Zhang Y, Xu L, et al. Comparison of language cortex reorganization patterns between cerebral arteriovenous malformations and gliomas: a functional MRI study. J Neurosurg. 2015;122(5):996–1003.
113. Liu Z, Deng X, Cao Y et al. Does right-sided language lateralization on BOLD-fMRI affect postoperative language outcome for AVM patients. Turk Neurosurg. 2016.
114. Pouratian N, Bookheimer SY, Rex DE, et al. Utility of preoperative functional magnetic resonance imaging for identifying language cortices in patients with vascular malformations. J Neurosurg. 2002;97:21–32.
115. Pallud J, Capelle L, Taillandier L, et al. The silent phase of diffuse low-grade gliomas. Is it when we missed the action. Acta Neurochir. 2013;155:2237–42.
116. Thiel A, Habedank B, Winhuisen L, et al. Essential language function of the right hemisphere in brain tumor patients. Ann Neurol. 2005;57:128–31.
117. Kristo G, Raemaekers M, Rutten GJ, et al. Inter-hemispheric language functional reorganization in low-grade glioma patients after tumour surgery. Cortex. 2015;64:235–48.
118. Benzagmout M, Gatignol P, Duffau H. Resection of World Health Organization Grade II gliomas involving Broca's area: methodological and functional considerations. Neurosurgery. 2007;61:741–52.

119. Goldstein K. Language and language disturbances. Aphasic symptoms complexes and their significance for medicine and theory of language. New York: Grune and Stratton; 1948.
120. Laurence S, Stein DG. Recovery of brain damage and the concept of localization of function. In: Finger S, editor. Recovery from brain damage. Research and theory. New York: Plenum Press; 1978. p. 369–407.
121. Nadeau SE. Neuroplastic mechanisms of language recovery after stroke. In: Tracy JI, Hampstead BM, Sathian K, editors. Cognitive plasticity in neurological disorders. New York: Oxford University Press; 2015. p. 61–84.
122. Papagno C, Vallar G. A plastic brain for a changing environment. Cortex. 2014;58:248–50.
123. Pascual-Leone A, Amedi A, Fregni F, Merabet LB. The plastic human brain cortex. Annu Rev Neurosci. 2005;28:377–401.
124. Doraiswamy PM, Agronin ME. Brain games: do they really work? Sci Am. 2009.
125. Brady MC, Kelly H, Godwin J, et al. Speech and language therapy for aphasia following stroke. Cochrane Database Syst Rev. 2016;6:CD000425.
126. Cappa SF, Benke T, Clarke S, et al. EFNS guidelines on cognitive rehabilitation: report of an EFNS task force. Eur J Neurol. 2005;12:665–80.
127. Greener J, Enderby P, Whurr R. Speech and language therapy for aphasia following stroke. Cochrane Database Syst Rev. 2000: CD000425.
128. Bowen A, Hesketh A, Patchick E, et al. Effectiveness of enhanced communication therapy in the first four months after stroke for aphasia and dysarthria: a randomised controlled trial. BMJ. 2012;345:e4407.
129. Studdert-Kennedy M, Mody M. Auditory temporal perception deficits in the reading-impaired: a critical review of the evidence. Psychon Bull Rev. 1995;2:508–14.
130. Snowling MJ. From language to reading and dyslexia. Dyslexia. 2001;7:37–46.
131. Schwartz JM, Begley S. The mind and the brain. 2002.
132. Merzenich MM, Jenkins WM, Johnston P, et al. Temporal processing deficits of language-learning impaired children ameliorated by training. Science. 1996;271:77–81.
133. Tallal P, Miller SL, Bedi G, et al. Language comprehension in language-learning impaired children improved with acoustically modified speech. Science. 1996;271:81–4.
134. Temple E, Deutsch GK, Poldrack RA, et al. Neural deficits in children with dyslexia ameliorated by behavioral remediation: evidence from functional MRI. Proc Natl Acad Sci U S A. 2003;100:2860–5.
135. Naeser MA, Martin PI, Nicholas M, et al. Improved picture naming in chronic aphasia after TMS to part of right Broca's area: an open-protocol study. Brain Lang. 2005;93:95–105.
136. Naeser MA, Martin PI, Ho M, et al. Transcranial magnetic stimulation and aphasia rehabilitation. Arch Phys Med Rehabil. 2012;93:S26–34.
137. Ren CL, Zhang GF, Xia N, et al. Effect of low-frequency rTMS on aphasia in stroke patients: a meta-analysis of randomized controlled trials. PLoS One. 2014;9:e102557.
138. Barcia JA, Sanz A, Balugo P, et al. High-frequency cortical subdural stimulation enhanced plasticity in surgery of a tumor in Broca's area. Neuroreport. 2012;23:304–9.
139. Takahashi S, Jussen D, Vajkoczy P, Picht T. Plastic relocation of motor cortex in a patient with LGG (low grade glioma) confirmed by NBS (navigated brain stimulation). Acta Neurochir. 2012;154:2003–8.
140. Duffau H, Capelle L, Denvil D, et al. Functional recovery after surgical resection of low grade gliomas in eloquent brain: hypothesis of brain compensation. J Neurol Neurosurg Psychiatry. 2003;74:901–7.
141. Rutten GJ, Ramsey NF, van Rijen PC, et al. fMRI-determined language lateralization in patients with unilateral or bilateral language dominance according to the Wada test. Neuroimage. 2002;17:447–60.

Epilogue

The intention to write this book was born out of amazement at the persistent clinical role of the classic language model over the past 150 years.[a] Countless arguments have been made for a more sophisticated and complex view of language in the brain, yet the model has prevailed in textbooks and medical schools. The names of Broca and Wernicke continue to be used in neurological and neurosurgical departments, as well as in the many scientific and clinical language papers published. We know now that the classic language model is not a good predictor for the different types of aphasia (Chaps. 1 and 2). It does not account for individual differences in cortical anatomy (Chap. 5) or function (Chap. 6) nor for the many different subcortical language pathways that have been described only more recently (Chaps. 6 and 7) or for functional rehabilitation after brain damage (Chap. 9). In addition to all of these shortcomings, it should be realized that many historical authors, in particular those that have been characterized as the diagram makers, have been misinterpreted to begin with. Wernicke, as we have seen in Chap. 2, was convinced that the whole perisylvian area and the insula were involved in language. He laid out a connectionist view that predated our current network approach to brain functions, but that was put aside in favour of a simpler model.

Particularly in the last 20 years, ideas of how language is implemented in the brain have changed dramatically. There is now broad consensus in the neuroscientific community that there are no well-demarcated language centres, that there is not a single hemisphere that does all the language processing and that language functions cannot strictly be decomposed into a handful of linguistic components. Instead, we have come to realize that information is processed in several different streams (decoding of speech, mapping of sound onto meaning, etc.) and that these various subnetworks are highly interrelated, as well as being involved in other cognitive and sensorimotor functions. Language functions, in fact, should be considered as emergent properties of very complex and interconnected systems.

In recent years, increasing numbers of authors have explicitly disqualified the classic language model and proposed alternative views for the neural basis of language.

[a]In various compositions, it is usually referred to as the Broca–Wernicke–Lichtheim–Geschwind model.

G.-J. Rutten, *The Broca-Wernicke Doctrine*, DOI 10.1007/978-3-319-54633-9

Some have even declared that 'Broca and Wernicke are dead'.[b] However, this is not something new but has occurred repeatedly over the course of history. Remember, for instance, Pierre Marie's fierce opposition against the 'dogma of the third convolution' and his denial of the existence of a frontal language area (Chap. 1), or Luria's description of functions as complex and dynamical systems that are never simply localized (Chap. 7): 'the cerebral cortex does not consist of separate, isolated centers and (…) the recovery of a function must not be attributed to transfer of the function to a new, vicarious center but rather, to a structural reorganization into a new, dynamic system widely dispersed in the cerebral cortex and lower formation'.[c]

Why then has the classic model retained such a large influence in our current clinical and scientific work? The philosopher and historian Thomas Kuhn argued that people and systems generally tend to resist change and that prevailing and universal scientific frameworks—that he named 'paradigms'—are not easily overthrown.[d] Even if contemporary theories have obvious flaws, he said that it will take great effort and several 'anomalies' before these errors are truly acknowledged.[e] Eventually, the accumulation of errors will result in a state of 'scientific crisis' that persists until a new paradigm is formed and adopted.

There is likely a more specific reason that the classic language model has remained so popular, and that is because it is so attractive from a conceptual and perhaps even aesthetic point of view. Models are meant to reduce the complex reality into something far more simple and understandable. We are all accustomed to models in our daily lives and usually experience their benefits: in weather forecasting, economy, education, etc.[f] However, these models are only valid when they have a certain predictive value for real-world phenomena, and—most importantly—their limitations are recognized and respected by their users. The latter is not always the case, especially when models are elegant and seem correct from an intuitive point of view.[g] There is a real danger that these models are interpreted too literally and

---

[b] Tremblay and Dick, *Brain and Language* (2016)

[c] Luria, *Higher Cortical Functions in Man—Second Edition* (1980)

[d] Classic examples of paradigm changes are those from classical to quantum mechanics, and from the cosmological view of Ptolemaeus to that of Copernicus. Scientific communities consider paradigms a foundation for further practice, whereas in Kuhn's opinion these paradigms do not leave much room for true novelty and innovation. At the time, Kuhn's propositions were considered quite shocking by many, as he in fact stated that normal science is working only on a few puzzles that are left open in a current field of knowledge. In his view, normal science is predicated on the assumption that the scientific community knows what the world is like, thereby often suppressing fundamental novelties; these novelties are only acknowledged after a shift of paradigm. See Kuhn, *The Structure of Scientific Revolutions* (1962).

[e] Somehow, people have a tendency to dismiss their errors. This is unfortunate, as errors play an important and necessary role in innovation. They help us to create a path out of our comfortable assumptions. See, for instance, Steven Johnson's *Where Good Ideas Come From* (2010).

[f] Our brains, of course, also use predictive models to interact with the environment.

[g] Scientific models and theories do not always match with intuition or personal observations. For example, many people still believe that a force is needed to keep an object in motion, because that is what they *think* they experience in daily life, whereas, in fact, an object will keep moving forever unless there is an opposing force that will stop it (Newton's first law).

that they begin to function as a new reality for many, thereby seriously impeding scientific progress. In brain research, a good example is the assignment of functions to either the left or the right hemisphere (linguistic vs. spatial, verbal vs. nonverbal, rational vs. irrational, ego vs. id, etc.). Other powerful concepts that withstood fierce criticism for a long time are the somatotopic representation of sensorimotor functions along the precentral gyrus and the different anatomical locations of perceptive and productive language functions. The schemes of Penfield and Wernicke still suffer from shallow and incorrect interpretations, despite explicit warnings of the authors from the beginning (Chaps. 2 and 6). Something alike can be said in our modern era for maps that contain information from functional MRI or fibre tractography (Chap. 8).

Are There Any Language-Critical Cortical Areas?

A frequent source of confusion in the description of language areas and their pathways is the terms 'supportive', 'involved' and 'critical'. Critical language regions can really only be defined post hoc, when damage has resulted in lasting and measurable deficits. As these critical regions are smaller than language-supportive regions, the language maps of clinicians and neuroscientists are different to begin with. The investigator is another factor that plays a role in how language deficits are qualified. A neurologist or neurosurgeon will often judge a language deficit differently than a linguist or a neuroscientist.

The restoration of function in response to a lesion is a complex process that leads to changes in the remaining—healthy—part of the brain (Chap. 9). When we want to investigate whether a part of the brain is critical, we are in fact assessing to what extent the brain can sustain (most of) its functionality without this part. Sometimes these inferences are fairly straightforward, for example, in case of a lesion of the optic radiation. If such a lesion consistently results in visual field deficits, it is clear that there is no redundancy within the system and that the damaged optic radiation is the cause of the functional impairment. In other instances, especially when a lesion destroys associative systems, it can be very difficult or even impossible to deduce where the affected functions are 'located'. This was realized by several authors in the past, in particular by Hughlings Jackson, who stated that 'to locate the damage which destroys speech and to locate speech are two different things' (Chap. 4).[h] Others would later add that the reaction of the organism as a whole and in particular its interaction with the environment also significantly contribute to functional recovery (Chap. 9).

'There is greater revelation in pathological phenomena', wrote Goldstein, and he meant that patient studies can greatly advance our understanding of normal phenomena. Evidence for the existence of critical language structures must therefore come from brain-lesioned patients. We should learn from these patients how damage affects the functional organization of the brain and use this information to

[h] Hughlings Jackson, *On the Nature of the Duality of the Brain* (1874)

develop methods to predict the compensatory potential of the brain prior to damage. There are currently no technologies that can do this very accurately, despite claims made for electrical stimulation mapping and the Wada test. New techniques, such as MRI-based tractography and functional MRI (fMRI), seem better candidates for this purpose but require further development and validation in order to become clinically relevant (see Chap. 8 for a discussion on fMRI). But Goldstein also warned us that inappropriate use of pathological material or insufficient analyses can have a 'fatal influence' in the field of normal psychology or for the theory of the structure of language.[i]

Before answering the proposed question of whether there are any critical cortical language areas, we should realize that lesions invariably affect both cortical and subcortical tissue. Broca knew that there was damage to the underlying white matter of the frontal lobe in his patients, but he considered only the cortical part relevant for their language problems (Chap. 1). White matter pathways have long played only a minor role in the neural basis of language and aphasia, but with the advent of diffusion-weighted imaging and intraoperative subcortical electrical stimulation, there is now overwhelming evidence that language is supported by multiple fibre pathways. As a consequence, it is difficult to judge the functional contribution of a single cortical area separately from the other brain regions with which it is connected (in a similar fashion that functions cannot be attributed to single fibre pathways). Let us assume, though, in an attempt to answer the question, that the classic language areas each encompass only a single anatomical region, respectively, the pars triangularis and opercularis of the left inferior frontal gyrus (Broca's area) and the posterior part of the left superior temporal gyrus (Wernicke's area). What will happen if any of these two areas is selectively damaged?

A lesion of Broca's area, or at least a part of it, is likely to induce only mild or transient language deficits, as long as there is not too much damage to adjacent subcortical pathways. There is reasonable evidence for this from stroke studies (Chap. 1) and from a selective number of surgical procedures that have used the left frontal operculum as a corridor to deeper-located pathological lesions (Chap. 6). In a clinical sense, Broca's area seems less critical than is usually advocated. This does not necessarily imply, however, that it plays only a minor role in language. Damage to the classic Wernicke's area also does not result in the deficits with which it is usually associated, because language comprehension is not a very localized function and involves large parts of at least both temporal and inferior parietal lobes (Chaps. 2, 7 and 8). More recent opinion claims a role for Wernicke's area in phonological processing (as previously also proposed by Luria; see Chap. 7). Lesions within this region have been associated with good comprehension but abundant phonological errors during speech production (i.e. conduction aphasia). Data from selective and acute-onset cortical lesions, however, are lacking.

In conclusion, Broca's aphasia is not caused by damage to Broca's area alone, and Wernicke's aphasia is not caused by damage to classic Wernicke's area. This has repeatedly been put forward by different authors over the past decades. As such,

---

[i]Goldstein, *The Organism* (1932)

these areas play a different and a more limited clinical role than classically advocated. Several other areas have been proposed for inclusion in various neoclassic language models, in particular the anterior part of the insula, the middle temporal gyrus and the anterior part of the temporal lobe. However, none of these cortical areas seems to fulfil the critical role that has previously—wrongly—been attributed to Broca's and Wernicke's areas.

When Will Clinicians Put Aside the Classic Model?

Use of the Broca–Wernicke terminology has led to much confusion and error. Consensus on the classic language territories and their functions was never reached, which is already an indication that the classic model is not a very realistic view of the neural basis of language. Still, the model remains widely used in patient care. Apparently, medical doctors rarely encounter 'evidence' in their practice that pleads against this model. In any case, the occasional exceptions (language impairments after surgery or a stroke in the right hemisphere, for instance) clearly never had enough momentum to overturn the old dogma.

What is changing this situation is the growing emphasis on subcortical pathways as crucial structures for normal language function, and the idea that functions are dynamically represented in networks. This started in the 1990s with the concept of large-scale networks to explain brain–behaviour relationships[j] and got support from clinical observations when electrical stimulation began to identify not one (arcuate fasciculus) but several different pathways that play a role in language (inferior fronto-occipital fasciculus, fronto-aslant tract, subcallosal fasciculus, uncinate fasciculus and inferior longitudinal fasciculus). The new cortical–subcortical concept is already changing the surgical approach to brain tumours and provides us with a better understanding of patients' language impairments in diseases like stroke or primary progressive aphasia.

As a final point in the discussion towards a new clinical view on language representation, it is worth mentioning that in my opinion our current invasive gold standards, in particular electrical stimulation mapping, stand in the way of progress of our knowledge of language in the brain. As was described in Chap. 6, electrical stimulation mapping usually finds only a few small cortical sites—and not the larger regions that are associated with the classic language view. These sites are highly respected by neurosurgeons as critical language areas. However, it seems very unlikely that a selective lesion of one of these small sites in Broca's or Wernicke's region will result in permanent and grave impairments. These sites are not critical language areas *on their own*, as this simply does not fit with the large-scale network view of distributed functions or with clinical experience that limited damage (small cortical lesions of 1–2 cm) in patients is generally asymptomatic. The exception, perhaps, is the left ventral premotor cortex; this relatively small unimodal area on

[j] Mesulam, *Large-Scale Neurocognitive Networks and Distributed Processing for Attention, Language, and Memory* (1990)

the precentral gyrus—which invariably yields a speech arrest when electrically stimulated—may serve as a final common pathway for speech production. Problem is that the hypothesis that all language sites that are found with electrical stimulation are critical, is notoriously hard to disprove as these sites are never deliberately resected. Still, clinicians should keep in mind that the effects of postlesional reorganization cannot reliably be predicted by electrical stimulation mapping and that the procedure itself also has a limited ability to test more complex language functions. Careful research is needed to resolve these issues, whereby patient outcome is the true gold standard against which the results of any technique (old or new) should be compared.

So, should we refrain of the names of Broca and Wernicke all together? Tremblay and Dick have a point when they write that maintaining the terminology for the classic model 'artificially maintains it as a legitimate model'. In practice, however, abandoning these terms will not be easy, as clinicians and neuroscientists are so familiar with the names of Broca and Wernicke. It is probably better to keep in mind that there was never much consensus on the classic language model and use Broca's and Wernicke's names to coarsely indicate the anterior and the posterior perisylvian regions, respectively (i.e. refer to these areas without much anatomical specifications or functional connotations). In this way, we can commemorate these men for their innovative ideas and discoveries, which—unfortunately—were not always judged appropriately.

The manufacturer's authorised representative in the EU is Springer Nature Customer Service Centre GmbH, Europaplatz 3, 69115 Heidelberg, Germany. If you have any concerns regarding our products, please contact ProductSafety@springernature.com

Printed and bound by CPI Group (UK) Ltd, Croydon, CR0 4YY
25/09/2026
02222863-0006